Advances in Intelligent Systems and Computing

Volume 225

Series Editor

J. Kacprzyk, Warsaw, Poland

For further volumes:
http://www.springer.com/series/11156

Dhinaharan Nagamalai · Ashok Kumar
Annamalai Annamalai
Editors

Advances in Computational Science, Engineering and Information Technology

Proceedings of the Third International Conference on Computational Science, Engineering and Information Technology (CCSEIT-2013), KTO Karatay University, June 7–9, 2013, Konya, Turkey – Volume 1

Springer

Editors

Dhinaharan Nagamalai
Dept. of Computer Engineering
Faculty of Engineering
KTO Karatay University
Konya
Turkey

Annamalai Annamalai
Dept. of Electrical and Computer Engineering
Prairie View A & M University
Texas
USA

Ashok Kumar
School of Computing and Informatics
University of Louisiana at Lafayett
Louisiana
USA

ISSN 2194-5357 ISSN 2194-5365 (electronic)
ISBN 978-3-319-00950-6 ISBN 978-3-319-00951-3 (eBook)
DOI 10.1007/978-3-319-00951-3
Springer Cham Heidelberg New York Dordrecht London

Library of Congress Control Number: 2013939955

Printed on acid-free paper

Springer is part of Springer Science+Business Media (www.springer.com)

Preface

The Third International Conference on Computational Science, Engineering and Information Technology (CCSEIT-2013) was held in Konya, Turkey, June 7–9, 2013. CCSEIT-2013 attracted many local and international delegates, presenting a balanced mixture of intellect from the East and from the West. The goal of this conference series is to bring together researchers and practitioners from academia and industry and share cutting-edge development in the field. The conference will provide an excellent international forum for sharing knowledge and results in theory, methodology and applications of Computational Science, Engineering and Information Technology. Authors were invited to contribute to the conference by submitting articles that illustrate research results, projects, survey work and industrial experiences describing significant advances in various areas of Computational Science, Engineering and Information Technology.

The CCSEIT-2013 Committees rigorously invited submissions for many months from researchers, scientists, engineers, students and practitioners related to the relevant themes and tracks of the conference. This effort guaranteed submissions from an unparalleled number of internationally recognized top-level researchers. All the submissions underwent a strenuous peer-review process which comprised expert reviewers. These reviewers were selected from a talented pool of Technical Committee members and external reviewers on the basis of their expertise. The papers were then reviewed based on their contributions, technical content, originality and clarity. The entire process, which includes the submission, review and acceptance processes, was done electronically. All these efforts undertaken by the Organizing and Technical Committees led to an exciting, rich and a high quality technical conference program, which featured high-impact presentations for all attendees to enjoy, appreciate and expand their expertise in the latest developments in computer network and communications research.

In closing, CCSEIT-2013 brought together researchers, scientists, engineers, students and practitioners to exchange and share their experiences, new ideas and research results in all aspects of the main workshop themes and tracks, and to discuss the practical challenges encountered and the solutions adopted. We would like to thank the General and Program Chairs, organization staff, the members of the Technical Program Committees and external reviewers for their excellent and tireless work. We sincerely wish that all attendees benefited scientifically from the conference and wish them every success in their research.

It is the humble wish of the conference organizers that the professional dialogue among the researchers, scientists, engineers, students and educators continues beyond the event and that the friendships and collaborations forged will linger and prosper for many years to come. We hope that you will benefit from the fine papers from the CCSEIT-2013 conference that are in this volume and will join us at the next CCSEIT conference.

Dhinaharan Nagamalai
Ashok Kumar
Annamalai Annamalai

Organization

General Chair

Ali Okatan	KTO Karatay University, Turkey
David C. Wyld	Southeastern Louisiana University, USA

General Co-Chairs

Jae Kwang Lee	Hannam University, South Korea
Michal Wozniak	Wroclaw University of Technology, Poland

Steering Committee

Natarajan Meghanathan	Jackson State University, USA
Chih-Lin Hu	National Central University, Taiwan
Dhinaharan Nagamalai	KTO Karatay University, Turkey
Chin-Chih Chang	Chung Hua University, Taiwan
Ashok Kumar	University of Louisiana at Lafayette, USA

Volume Editor

Annamalai A	Prairie View A & M University, USA

Publicity Chairs

Kasym Oztoprak	KTO Karatay University, Konya, Turkey
Ali Ozturk	KTO Karatay University, Konya, Turkey
Refik Caglar Kyzylyrmak	KTO Karatay University, Konya, Turkey

Program Committee Members

Piotr Zwierzykowski	Poznan University of Technology, Poland
Luca Caviglione	National Research Council of Italy (CNR) - ISSIA
Danda B. Rawat	Eastern Kentucky University, USA
Gregorio Martinez	University of Murcia, Spain
Antonio Ruiz-Martinez	University of Murcia, Spain
Emilio Jimenez Macias	University of La Rioja, Spain
Eugenia Moreira Bernardino	School of Technology and Management, Polytechnic Institute of Leiria, Portugal
Kamran Arshad	University of Greenwich, United Kingdom
Josip Lorincz	FESB, University of Split, Croatia
Carlos Miguel Tavares Calafate	Universidad Politecnica de Valencia, Spain
Namje Park	Jeju National University, Korea
Praveen Ranjan Srivastava	BITS Pilani, Rajasthan, India
Chitra	PSG College of Technology-Coimbatore, India
A.Velumani	Avinashilingam University-Coimbatore, India
Aarti M. Karande	Mumbai University, India
Abdul Aziz	University of Central Punjab, Pakistan
Abdul Kadir Ozcan	KTO Karatay University, Turkey
Acharyulu K.V.L.N	Bapthala Engineering College, India
Aditi Sharan JNU	New Delhi, India
Ajay K Sharma	Dr B R Ambedkar National Institute of Technology, India
Amandeep Singh Thethi	Guru Nanak Dev University Amritsar, India
Amit Kumar Singh Sanger	Hindustan College of Science and Technology, India
Andreas Riener	Johannes Kepler University Linz, Austria
Andy Seddon	Asia Pacific Institute of Information Technology, Malaysia
Ankit Chaudhary	BITS Pilani, India
Antelin vijila	Manonmaniam Sundaranar University, India
Arokiasamy A	Eastern Mediterranean University, Cyprus
Arunadevi J	Thiagarajar college, Madurai, India
Arvinth kumar	M.S. University, India
Ashok kumar Sharma	YMCA Institute of Engineering, India
Athanasios Vasilakos	University of Western Macedonia, Greece
Atilla Elci	Eastern Mediterranean University, Cyprus
Azween Bin Abdullah	Universiti Teknologi Petronas, Malaysia
Kalpana	Avinashilingam University-Coimbatore, India

Hamid Reza Karimi	University of Agder, Norway
Hao Shi	Victoria University, Australia
Hao-En Chueh	Yuanpei University, Taiwan, R.O.C.
Hariharan S	B.S.Abdur Rahman University, India
Hariharan S	Pavendar Bharathidasan College of Engineering and Technology, India
He Xu	Nanjing University of Post and Telecommunications, China
Henrique J.A. Holanda	UERN - Universidade do Estado do Rio Grande do Norte
Henrique Joao Lopes Domingos	University of Lisbon, Portugal
Ho Dac Tu	Waseda University, Japan
Hoang	Huu Hanh, Hue University, Vietnam
Hwangjun Song	Pohang Univ of Science and Technology, South Korea
I. Elizabeth Shanthi	Avinashilingam University-Coimbatore, India
Indrajit Bhattacharya	Kalyani Govt. Engg. College, India
Jacques Demerjian	Communication & Systems, Homeland Security, France
Jae Kwang Lee	Hannam University, South Korea
Jalel Akaichi	University of Tunis, Tunisia
Jansirani	Manonmaniam Sundaranar University, India
Jeong-Hyun Park	Electronics Telecommunication Research Institute, South Korea
Jitendra Prithviraj	GGITS, Jabalpur, India
Jivesh Govil	Cisco Systems Inc - CA, USA
Johann Groschdl	University of Bristol, UK
Johnson Kuruvila	Dalhousie University, Halifax Canada
Jose Enrique Armendariz-Inigo	Universidad Publica de Navarra, Spain
Jung-Gil Song	Hannam University, South Korea
Jungwook Song	Konkuk University, South Korea
Jyoti Singhai	Electronics and Communication Deptt-MANIT, India
K.Somasundaram	Gandhigram Rural University, India
K.Thangavel	Periyar University-Salem, India
Kamaljit I Lakhtaria	Atmiya Institute of Technology & Science, India
Kami Makki S	Lamar University, USA
Kannan A	K.L.N. College of Engineering, India
Kannan	Anna University, Chennai, India
Khamish Malhotra	University of Glamorgan, UK
Khoa N. Le	Griffith University, Australia
Krishnan N	Manonmaniam Sundaranar University, India

Krzysztof Walkowiak	Wroclaw University of Technology, Poland
L.Ganesan	Alagappa University, India
Lakshmi Rajamani	Osmania University, India
Lu s Veiga	Technical University of Lisbon, Portugal
Lu Yan	University Of Hertfordshire, UK
Lylia Abrouk	University of Burgundy, France
M.Mayilvaganam	PSG College of Arts and Science-Coimbatore, India
M.Thangaraj	Madurai Kamaraj University-Madurai, India
Madhan KS	Infosys Technologies Limited, India
Manik Gupta	LSI Research and Development, India
Maniza Hijab	Osmania University, India
Maode Ma	Nanyang Technological University, Singapore
Marco Roccetti	Universty of Bologna, Italy
Masaki Murakami	Okayama University, Japan
Michael Peterson	University of Hawaii at Hilo, United States
Mohsen Sharifi	Iran University of Science and Technology, Iran
Murali R	Dr. Ambedkar Institute of Technology, Bangalore
Murugan D	Manonmaniam Sundaranar University, India
Murugeswari	M.S.University, India
Muthu Selvi R	Thiagarajar College of Engineering, India
Muthulakshmi	M.S.University, India
N.Ganesan	Director Regional Institute of Cooperative Management-Bangalore, India
N.Valliammal	Avinashilingam University-Coimbatore, India
Nabendu Chaki	University of Calcutta, India
Neerajkumar	Madha Vaisnavi Devi University, India
Nicolas Sklavos	Technological Educational Institute of Patras, Greece
Nidaa Abdual Muhsin	Abbas- University of Babylon, Iraq
Nitiket N Mhala	B.D. College of Engineering - Sewagram, India
P. Narayanasamy	Anna University-Chennai, India
P. Ilango	VIT, Chennai, India
P. Subashini	Avinashilingam University-Coimbatore, India
Patrick Seeling	University of Wisconsin - Stevens Point, USA
Paul D. Manuel	Kuwait University, Kuwait
Phan Cong Vinh	London South Bank University, United Kingdom
Ponpit Wongthongtham	Curtin University of Technology, Australia
Prabu Dorairaj	NetApp Inc, India
Pravin P. Karde	HVPM's College of Engineering & Technology - Amravati, India
Premanand K. Kadbe	Vidya Pratishthan's College of Engineering, India

Shin-ichi Kuribayashi	Seikei University, Japan
Shivan Haran	Arizona State University, USA
Shrirang.Ambaji.Kulkarni	National Institute of Engineering, India
Singh M.P	National Institute of Technology Patna
Somitra Sanadhya	IIT-Delhi, India
Soodamani Ramalingam	Uuniversity of Hertfordshire, UK
Sriman Narayana Iyengar	VIT University, India
Srinivas Acharyulu P.V	NTPC Limited, India
Srinivasan B	Monash University, Australia
Subashini Parthasarathy	Avinashilingam University for Women, India
Subha	M.S. University, India
Sudip Misra	Indian Institute of Technology-Kharagpur, India
Sundaresan	Manonmaniam Sundaranar University, India
SunYoung Han	Konkuk University, South Korea
Suruliandi	Manonmaniam Sundaranar University, India
Susana Sargento	University of Aveiro, Portugal
Syed Abdul Sattar	Royal Institute of Technology and Science, India
Syed Rahman	University of Hawaii-Hilo, United States
Syed Rizvi	University of Bridgeport, United States
T Venkat Narayana Rao	Hyderabad Institution of Technology and Management, India
T. Santhanam	D.G. Vaishnav College-Chennai, India
T. Devi	Bharathiyar University-Coimbatore, India
T.K.S. Lakshmipriya	Avinashilingam University-Coimbatore, India
T.V. Geetha	Anna University-Chennai, India
Tejbanta Chingtham	Sikkim Manipal Institute of Technology, India
Thamaraiselvi	Madras Institute of Technology, India
Thambidurai	Pondicherry University, India
Thooyamani K.P	Bharath University, India
Utpal Biswas	University of Kalyani, India
V. Srividhya	Avinashilingam University-Coimbatore, India
Vasantha kalyani David	Avinashilingam University-Coimbatore, India
Velmurugan Ayyadurai	Center for Communication Systems, United Kingdom
Vijayanandh R	ICIT, India
Vikram Jain	Shri Ram Institute of Technology Jabalpur, India
Vishal Sharma	Metanoia Inc, USA
Wei Jie	University of Manchester, UK
Wided oueslati	l'institut superieur de gestion de tunis, Tunisia
Xiaofeng Liao	Chongking University, China
Yan Luo	University of Massachusetts Lowell, United States
Yannick Le Moullec	Aalborg University, Denmark

Yao-Nan Lien	National Chengchi University, Taiwan
Yazid Saman	Universiti Malaysia Terengganu, Malaysia
Yeong Deok Kim	Woosong University, South Korea
Yuh-Shyan Chen	National Taipei University, TAIWAN
Yung-Fa Huang	Chaoyang University of Technology, Taiwan
Wael M El-Medany	IT College,University Of Bahrain
Raja Kumar M	National Advanced IPv6 Center (NAv6), Universiti Sains Malaysia
Sevil Sen	Hacettepe University, Turkey
Aysel Safak	Baskent University, Turkey
Jalil Jabari Lotf	Islamic Azad University, Iran
Mohammad Saleem	DERI Ireland, Ireland
Ankit Thakkar	Nirma University, India
Gagan Singla	Aryabhatta college of engineering and technology, India
Gaurav Kumar Tak	Lovely Professional University, India
J. Indumathi	Anna University, India
K. Ganesh Reddy	NITK surathkal, India
M. Sandhya	B.S. Abdur Rahman University, India
Meenu Chawla	MANIT Bhopal, India
R K Mishra	Feeroze Gandhi Institute of Engineering and Technology, India
Rama Krishna C	NITTTR Chandigarh, India
Ramesh Babu.H.S	Acharya Institute of Technology, India
Sanjay Kumar Sharma	Oriental Institute of Science & Technology, India
Santhi Thilagam	N.I.T.K-Surathkal, India
Sarbani Roy	Jadavpur University, India
T. Meyyappan	Alagappa University, India
Seckin Tunalilar	Aselsan, Turkey

Program Organizers

Chief Patron

Omer TORLAK	Rector, KTO Karatay University, Konya, Turkey

Patron

Ali Okatan	KTO Karatay University, Konya, Turkey

President

Mehmet CELIK	KTO Karatay University, Konya, Turkey

Vice - President

Hasan Fehmi ATASAGUN	KTO Karatay University, Konya, Turkey

Committee Chairpersons

Kasym Oztoprak	KTO Karatay University, Konya, Turkey
Ali Ozturk	KTO Karatay University, Konya, Turkey
Refik Caglar Kızılırmak	KTO Karatay University, Konya, Turkey
Ayben Karasu Uysal	KTO Karatay University, Konya, Turkey
Hulusi Acıkgoz	KTO Karatay University, Konya, Turkey
Mehmet Ozbay	KTO Karatay University, Konya, Turkey
Semih Yumusak	KTO Karatay University, Konya, Turkey
Adem Yılmaz	KTO Karatay University, Konya, Turkey
Mastaneh Torkamani Azar	KTO Karatay University, Konya, Turkey

Convener

Abdul Kadir Ozcan	KTO Karatay University, Turkey

Contents

Urban Traffic Management System by Videomonitoring

José Raniery Ferreira Junior

ZUQ - Intelligent Transportation
Lourival Melo Mota Ave, 12 Building, 107 Room, Federal University of Alagoas
Tabuleiro dos Martins, Maceió, Alagoas, Brasil, 57072-970
Institute of Computing, Master's Program in Informatics
Federal University of Alagoas
jose.raniery@gmail.com
http://www.zuq.com.br

Abstract. As the big cities grow, it's more necessary the use of cameras for urban traffic monitoring. The increase in the number of vehicles on the streets makes the traffic congestion, one of the largest metropolis problems, even more often to happen. To avoid this kind of issue, this paper proposes a management system by videomonitoring for the urban traffic. And the goal is to identify the vehicles e count them in period of time using Computer Vision and Image Processing techniques.

Keywords: videomonitoring, urban traffic, computer vision.

1 Introduction

Vehicle detection through videomonitoring is a important tool for real time traffic management systems. It offers some advantages against the traditional methods, such as loop detectors. Besides vehicle count, video images can provide more informations about the traffic: speed and vehicle classification [3].

So it's expectated the decrease of the congestion in the big cities and the number of accidents in the urban roads, big issues that metropolis in the whole world have to face.

The presence of Graphics Computer in the human day-by-day is increasing, and even more Computer Vision and Image Processing. The computers and cameras prices are the lowest ever seen and it gets more viable to equip the roads with an artificial vision system that is able to detect movements, follow vehicle track and extract informations, such as speed, dimensions and traffic density [1].

Some useful algorithms for vehicle detection in the literature were: background classification learning [3], image segmentation [1], application of mathematical morphology operations, such eroding and dilation [4], others [5] [2].

The goal of this paper is to develop a system that aids the traffic management of main urban roads and aim to decrease the number of vehicles. This way, the expectation is the congestions end by means of vehicle detection and count.

D. Nagamalai et al. (Eds.): *Adv. in Comput. Sci., Eng. & Inf. Technol.*, AISC 225, pp. 1–9.
DOI: 10.1007/978-3-319-00951-3_1

As secondary goal, it expects that the algorithm has low processing cost, using opensource tools.

2 Methods

All video samples used in this system had speed rate of 30 frames per seconds and dimensions 640x480 coloured pixels.

The system was developed in Java 1.7 programming language, along with OpenCV (Open Source Computer Vision) version 2.4.0 graphic library with a wrapper to Java called JavaCV. OpenCV is an opensource library useful in several areas and techniques, as Computer Vision, Image Processing and Segmentation, Machine Learning, and others [6].

The system runs some steps, listed below. Bradski [7] affirms that the Warping operation is a geometric transformation based on image not uniform resizing. The Figure 1 shows a Warping operation example. In the system algorithm case, four points of the image are selected to form a trapezoid. This trapezoid corresponds to the frame's region of interest (ROI). The ROI is the image area where vehicle recognition happens. In the system case, only part of the street is the region of interest (Figure 2(a)). Vehicles that are far away are harder to identify, because of the proximity of itself and other vehicle. The areas that correspond to the sidewalk and the begin of the street are not of interest for processing. After the trapezoid's definition, the Warping operation is done and the image becomes the whole area corresponding to the trapezoid (Figure 2(b)).

1. Warping of the region of interest of the frame
2. Image conversion in grayscale
3. Background removal
4. Image binarization
5. Eroding and Dilation application
6. Contour areas detection
7. Count of the areas

Following, the image conversion in grayscale is performed, a necessary step to achieve better results in posterior steps (binarization e.g.). After that, the image background is removed, to identify only the moving objects (vehicles). This operation is performed by the OpenCV's function *cvAbsDiff*. This function performs the subtraction of two images (original e background), resulting the foreground. After that, the binarization operation is performed, so the image is segmented with 128 threshold, resulting in a black and white image (Figure 3(a)).

The mathematical morphology operations, eroding e dilation, are executed in sequence to eliminate isolated "spots" that do not make part of a vehicle and to group the "spots", respectively, that make part of the same vehicle, but somehow are separated (Figure 3(b)).

The localization of white "spots" contours is the next step of the algorithm. It is done by the OpenCV's function *cvFindContours*. As it was locating the "spots" frame by frame, it was able to perform the vehicle count by following:

(a) Original Image (b) Image after Warping operation

Fig. 1. Warping Operation

(a) ROI marked by red trapezoid (b) Image after Warping operation of the ROI

Fig. 2. Warping of the region of interest of the frame

when a new vehicle "is born" in the frame, a rectangle is designed and the vehicle is put within it. After that, the vehicle is followed until the last frame which the "spot" appears. By the moment the rectangle is designed, it is calculated the rectangle centroid or its central point (see Formula 1). When a centroid crosses the counting line, a new vehicle is registered and counted.

$$xCentroid = \frac{w}{2} + x;$$

$$yCentroid = \frac{h}{2} + y;$$

Formula 1. Point (xCentroid, yCentroid) of the rectangle's centroid.

The centroid is a point (xCentroide, yCentroide), and this point is calculated by the formula above, where (x,y) is the far left and high point of the "spots" rectangle, w is the rectangle width and h is the height.

(a) Binary image of moving vehicles (b) Binary image after the morphological operations

Fig. 3. Result of morphological operations in a binary image

After the "birth" of a "spot", it is attributed an id to it. In the next frame, it is checked if this centroid is located in the rectangle interior of some "spot" in the previous frame. If it is, that's because it's the same "spot", than it is the same id. If it's not, than it's a new one, than it's attributed a new id for it.

2.1 Shadow Removal

Shadows can cause some problems in background subtraction and vehicle detection, as you can see in Section 3 (Results and Discussion) of this paper. The detection of cast shadows as foreground objects is very common, producing undesirable consequences: shadows can connect different people walking in a group, generating a single object as output of background subtraction [8], but there's a technique in literature that can eliminate them of the grayscale image. Jacques [8] affirms that the normalized cross-correlation (NCC) can be useful to detect shadow pixel candidates.

As Jacques [8] described, $B(i,j)$ is the background image and $I(i,j)$ is an image of the video sequence. For each pixel (i,j) belonging to the foreground, consider a $(2N + 1)$ X $(2N + 1)$ template T_{ij} such that $T_{ij}(n, m) = I(i + n, j + m)$, for $-N \leq n \leq N$, $-N \leq m \leq N$ (T_{ij} corresponds to a neighborhood of pixel (i,j)). Then, the NCC between template T_{ij} and image B at pixel (i,j) is given by Formula 2:

$$NCC(i,j) = \frac{ER(i,j)}{E_B(i,j)E_{Tij}},$$

where

$$ER(i,j) = \sum_{n=-N}^{N} \sum_{m=-N}^{N} B(i+n, j+m)T_{ij}(n,m),$$

$$E_B(i,j) = \sqrt{\sum_{n=-N}^{N} \sum_{m=-N}^{N} B(i+n, j+m)^2}, and$$

$$E_{Tij} = \sqrt{\sum_{n=-N}^{N} \sum_{m=-N}^{N} T_{ij}(n,m)^2}.$$

Formula 2. Normalized cross-correlation in a point (i, j).

A pixel *(i,j)* is pre-considered shadow if:

$$NCC(i,j) \geq L_{ncc}$$

and

$$E_{Tij} < E_B(i,j)$$

Formula 3. Condition for a point (i, j) to be pre-considered shadow.

where L_{ncc} is a fixed threshold. In this system, $N = 4$ and $L_{ncc} = 0.999$. Then, pixels that are false positive are eliminated. This stage consists of verifying if the ratio I(i, j)/B(i, j) in a neighborhood around each shadow pixel candidate is approximately constant, by computing the standard deviation of I(i, j)/B(i, j) within this neighborhood. More specifically, we consider a region R with $(2M+1)(2M+1)$ pixels (it was used $M = 1$ in all experiments) centered at each shadow pixel candidate *(i, j)*, and classify it as a shadow pixel if:

$$std_R(\frac{I(i,j)}{B(i,j)}) < L_{std}$$

and

$$L_{low} \leq (\frac{I(i,j)}{B(i,j)}) < 1$$

Formula 4. Condition to a point (i, j) to be a shadow pixel.

where $\mathrm{std}_R(\frac{I(i,j)}{B(i,j)})$ is the standard deviation of quantities I(i,j)/B(i,j) over the region R, and L_{std}, L_{low} are thresholds. In this system, it was used $L_{std} = 0.05$ and $L_{low} = 0.3$. Then it was applied morphological operations to enhance the results. As experiment, it was used a frame with three "spots", and after the shadow removal algorithm has been applied, the "spots" decreased the area in 25%, 18% and 19%, as it can be seen in Figure 4.

Despite this results, the shadow removal algorithm was not used in the main one because it demands much processing time (it takes some seconds to process a single frame), and one of the goals of the system is to have a low processing cost. So the algorithm will be enhanced, so in the future it can be used in the main algorithm.

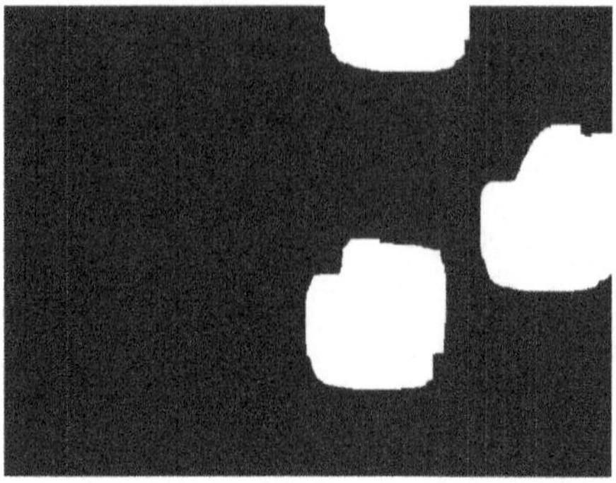

(a) "Spots" before shadow removal algorithm

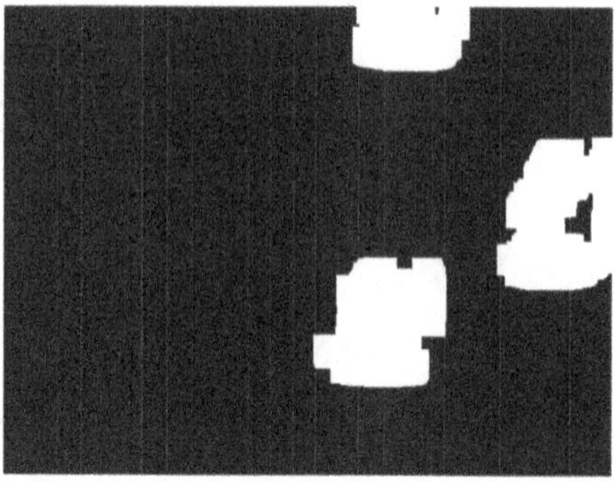

(b) "Spots" after shadow removal algorithm

Fig. 4. Result of shadow removal

3 Results and Discussion

Figure 5 shows a frame completely processed, so it was performed all the described operations before, needed to recognize vehicles. In the Figure 5, six vehicles are shown, all identified (with rectangle and centroid), but only two were counted, both inferior, as only their centroids were below the counting line (green colour). In the next frame, if these same vehicles are still in the frame, they are not counted again. If other vehicle's centroid crosses the line, it will be counted one more vehicle.

It was performed tests with three video sample. The results are in the Table 1. It was computed the processing times of vehicle detection for each frame. The PMAI column of the table shows the biggest time taken by a frame to detect all vehicles. PMEN is the lowest time and PMED, the mean time of all frames of the sample. The NVT column of the table is the total number of vehicles the is shown in the video sample, independent if the vehicle crossed the counting line or not. NVM is the total number of vehicles that crossed the counting line, under manual count. The number of vehicles that crossed the green line that were counted by the algorithm can be seen by the NVA column.

Table 1. Sample results

Sample	Duration	Number of frames	PMAI	PMEN	PMED	NVT	NVM	NVA
1	15 s	450	41 ms	5 ms	9 ms	23	21	21
2	30 s	900	32 ms	5 ms	5 ms	27	22	23
3	60 s	1800	49 ms	5 ms	7 ms	74	69	63

The sample 1 showed a processing mean bigger than all samples, but all the vehicles that crossed the counting line were counted. Just two vehicles stayed

Fig. 5. Recognition of vehicles in a frame

above the counting line, so that they were not registered. In sample 2, video duration was bigger, but the processing time for the recognition of the vehicles was lower, considering that its mean was 5 milliseconds, against 9 milliseconds for the first sample. It was counted 27 vehicles in total, but only 22 crossed the counting line.

The sample 3 had bigger video duration, and still a processing time mean lower than the sample 1. That means that sample 1 reached a bigger number of "spots" by frame than all others samples, but by the time the time duration was lower than the others, the number of vehicles also was lower.

Still about the sample 3, it was counted 74 vehicles manually, but only 69 could be counted because they were below the counting line. The algorithm counted 63 vehicles below the line.

Some error cases in the algorithm were found, such:

- "Spots" inside others are not considered;
- "Spots" with small areas, that are not vehicles, are not counted;
- "Spots" with large width (likely two cars side by side that were not separated), the area was divided in two, so it will count two vehicles;
- Vehicles that "are born" with two "spots" and than become one, is counted only once (the algorithm counts two times because of the two initial "spots");
- Vehicles that "are born" with one spot and then becomes two, are counted only once.

Nevertheless, in this last case, there is a problem: when two vehicles are too close to each other in the "birth", only one "spot" is considered because of the shadow of both. After a while, much times the "spot" of two vehicles splits. So, by the previous rule, it will not count the second car as a new one, considering that only one "spot" in the beginning.

3.1 Tests

To test the accuracy of the algorithm, it was created a comparison mechanism of the manual and the algorithm count. Given a short interval of time, five seconds e.g., it is counted the number of vehicles, both manually and by the algorithm, that crosses the counting line. Then these counts are compared and it is obtained some indicators, like: difference between the manual and the algorithm counts, the total number of errors of the algorithm, the interval of time that has the larger number of errors, absolut mean, relative mean, variance and standard deviation of the errors, among others.

Some of these results and indicators are listed below in Table 2:

Table 2. Tests results

Sample	Duration	Number of Frames	NIE	TNE	AME	RME	Variance	SD
1	60 s	1800	7	8	1.14	0.51	0.12	0.35
2	160 s	4800	27	78	2.89	0.64	4.54	2.13

NIE column represents the number of intervals that have errors in the whole video sample. TNE is total number of errors in all intervals. AME and RME are the absolute and the relative mean of errors that a interval has, respectively. SD is the standard deviation of all errors in the video sample.

4 Conclusion

This paper presents a systems with low processing cost algorithm and success rate of 90% like showed in the three video samples in Table 1. Also it has good perspectives for the future to enhance the algorithm, considering that shadow removal technique achieved good initial results.

To obtain a bigger success rate, the ideal view would be a camera in the vertical position or 90 degrees angle towards the street and not diagonal, like presented in the video samples. So that, vehicles very close to each other would not be connected in a non appropriated way.

The disadvantages of the algorithms are: the need to obtain a background image manually and the difficulty to separate vehicles that are very close to each other. For future works, it is proposed an algorithm to eliminate the image background automatically and to integrate the shadow algorithm to the main one.

References

1. Haupt, A.G.: Detecção de Movimento, Acompanhamento e Extração de Informações de Objetos Móveis. Universidade Federal do Rio Grande do Sul (2004)
2. Gupte, S., Masoud, O., Martin, R.F.K., Papanikolopoulos, N.P.: Detection and Classification of Vehicles. IEEE Transactions on Intelligent Transportation Systems (2002)
3. Tan, X., Li, J., Liu, C.: A video-based real-time vehicle detection method by classified background learning. World Transactions on Engineering and Technology Education (2002)
4. Purnama, I.K.E., Zaini, A., Putra, B.N., Hariadi, M.: Real Time Vehicle Counter System for Intelligent Transportation System. In: International Conference on Instrumentation, Communications, Information Technology, and Biomedical Engineering (2009)
5. Daigavane, P.M., Bajaj, P.R.: Real Time Vehicle Detection and Counting Method for Unsupervised Traffic Video on Highways. International Journal of Computer Science and Network Security (2010)
6. OpenCV Official Homepage, `http://opencv.willowgarage.com/wiki/` (last access: July 26, 2012)
7. Bradski, G., Kaehler, A.: Learning OpenCV - Computer Vision with the OpenCV Library. O'Reilly Publisher (2008)
8. Jacques Jr., J.C.S., Jung, C.R.: Background Subtraction and Shadow Detection in Grayscale Video Sequences. In: Proceedings of the XVIII Brazilian Symposium on Computer Graphics and Image Processing (2005)

Symbolic Classification of Traffic Video Shots

Elham Dallalzadeh[1], D.S. Guru[2], and B.S. Harish[3]

[1] Department of Computer, Marvdasht Branch, Islamic Azad University, Marvdasht, Iran
elhamdallalzadeh@gmail.com
[2] Department of Studies in Computer Science, University of Mysore, Manasagangothri, Mysore - 570 006, Karnataka, India
dsg@compsci.uni-mysore.ac.in
[3] Department of Information Science and Engineering, S J College of Engineering, Mysore, Karnataka, India
bharish@ymail.com

Abstract. In this paper, we propose a symbolic approach for classification of traffic video shots into light, medium, and heavy classes based on their content (congestion). We propose to represent a traffic video shot by an interval valued features. Unlike the conventional methods, the interval valued feature representation is able to preserve the variations existing among the extracted features of a traffic video shot. Based on the proposed symbolic representation, we present a symbolic method of classifying traffic video shots. The symbolic classification method makes use of a symbolic similarity measure for classification. An experimentation is carried out on a benchmark traffic video database. Experimental results reveal the efficacy of the proposed symbolic classification model. Moreover, it achieves classification within negligible time as it is based on a simple matching scheme.

Keywords: Traffic congestion, classification of traffic video shots, symbolic representation, interval valued features, symbolic similarity measure, symbolic classifier.

1 Introduction

Traffic congestion is a serious issue in many urban streets and highways. In order to reduce the traffic congestion, traditional solutions have been employed that are based on increasing the supply of roads. However, such solutions are costly. Furthermore, it is not always possible to employ because of the lack of suitable lands. Recently, interest has been grown up in optimizing the throughput of the existing roads using vision-based traffic monitoring systems.

Vision-based traffic video monitoring systems help us in gathering statistical data on traffic activity through monitoring the density of vehicles. Furthermore, it assists taking intelligent decisions and actions in any abnormal conditions by analyzing the traffic information. Hence, an intelligent automated monitoring system can detect

D. Nagamalai et al. (Eds.): *Adv. in Comput. Sci., Eng. & Inf. Technol.*, AISC 225, pp. 11–22.
DOI: 10.1007/978-3-319-00951-3_2

traffic jams and divert vehicles from congested roads to less crowded ones. On the other hand, it is a key task to collect information on traffic flows in real-time.

Most of the information required for classification of traffic video shots can be based on motion that each traffic video contains. Thus, a holistic representation can be used to capture the variability of the motion without the need for segmenting or tracking individual components.

Yu et al., [2] presented a novel algorithm to directly extract highway traffic information such as average vehicle speed and density from MPEG compressed Skycam video. The MPEG motion vector field is filtered to remove vectors that are not consistent with vehicle motion. The traffic flow is then estimated by averaging the remaining motion vectors.

Porikli and Li [3] proposed an unsupervised, low-latency traffic congestion estimation algorithm that operates on the MPEG video data. Congestion features are directly extracted from the compressed domain. Gaussian Mixture Hidden Markov Models (GM-HMM) is employed to detect traffic congestion. Traffic patterns are detected as five traffic patterns (empty, open flow, mild congestion, heavy congestion, and stopped).

A framework to collect traffic flow information from urban traffic scenes is proposed by Lee and Bovik [1]. The traffic flow is presented by defining traffic regions. Basic statistics of traffic flow vectors inside the traffic regions are then computed.However, extracting reliable measurements of flow is difficult in traffic scenarios due to environmental conditions. Moreover, the extracted measurements are subjected to noise.

Chan and Vasconcelos [4] proposed an approach to model the entire motion field as a dynamic texture. It is an auto-regressive stochastic process with both spatial and temporal components. The auto-regressive stochastic process encodes the appearance and the underlying motion separately into two probability distributions. Distances among dynamic textures are computed using information theoretical measures of divergence (Kullback-Leibler divergence) between the associated probability distributions or through geometry measures based on their observable space (Martin distance). With these distance measures, the traffic congestion is classified using a Nearest Neighbor (NN) classifier or by the use of a Support Vector Machine (SVM) classifier with the Kullback-Leibler kernel. However, the proposed model is computationally complex. It cannot be well-adopted for on-line and real-time estimation of traffic congestion.

Derpanis and Wildes [5] described a system based on 3D spatio-temporal oriented energy model of dynamic texture. Traffic patterns are classified in terms of their dynamics measures aggregated over regions of image space-time. For such purposes, local spatio-temporal orientation is derived. Each traffic scene is then associated with a distribution (histogram) of measurements. Classification is performed based on matching such distributions (or histograms) of space-time orientation structure. However, heavy traffic in which the traffic stopped completely cannot be distinguished from light traffic with almost an empty (stationary) roadway.

The above mentioned issues motivate us to propose a simple yet efficient model to classify traffic video shots based on their content (congestion). We outline a novel method of representing and classifying traffic video shots using symbolic data analysis concepts.

The remaining part of the paper is organized as follows. In section 2, we discuss the proposed model for symbolic representation and classification of traffic video shots. The details of an experimentation conducted to demonstrate our proposed model on UCSD benchmark traffic video database are given in section 3. The paper is concluded in section 4.

2 Proposed Model

In this section, we outline a novel method of representation and classification of traffic video shots based on their content (congestion) using symbolic data concepts. Initially, each traffic video shot is filtered using Gabor filter to segment the texture content of its frame sequence. We make use of motion, appearance, and texture features as the conventional features to construct feature matrices for the frame sequence of a traffic video shot.

However, the extracted feature matrices contain lots of intra-class variations. On the other hand, there exists low inter-class variations among such feature matrices extracted from various traffic video shots of different classes. To capture the intra-class variations as well as representing traffic video shots of different classes effectively, an interval valued feature vector representation is formulated to represent each traffic video shot. Traffic video shots are then classified based on symbolic similarity measure.

2.1 Pre-processing

Given a frame sequence of a traffic video shot, the pre-processing step is done to segment the content of the frames by their texture. A Gabor filter [8] with different frequencies along with predefined orientations is applied on the RGB channels of each of the frames of a traffic video shot. Each of these filters gives an image as a two-dimensional array of the same size as the input frame. The magnitude of each of the Gabor filtered images is taken and added to each other to get a final filtered image for a given frame. The obtained filtered image is then mapped on the given frame to segment its texture content. The margin of the segmented frame is then cropped to have only the content of the frame.

The above mentioned pre-processing step is employed to segment the content of the frame sequences of traffic video shots. The segmented texture content of the sample frames of three different traffic video shots are shown in Fig. 1.

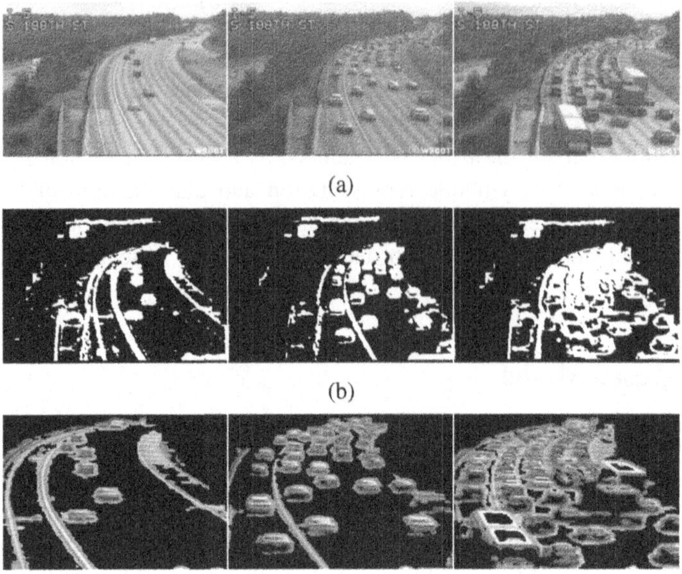

(a) (b) (c)

Fig. 1. Frame segmentation of traffic video shots. (a) Sample frames of light, medium, and heavy traffic, (b) Final Gabor filtered images of the frames, (c) Segmented texture content of the frames after mapping the filtered images on the sample frames and cropped the margin of the segmented frames.

2.2 Symbolic Feature Representation

To gather the statistical data on a traffic activity, the content of traffic video shots can be represented by motion. However, the motion of traffic video shots has a limited ability to distinguish between heavy traffic (almoststopped traffic) and lighttraffic(almost empty/stationary roadway). The estimated motion of such traffics is nearly the same. Moreover, the estimated motion of heavy traffic with slight speed cannot be well classified from the medium traffic at reduced speed.

Therefore, we propose to represent the content of a traffic video shot by appearance and texture in addition to motion. Hence, motion, appearance, and texture features of the content of traffic video shots are extracted.

Features Based on Motion

Motion estimation is the process of determining motion vectors that describe the transformation from one 2D image to another. We propose to estimate the motion of a traffic video shot using a global pixel-based method by considering the whole frame as one block. Sum of absolute differences (SAD) is considered as an evaluation metric to compute the motion of a frame with respect to its previous one.

Let $\{F_1, F_2, F_3, \ldots, F_R\}$ be a set of R frames of a traffic video shot. Considering the kth frame, say F_k, the motion of a frame F_k with respect to the previous frame F_{k-1} is calculated as,

$$MF_k = \sum_{x=1}^{u} \sum_{y=1}^{v} \left| F_k(x,y) - F_{k-1}(x,y) \right| \tag{1}$$

where, $F_k(x,y)$ and $F_{k-1}(x,y)$ are the intensity values of the pixels at (x,y) of the frames F_k and F_{k-1} respectively. Also, u and v define the size of the frames.

Therefore, a column vector called motion feature vectorsay MFof size R-1, considering the adjacent frames, is obtained.

In order to have a compact representation of the obtained motion feature vector in addition to capturing the variations of the motion feature values, we propose to represent the motion feature vector in the form of an interval valued feature vector. Though, it shall be noted that motion is a dominant feature for content representation of a traffic video shot.

Hence, we propose to symbolically represent the motion feature vector by preserving the variations of the features as well as conserving the domain of the features. So, we end up having an interval valued feature vector of dimension 2 representing motion of a traffic video shot.

An interval valued motion feature preserving the variations among the motion feature values is represented as,

$$\text{Motion\_Interval1} = \left[MF^-, MF^+ \right] \tag{2}$$

where,

$$MF^- = \mu(MF) - \tau(MF) \quad \text{and} \quad MF^+ = \mu(MF) + \tau(MF) \tag{3}$$

where, $\mu(MF)$ is the mean of the motion feature values, $\sigma(MF)$is the standard deviation of the motion feature values, and $\tau(MF)$is the function of $\sigma(MF)$given by $\tau(MF) = \alpha \times \tau(MF)$ for some scalar value αthat is set empirically.

Similarly, another interval valued motion feature capable of conserving the domain of the motion feature values is obtained as,

$$\text{Motion\_Interval 2} = \left[MF^-, MF^+ \right] \tag{4}$$

where,

$$MF^- = \min(MF) \quad \text{and} \quad MF^+ = \max(MF) \tag{5}$$

where, min(MF)andmax(MF)are the minimum and maximum of the motion feature values.

Hence, the motion feature vector is symbolically represented by two interval valued features named as Motion_Interval1 and Motion_Interval2.

Features Based on Appearance

To represent the distribution of the data values in a traffic video shot, a histogram of each of the frame is extracted. Considering the kth frameF_k, the RGB histograms of the frame F_k are represented by $\mathcal{H}$ number of gray levels, in here 256 gray levels, as the number of bins. The represented color channel histograms are assimilated and represented as a single appearance distribution vector for the frame F_k by considering the maximum frequency of the color channels.

A matrix called appearance feature matrix, say AFof size R×$\mathcal{H}$ (here R×256), is obtained. Each row is an appearance distribution vector representing its respective frame.

Instead of keeping this huge feature matrix, we recommend capturing the variation of the entire matrix in the form of an interval valued. Thus, we end up having an interval valued feature representing the appearance of a traffic video shotgiven by,

$$\text{Appearance\_Interval} = \left[AF^{-}, AF^{+} \right] \tag{6}$$

where, AF^{-} and AF^{+} are the limits of the interval computed as in Equation (3).

Hence, the appearance feature matrix is symbolically represented by an interval valued feature called Appearance_Interval.

Features Based on Texture

A statistical method of examining textures that consider the spatial relationship of pixels is the gray level co-occurrence matrix (GLCM), also known as the gray level spatial dependence matrix. The GLCM matrix can reveal certain properties about the spatial distribution of the gray levels in the texture image. The GLCM function calculates how often a pair of pixels with specific intensity (gray level) values and in a specified spatial relationship occur in an image [9].

We use a GLCM functionwhich creates multiple GLCMs for a single input frame. Thus, an array of offsets for the GLCM function is identified. These offsets define pixel relationships of varying directions and distances. After generating the GLCMs, four statistical textural features [9] such as contrast, correlation, energy, and homogeneity are derived.

Considering the kth frameF_k, the four GLCMs of the frame F_k are obtained by defining an array of offsets specifying four directions (0°,45°, 90°, 135°) with distance 1. For each of the obtained GLCMs, the statistical textural features such as contrast, correlation, energy, and homogeneity are calculated. In addition to these statistical textural features, the entropy of the frame F_k is also estimated as an added feature, thereby totally 17 features.

In general, $\mathcal{T}$number of statistical texture features are extracted for the frame F_k. So, a texture vector of size $\mathcal{T}$ is created (here $\mathcal{T}$=17). Furthermore, it has to be noticed that all the 17 statistical texture features are of different nature.

A matrix called texture feature matrix, say TFof size R×$\mathcal{T}$ (here R×17), is obtained.

To compress the representation of this matrix and to capture the variations of the extracted features, we recommend capturing the variations in each column in the form of an interval valued. Thus, we end up having 17 interval valued features given by,

$$\text{Texture\_Interval\_Vecor} = \left[TF_1, TF_2, ..., TF_{17} \right] \tag{7}$$

where, each TF_i, i =1, 2, ..., 17, is an interval valued feature computed as in Equation (3) for the respective column of TF.

2.3 Traffic Video Shots Representation and Classification

We propose to represent the traffic video shots in the knowledgebase by their reference feature vectors. The reference feature vectors are represented by assimilating the obtained interval valued features discussed in subsection 2.2. The reference feature vectors are used in classification.

Given a test sample traffic video shot, its similarity values with respect to all the reference feature vectors stored in the knowledgebase are computed. The test sample traffic video shot is then labeled by the class label of a traffic video shot with a maximum similarity.

Representation

Let there be O number of traffic video shots to be represented in the knowledgebase. A traffic video shot, say T_j (j =1, 2, 3, ..., O), is represented by a reference feature vector, say Ref_jgiven by,

$$Ref_j = \left(\text{Motion\_Interval 1, Motion\_Interval 2, Appearance\_Interval, Texture\_Interval\_Vector} \right) \tag{8}$$

Hence, a vector of interval valued features of dimension 20 is formed. Similarly, each traffic video shot is represented by its reference feature vector. Thus, O number of reference feature vectors are created and stored in the knowledgebase.

Classification

We propose to compute the similarity of a given test sample traffic video shot, say T_t, with respect to all the reference feature vectors stored in the knowledgebase by the use of symbolic similarity measure [6]. The similarity is estimated between the feature vector of a test sample and the reference feature vectors stored in the knowledgebase.

Thus, it is proposed to represent a test sample traffic video shot T_t by a crisp feature vector, say VF.

$$VF = \left(\text{Motion\_mean, Motion\_median, Appearance\_mean, Terxture\_mean\_Vector} \right) \tag{9}$$

where, Motion_mean is the mean of the motion feature vector, Motion_meadian is the median of the motion feature vector, Appearance_mean is the mean of the appearance feature matrix, and Texture_mean_vector is a vector obtained by computing the mean of each column of the texture feature matrix of the test sample traffic video shot T_t.

We use the symbolic similarity measure (Equation (10) and Equation (11)) to compute the similarity between the feature vector VF and all the reference feature vectors stored in the knowledgebase. The test sample traffic video shot T_t is then labeled by the class label of a traffic video shot with a maximum similarity.

$$\text{Total\_Sim}\left(\text{VF}, \text{Ref}_j\right) = \sum_{k=1}^{20} \text{Sim}\left(f_k, \left[f_{jk}^-, f_{jk}^+\right]\right) \tag{10}$$

$$\text{Sim}\left(f_k, \left[f_{jk}^-, f_{jk}^+\right]\right) = \begin{cases} 1 & \text{if } f_k \geq f_{jk}^- \text{ and } f_k \leq f_{jk}^+ \\ \max\left(\dfrac{1}{1+\left|f_k - f_{jk}^-\right|}, \dfrac{1}{1+\left|f_k - f_{jk}^+\right|}\right) & \text{otherwise} \end{cases} \tag{11}$$

where, f_k defines the kth feature value of VF, and $[f_{jk}^-, f_{jk}^+]$ defines the kth interval valued feature of the jth reference feature vector Ref_j.

3 Experimentation

In this section, we present the details of an experimentation conducted to demonstrate our proposed model for classification of traffic video shots on UCSD benchmark traffic video database. The traffic video shots are classified into light, medium, and heavy classes based on their content (congestion). Results of classification of traffic video shots using our proposed model are also presented.

3.1 Dataset

The UCSD traffic video database is used in this paper [7]. It consists of video sequences of daytime highway traffic in Seattle, Washington, totaling 20 minutes of video shots The traffic video shots were collected from a single stationary traffic camera over two days. The video shots contain a variety of traffic congestion patterns and weather conditions (e.g., raining, overcast, and sunny). Each video shot has a color resolution of 320×240 pixels with 42 to 52 frames captured at 10 frames per second. Also, hand-labeled ground truth is provided that describes the amount of traffic congestion in each sequence. In total, there are 254 video sequences, grouped into three classes of traffic congestion, i.e., light, medium, and heavy; see Table 1 for summary. Example frames from the database are as shown in Fig. 2.

Table 1. The UCSD Benchmark Traffic Video Database Summary

Traffic Condition	Description	Total No. of Traffic Video Shots
Light	Traffic around the speed limit; Free flowing traffic	165
Medium	Traffic at reduced speed	45
Heavy	Traffic at stopped and go or very slow speed	44

Fig. 2. Example frames from the UCSD benchmark traffic video database. The sample frames depict various traffic congestion categorized as light (top row), medium (middle row), and heavy traffic (bottom row).

3.2 Experimental Results

During experimentation, we conducted five different sets of experiments. In the first set of experiments, we used 40% of traffic video shots of each class to create training shot samples and the remaining 60% of the traffic video shots of each class for testing purpose. On the other hand, in the second set of experiments, the number of training and testing traffic video shots of each class are in the ratio of 50:50. Further, in the third set of experiments, we used 60% for training and 40% for testing. In the fourth set of experiments, the number of training and testing traffic video shots of each class are in the ratio of 70:30, and they are in the ratio of 75:25 in the fifth set of experiments respectively.

All the experiments are run for 5 trials by choosing the training sample traffic video shots randomly. Moreover, in each set of experiments, we used the proposed classification model to classify test sample traffic video shots. As measures of goodness of the proposed model, we computed accuracy, precision, recall, and F-measure.

The minimum, the maximum, and the average values of the classification accuracies of all the 5 trials are shown in Fig. 3. Further, the computed precision, recall, and F-measure of the trial responsible for maximum classification accuracy for all the sets of experiments are presented in Fig.4(a) and Fig.4(b) respectively.

Fig. 5shows several classification examples when the system is trained by experiment 5 (75:25). In Fig. 5(a), the first row shows several frames from the given test sample traffic video shots due to light traffic video shot classes, and the second row shows the frames of the corresponding nearest traffic video shots obtained. Similarly, in Fig. 5 (b) and Fig. 5 (c), the first rows present several frames from the given test sample traffic video shots due to medium and heavy traffic video shot classes, and the second rows present the frames of the corresponding nearest traffic video shots obtained.

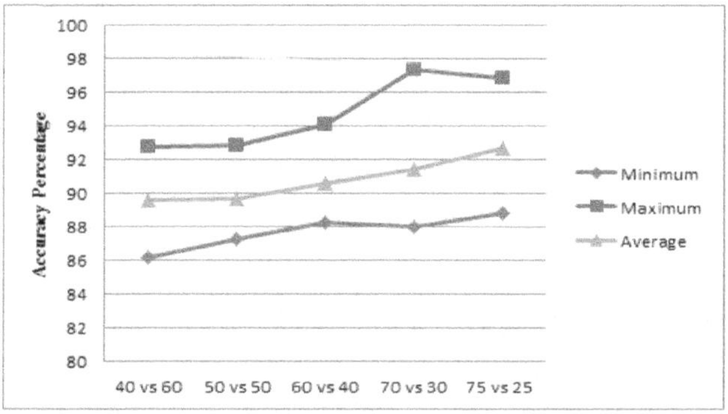

Fig. 3. Minimum, maximum, and average classification accuracy values on five different sets of experiments by the proposed model for classification of traffic video shots

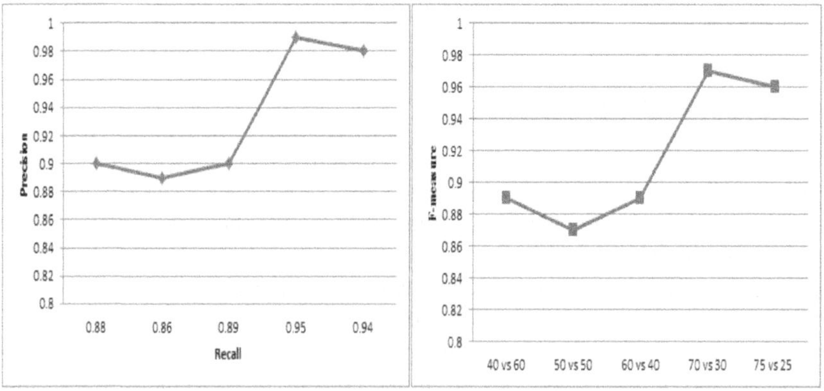

(a) (b)

Fig. 4. Precision-recall curve and F-measure of the trials responsible for maximum classification accuracies by the proposed modelfor classification of traffic video shots. (a) Precision-recall curve, (b) F-measure.

(a)

(b)

(c)

Fig. 5. Classification examples of traffic video shots using the proposed model for classification of traffic video shots on experiment 5 (75:25). (a) Light traffic, (b) Medium traffic, (c) Heavy traffic.

4 Conclusion

In this paper, we presented a classification model to classify traffic video shots based on traffic congestion. After feature extraction, simple yet effective representation and classification schemes are proposed. We explored a novel interval valued feature representation to represent the content of traffic video shots efficiently. We proposed to capture the variation of the features extracted from the corresponding frame sequence of a traffic video shot through feature assimilation. After feature assimilation and symbolic feature representation, a symbolic similarity measure is applied for classification of traffic video shots.

To corroborate the effectiveness and robustness of the proposed model, experimentation is conducted on the benchmark database. The obtained results reveal the efficiency of our proposed model for classification of traffic video shots based on their content (congestion).

The proposed model is a stand-alone system and first of its kind in literature for symbolic classification of traffic video shots. It can be expected to open up a new dimension for further research in the field of on-line and real-time vision-based traffic monitoring system by the notions of symbolic data representation.

References

1. Lee, J., Bovik, A.: Estimation and Analysis of Urban Traffic Flow. In: Proceedings of the 16th International Conference on Image Processing, pp. 1157–1160 (November 2009)
2. Yu, X.-D., Duan, L.-Y., Tian, Q.: Highway Traffic Information Extraction from Skycam MPEG Video. In: Proceedings of the 5th International Conference on Intelligent Transportation Systems, pp. 37–42 (2002)
3. Porikli, F., Li, X.: Traffic Congestion Estimation using HMM Models without Vehicle Tracking. In: Proceedings of the Intelligent Vehicles Symposium, pp. 188–193 (June 2004)
4. Chan, A.B., Vasconcelos, N.: Classification and Retrieval of Traffic Video Using Auto-Regressive Stochastic Processes. In: Proceedings of the Intelligent Vehicles Symposium, pp. 771–776 (June 2005)
5. Derpanis, K.G., Wildes, R.P.: Classification of Traffic Video Based on a Spatiotemporal Orientation Analysis. In: Proceedings of the Workshop on Applications of Computer Vision, pp. 606–613 (January 2011)
6. Nagendraswamy, H.S.: Fuzzy-Symbolic Approaches for Representation and Classification of Two dimensional Shapes. Ph.D. Thesis, University of Mysore, India (2006)
7. `http://www.wsdot.wa.gov`
8. Movellan, J.R.: Tutorial on Gabor Filters (2002)
9. Haralick, R.M., Shanmugam, K., Dinstein, I.: Textural Features for Image Classification. IEEE Transactions on Systems, Man, and Cybernetics, 610–621 (1973)

Implementing IT Security for Small Businesses within Limited Resources

Margie S. Todd[1], Syed (Shawon) M. Rahman[2], and Sevki Erdogan[3]

[1] School of Business & Technology, Capella University, Minneapolis, MN, USA
mtodd11@capellauniversity.edu
[2] University of Hawaii-Hilo, Hilo, USA and Capella University, Minneapolis, USA
SRahman@Hawaii.edu
[3] Dept. of Computer Science, University of Hawaii-Hilo, Hilo, HI 96720
Sevki@Hawaii.edu

Abstract. The purpose of this paper is to present a comprehensive budget conscious security plan for smaller enterprises that lack of security guidelines. We believe this paper will assist users to write an individualized security plan. We have also provided the top ten tools that are either free or affordable to get some sort semblance of security implemented.

Keywords: IT security, small business, security plan, risk planning.

1 Introduction

Enterprise security has grown in complexity over the last decade. The goal of this publication is to show the proper methods of creating, implementing, and enforcing an enterprise information security plan. During the initial phase of the plan, the assessment, it is imperative that the shareholders of the enterprise identify the most valuable assets that need protection. The information technology department must be part of this determination so that strategies for network vulnerability mitigation can begin to emerge. Once there is a clear image of what it is that needs the most protection then costing analysis can start. That is why identifying all possible vulnerabilities is critical so that if cost is an issue we can start prioritizing those assets in terms of what risks can the business afford versus what policies must be in place to avoid the exposure of the asset to the risk by curbing employees' current behaviour.

Once the list of assets is solidified, begin conducting real time threat assessments keeping in mind that there are external (via Internet) as well as internal (employees) threats that must be considered. Each threat to our assets comes with its own vulnerability areas. Once the threats are identified we almost have to become criminal-minded to identify how much vulnerability accompanies each threat. Share the list with the shareholders and our management staff so that different perspectives are taken into consideration. Some of the vulnerabilities may be averted by creating policies for permissible network usage.

Write a comprehensive security plan for the enterprise and add it to our Business Continuity Plan and our Disaster Recovery Strategy. A comprehensive plan

D. Nagamalai et al. (Eds.): *Adv. in Comput. Sci., Eng. & Inf. Technol.*, AISC 225, pp. 23–32.
DOI: 10.1007/978-3-319-00951-3_3

encompasses a clearly identified risk list with priorities and the impact of each security break in terms of how it affects the business. Supplement our security plan with a logging mechanism for all security events (real and perceived). If we create an event ticket tracking system it will encourage all employees to participate in the process thus reducing the time from reporting to containment. Be prepared to explain to the shareholders our rationale for our recommendations of one tool versus the other. The top ten affordable solutions are offered to minimize the initial cost of implementing some of the security risks by utilizing available tools from trustworthy sources to assist in software and hardware risk mitigation.

The next phase involves creating policies and procedures that support the security plan to be implemented. The policies must be polished and scrutinized so that we adhere to federal and state laws to ensure employee's rights are not violated; a feat somewhat easier to accomplish on the private sector. Once the policies are written and approved, then procedures can be written on carrying out the plan, the policies, and the consequences employees will face for failure to comply with the new secure plan. Once we are ready to implement this plan, and realizing that change is rather difficult, begin by conducting small trial runs and modify the plan and procedures accordingly. A proper implementation plan must be accompanied by a thorough training curriculum, although our goal is to devise a plan that is comprehensive yet simple enough that anyone in the corporation can easily follow it.

2 Top 10 Free or Nearly Free Security Measures to Implement Right Now

Many companies think of dollars signs when anyone mentions the word "security." If companies knew that the top 10 steps of basic security cost nothing at all – it costs nothing at all because it is something that, although in limited supply, still exists: common sense. Allow us to elaborate by using the top 10 security audit practices provided by itsecurity.com as a checklist to secure our small business.

1) Know our equipment – we do not know what we need to secure and how unless we have a list of all our assets (including port numbers, ip addresses, printers, scanners, etc.) Name our assets so that they make sense to us and make auditing and inventorying easy. For instance, we are currently migrating from XP to Windows 7. One clever idea that we have always used is to name machines with the user's initials followed by the operating system and the year it was built. For example, MT711. Simply by looking at our network neighbourhood we can immediately tell what operating system they are on as well as the age of the device. In this case our machine is a Windows 7 machine and it was just replaced this year. Why is this important? First of all we can keep track of the migration status as well as we can tell when something does not belong on that list.

2) Contain our threats by staying a step ahead – identify all possible ways our network can be compromised and develop a plan to combat it and in the worst case scenario make sure our crisis management plan includes a plan of action should our security is compromised.

3) Study our past so we will not make the same mistakes in the future – studying past threats can help us to predict future ones. Use past mistakes as a training tool for our own users and our own support staff. Do not spend resources guarding the wrong assets. For instance, we get no walk-in traffic. All traffic is signed on and by appointment only. The computer room is not even in site of visitors. Why spend a ridiculous amount of money on cardkey access to the computer room if the threat is not there. We do spend money on a great e-mail filtering system since we handle close to 1 million e-mails a year so phishing and trojans are a definite concern for us.

4) Prioritize our security concerns – as mentioned above physical security is not a threat, however, is there a slight chance an employee has access to the computer and may take a spare monitor out of the room? Of course. However, the threat is so small that on the list of concerns this is probably near or at the bottom. All servers are locked so there is no chance they would sit at a server and use it. It does not mean that the computer is a free for all and everyone comes in and out but the real threat of a security incident is near zero.

5) Control access to our network – another free tool since we can design our own access control list for free. Make good use of groups and permissions. Any employee requiring access via Internet must pass one of our IT audits in order to connect. It must be done from an employer provided machine, must have the proper Cisco software and certificates installed and must abode by our strict computer and network usage policy. We do audit who logs on remotely and we record transactional logs of activity. We have a gentleman in upstate New York who has trouble sleeping and he is the only one we see logged on at 1 or 2 am. We did advise him that once 3:00 am comes he must be off the system as that's when backups and antivirus upgrades occur. When we looked at the logs we did call him the first time we saw him logged on at 1:00 am to make sure it was him in fact and not someone logging on as him. He did sound puzzled at the fact that we knew he was logged on – we told him we see everything!

6) Test our firewall – there are some great freebies from our own software vendors that can test our firewalls for intrusions. Some utilities simply log the intrusions while other more advanced ones will react based on the threat.

7) Give only what is necessary – employees will be curious if we let them be. Some are not malicious they are just too curious for their own good. Only give them access to what is required. We do employ NTFS so access to folders is on a per-employee basis. We take it a step further. We have an entire drive that is for manager content only. It simply just does not exist (it is not shared to authenticated users) nor is it mapped to everyone. One recommendation is to have our managers' get into the good habit of locking their screens when they walk away from their desks. This minimizes the threat of someone sitting at a manager's desk and gaining access to some of the sensitive information.

8) Backups, backups, backups – this is so important for security purposes. There is a little dirty secret about backups though that we do not see discuss in any of the material we have read. SQL database backups are fantastic, however, when the software versions are upgraded those backup are no longer compatible with the software until and unless they are converted to the latest database version. Whenever

there is a software upgrade those backups should be restored to a play database, converted, and then archived in the new format again. Why should this matter? A backup is only as good as being able to read and restore the data in it. An interesting suggestion for further research would be to make a study nationwide of how many companies do test their backups and of those, how many are actually able to use them right out of the box?

9) All e-mail is evil – teach users to suspect everyone. It is estimated that each day 55,000,000,000 spam e-mails circulate the cyberglobe (itsecurity.com, 2007). We have a sales manager who loved to read internet articles (for "research") and had the bad habit of sending hyperlinks with nothing else but his signature. He ended up sending a bad link to two employees and thank goodness for anti malware software both of them had the threat stopped. Needless to say we had a conversation with him. Any technical person would have looked at the link and new it was not a valid hyperlink but that's just it – a technical person not an everyday Joe. Definitely it is worthy but the free version works well for a budget conscious company.

10) The Tele-Worker Factor –kids and four legged companions are a big threat to computer equipment (particularly those users who take laptops home). Stress the fact that the company issued laptop is not to be used for Spongebob's revenge, nor is the warm top of the laptop a warm bed for a feline or canine companion. Enforce passwords on all portable devices. Yearly we send out reminders to our home laptop users on proper care of their equipment. We have quarterly company meetings in which all tele-workers come in to the office with their laptops. It is a two day event and while they are meeting we are busy maintaining their equipment. We have it down to a science but we also take the opportunity to visually inspect the laptop for scratches, dents, dust, humidity problems, etc and we help the employee remediate the problem.

3 Physical Security Measures & Rationale

Data theft can occur by electronic means as previously mentioned; however the physical threats to information can arise from within the enterprise. As mentioned in the prior section's table, XYZ Company's physical threats are defined as follows:

1) Natural disasters – although our office is located on the East Coast, far from water and in a traditionally non-tornado zone, last year brought us the reminder that mother nature listens to no one and follows no rules. We had over 70 inches of snow, a tornado touchdown less than 10 miles from us and if that was not enough we even had an earthquake. Needless to say disaster planning is a must. As such, the business is equipped with an A.D.T. alarm system so that if anything were to occur after the building is locked, three of us will be notified immediately. We will immediately deploy the Crisis Management procedures (nowadays known as Disaster Recovery and Business Continuity Plan).

2) Building Break In – our office building is located inside a business park. It is safe to assume that non-business traffic is minimal, and it is the type of business where most (if not all) of us know who drives what type and model of car by simply walking out into the parking lot. Needless to say visitors stick out like a sore thumb.

Our building is closed to the public and there are signs everywhere that visitors must check in and check out at the front desk. As such, there is no reason for a company our size to invest in closed circuit video equipment, or key card access to the building. Our distribution industry (and the type of complex products we distribute) does not lend itself to walk-in business. All visitors must be announced and there is a clearly displayed door sign reading "NO SOLICITORS."

3) Computer Break In- these types can occur in the form of theft. One cannot prevent theft with certainty. Mobile and hand held devices can be forgotten, left behind, stolen, and even destroyed by accident. It is imperative that employees understand the role they play in securing these devices electronically and physically. It is so critical that we have made it a part of the Associate Handbook. Set up strong passwords, provide laptop carrying cases that are easy but secure and encourage employees to always carry their devices in the proper luggage. (Hint: emphasize the potential dangers of having radioactive devices in their pockets).

4) Power Failure – as a smaller company we cannot afford to have all computers connected to Uninterrupted Power Source (UPS) units. Rationally we have critical systems plugged into UPS systems (i.e. servers, firewall, phone system, T1s, packaging machines). We force all users to save their documents on employee shares on the network and we have auto save options set to ON for all PCs. When power failures occur we have the benefit to ascertain whether it is a long term power outage (by contacting the utility company) or if it is just a brown out. If it is a long term outage we then have the benefit of conducting a differential backup and shutting down through proper procedures. Our UPS boxes can sustain backup power for two hours which is ample time to conduct the necessary preparations in the event of a long term down time. Should the power outage occur after hours, ADT will contact all three individuals on the list and we will act accordingly. We have access to the entire network remotely and can shut the system down completely.

4 Information Security Policies and Procedures

The biggest hurdle is creating all encompassing security policies and procedures that will address every situation imaginable. Realizing that it is impossible to be 100% safe all the time, the task will actually become somewhat easier. Start with the most obvious and build on it. We do not want to be the target of every attack before developing a comprehensive plan but we have to start somewhere. Both Microsoft (MCS, 2007) and the C.I.A. have templates of policies that can be adopted and modified to fit each enterprise. The policies will in turn dictate the procedures as these are unique to each company. Our security policies are clearly stated in our Associate Employment Handbook. The procedures are then governed and reviewed by our ISO certified quality program.

Information security policies must be first discussed and agreed to by the entire management staff. The policies set forth must be in tune with the current industry. For instance, let's analyse the following social engineering policy. Even though we say:

"XYZ Company, Inc. is increasingly exploring how online discourse through Social Media can empower XYZ Company, Inc. as an industry leader. It is very much in the XYZ Company, Inc interest to be aware of and participate in this sphere of information, interaction and idea exchange.

The same principles and guidelines that apply to XYZ Company, Inc associates' activities in general also apply to employees' Social Media activities via Facebook, MySpace, LinkedIn, YouTube, Plaxo, Twitter, etc. Social media describes the online technologies and practices that people use to share opinions, insights, experiences, and perspectives. Social media can take many different forms, including text, images, audio, and video. Social Media sites typically use technologies such as websites, blogs, message boards, podcasts, wikis, and blogs to allow users to interact" (XYZ Company, 2012).

On the same policy we explicitly state:

"Personal Social Media activities must not take place during work hours or using XYZ Company, Inc equipment. Refer to the XYZ Company, Inc Computer Policy" (XYZ Company, 2012).

We are not Neanderthals, we obviously understand that social media is here to stay. However, our industry is not well served by social interaction with the public. We provide sealing solutions to complex environmental and chemical problems. In addition to the information not being highly exciting and complicated, there is no value to "liking" XYZ Company on Facebook as it brings no value to the general consumer. If we were in the retail business perhaps then it would benefit us from participating in such activities. There is an inherent danger in social engineering and the spread of worms, trojans, and viruses through participating n them with friends. Why take the risk if the reward is zero?

Notice that we did not mention a social engineering site such as LinkedIn. We do actively participate in LinkedIn as it is a business directory. This particular site is useful to us as we may be able to gain accessibility to a certain business through an associate. LinkedIn has very strict participation rules and no risky web browsing is involved, at least for the time being. The main purpose is that the policy that is developed supports the security plan being implemented. If the enterprise cannot afford the risk then make it a part of the security policy. Another critical portion of said policy is that it states consequences for non-compliance. Employees have to understand that the cost of the risk and potential security break can be detrimental to the business.

5 Security Plan Implementation

It is great to have a plan drafted on paper but it is worth its weight in gold if it is not properly implemented. The biggest hurdle to get over is answering the two questions that burn in everyone's mind when the issue of security policy implementation comes up: Why and How much? It is bothersome how in a post 9/11 world there are still people that ask the why question. In a world riddled with wars, conflicts, never before seen weather phenomenon, an overactive sun, aging equipment, etc we must not question if we will have to deal with a major security event in our lifetime but rather when it will occur. We love answering one question with another. What would happen if a pay check did not come during the next pay period? How about for the next six pay periods? Regardless of who is asked, the answer is probably the same: We don't know. What if there was a plan developed so that mitigation was in place ahead of time should the situation arrive? Think of equating protecting the enterprise to protecting a pay check (in reality that is what it is all about). How much will it cost? We presented information earlier that showed how a basic security plan does not have to be expensive. Most of it can be done with existing resources and relies heavily on common sense and education.

6 Training

As employees continue to be the highest information security threat, we make it a departmental goal to conduct a set number of training sessions per year. Some training is mandatory (such as security) and some is voluntary. The year of 2001 was very defining for many companies as up until then we lived in denial. The September 11th attacks brought about major changes in the industry. Training and awareness became the focal point and continues to be today. Back then we started developing training guides for XYZ Company that encompassed all areas of the business. We had some training in place but we sadly realized we were doing a poor job. We began putting together an IT curriculum that continues in existence to this day and gets reviewed on a constant basis. Every other year has become a training year for every department.

During the year of 2010 we conducted 100 employee training sessions – we spent almost half a business year training as well as dealing with regular IT issues. The year of 2011 became the train-the-trainer year. We partnered with a local IT training school (New Horizons Computer Learning Centers) with a special pricing program that allowed any of our IT employees to attend any online classes any time of the year. We were going to begin migrating to Windows 7 and Windows 2008 R2 software so we took the opportunity to upgrade our skills before deployment. The amount of confidence an employee gains from proper training is priceless. The work gets done more efficiently and deployment becomes a very smooth process. It also brings light to the new security threats and countermeasures based on all the new software changes and the improvement in monitoring that occur with upgrades. It also forces the company to review existing processes to enhance them with newer updated information.

In order to develop ongoing training the first item is to create a matrix in either Microsoft Excel or Microsoft Access. The querying reporting capabilities in Access make tracking XYZ Company employee training a breeze. Comparative reports between departments were created to track progress and ensure compliance. A list of information security training materials at XYZ Company for the Customer Service Department is as follows:

1) Introduction to Computers – a book we wrote on CD format available for employees only that gives then an overview of computers including a hands on lab in which we take a junked computer apart and show them the components and liken them to human organs so that employees are not intimidated by them and learn some of the basics – ultimately humans fear the unknown.

2) Email Etiquette – a comprehensive e-mail training power point presentation in a classroom setting where topics are analysed and demonstrated (i.e. identifying scam e-mails where we show them how hovering over a link without clicking on it can show them the likelihood an e-mail is bogus: We show them an actual junk e-mail that purports to come from Chase Bank when the hyperlink points to some suspected website.

7 Conclusions

The overall goal of this article was to create a budget conscious security plan after a thorough analysis of the enterprise. We believe readers will be able to draft, organize and create a comprehensive security plan by following the recommendations presented. The plan will be comprised of all the necessary components of a thorough enterprise analysis such as: preliminary security assessment, security requirements, security plan, security plan policies, and security procedures. Basic and affordable security monitoring recommendations are also presented to get an enterprise headed in the proper direction to create a culture of security minded employees.

References

1. Byrne, B.: Voip hacking about to hurt small businesses? (June 3, 2011), `http://meship.com/Blog/2011/06/03/voip-hacking-about-to-hurt-businesses/` (retrieved)
2. DePasquale, S.: How to analyze security needs (May 1, 2004), `http://securitysolutions.com/mag/security_analyze_security_needs/index1.html` (retrieved)
3. Harris, S.: CISSP all-in-one exam guide, 5th edn. McGraw-Hill, New York (2010)
4. itsecurity.com. Create your own security audit (January 4, 2007), `http://www.itsecurity.com/features/it-security-audit-010407/` (retrieved)
5. MCS, Musil, S.: Business continuity planning: best practices for your organization (2007), `http://www.mcsmanagement.com/WhitepapersUpload` (retrieved)
6. Microsoft. Access control best practices (2011), `http://technet.microsoft.com/en-us/library/cc778399(WS.10).aspx` (retrieved)
7. Microsoft. Auditing security events best practices (2011), `http://technet.microsoft.com/en-us/library/cc778162(WS.10).aspx` (retrieved)

8. Musil, S.: 'socialbots' steal 250gb of user data in Facebook invasion. Cnet.com (November 1, 2011), http://news.cnet.com/8301-1009_3-20128808-83/socialbots-steal-250gb-of-user-data-in-facebook-invasion/?tag=txt;title (retrieved)
9. onpoint. (n.d.). Incorporating security into the system development life cycle(SDLC), http://www.onpointcorp.com/documents/Security_in_the_SDLC.pdf (retrieved)
10. Xavier, G.: Specifications for cat5, cat6 and cat6e cables (September 18, 2011), http://www.ehow.com/about_6521486_specifications-cat5_-cat6-cat6e-cables.html (retrieved)
11. Whitman, M.E., Mattord, H.J.: Principles of information security, 4th edn. Course Technology Ptr. (2011)
12. Henderson, J., Rahman, S.(Shawon): Working Virtually and Challenges that must be overcome in today's Economic Downturn. International Journal of Managing Information Technology (IJMIT) ISSN : 0975-5586 (Online), 0975-5926 (Print)
13. Dreelin, S., Gregory, Rahman, S.(Shawon): Enterprise Security Risk Plan for Small Business. International Journal of Computer Networks & Communications (IJCNC) ISSN : 0974 – 9322 (Online), 0975- 2293 (Print)
14. Donahue, K., Rahman, S.(Shawon): Healthcare IT: Is your Information at Risk? International Journal of Network Security & Its Applications (IJNSA) 4(5) (September 2012) ISSN:0974-9330 (online), 0975-
15. Rice, L., Rahman, S.(Shawon): Non-Profit Organizations' need to Address Security for Effective Government Contracting. International Journal of Network Security & Its Applications (IJNSA) 4(4) (July 2012)
16. Neal, D., Rahman, S.(Shawon): Video Surveillance in the Cloud? The International Journal of Cryptography and Information Security (IJCIS) 2(3) (September 2012)
17. Halton, M., Rahman, S.(Shawon): The Top 10 Best Cloud-Security Practices in Next-Generation Networking. International Journal of Communication Networks and Distributed Systems (IJCNDS); Special Issue on: Recent Advances in Next-Generation and Resource-Constrained Converged Networks 8, 70–84 (2012) ISSN: 1754-3916
18. Amin, S., Pathan, A.-S., Rahman, S.(Shawon): Special Issue on Recent Advances in Next-Generation and Resource-Constrained Converged Networks. International Journal of Communication Networks and Distributed Systems (IJCNDS)
19. Mohr, S., Rahman, S.(Shawon): IT Security Issues within the Video Game Industry. International Journal of Computer Science & Information Technology (IJCSIT) 3(5) (October 2011) ISSN:0975-3826
20. Dees, K., Rahman, S.(Shawon): Enhancing Infrastructure Security in Real Estate. International Journal of Computer Networks & Communications (IJCNC) 3(6) (November 2011)
21. Hood, D., Rahman, S.(Shawon): IT Security Plan for Flight Simulation Program. International Journal of Computer Science, Engineering and Applications (IJCSEA) 1(5) (October 2011)
22. Schuett, M., Rahman, S.(Shawon): Information Security Synthesis in Online Universities. International Journal of Network Security & Its Applications (IJNSA) 3(5) (September 2011)
23. Jungck, K., Rahman, S.(Shawon): Cloud Computing Avoids Downfall of Application Service Providers. International Journal of Information Technology Convergence and services (IJITCS) 1(3) (June 2011)

24. Slaughter, J., Rahman, S.(Shawon): Information Security Plan for Flight Simulator Applications. International Journal of Computer Science & Information Technology (IJCSIT) 3(3) (June 2011)
25. Benson, K., Rahman, S.(Shawon): Security Risks in Mechanical Engineering Industries. International Journal of Computer Science and Engineering Survey (IJCSES) 2(3) (August 2011)
26. Bisong, A., Rahman, S.(Shawon): An Overview of the Security Concerns in Enterprise Cloud Computing. International journal of Network Security & Its Applications (IJNSA) 3(1) (January 2011)
27. Mullikin, A., Rahman, S.(Shawon): The Ethical Dilemma of the USA Government Wiretapping. International Journal of Managing Information Technology (IJMIT) 2(4) (November 2010)
28. Rahman, S.(Shawon), Donahue, S.: Convergence of Corporate and Information Security. International Journal of Computer Science and Information Security (IJCSIS) 7(1) (2010)
29. Hailu, A., Rahman, S.(Shawon): Protection, Motivation, and Deterrence: Key Drivers and Barriers of Organizational Adoption of Security Practices. In: IEEE the 7th International Conference on Electrical and Computer Engineering (ICECE), Dhaka, Bangladesh, December 20-22 (2012)
30. Hailu, A., Rahman, S.(Shawon): Security Concerns for Web-based Research Survey. In: IEEE the 7th International Conference on Electrical and Computer Engineering (ICECE), Dhaka, Bangladesh, December 20-22 (2012)
31. Neal, D., Rahman, S.(Shawon): Consider video surveillance in the cloud-computing. In: IEEE the 7th International Conference on Electrical and Computer Engineering (ICECE), Dhaka, Bangladesh, December 20-22 (2012)
32. Neal, D., Rahman, S.(Shawon): Securing Systems after Deployment. In: The Third International Conference on Communications Security & Information Assurance (CSIA 2012), Delhi, India, May 25-27 (2012)
33. Johnson, M., Rahman, S.(Shawon): Healthcare System's Operational Security. In: IEEE 2011 14th International Conference on Computer and Information Technology (ICCIT), Dhaka, Bangladesh, December 22-24 (2011)
34. Rahman, S.(Shawon): System Security Specifications for a Multi-disciplinary Research Project. In: 7th International Workshop on Software Engineering for Secure Systems conjunction with The 33rd IEEE/ACM International Conference on Software Engineering (ICSE 2011), Honolulu, Hawaii, May 21-28 (2011)
35. Rahman, S.(Shawon), Donahue, S.: Converging Physical and Information Security Risk Management. In: Executive Action Series, The Conference Board, Inc. 845 Third Avenue, New York, New York 10022-6679, United States,
36. Rahman, S.(Shawon), Peterson, M.: Security Specifications for a Multi-disciplinary Research Project. In: The 2011 International Conference on Software Engineering Research and Practice (SERP 2011), Las Vegas, Nevada, USA, July 18-21 (2011)
37. Jungck, K., Rahman, S.(Shawon): Information Security Policy Concerns as Case Law Shifts toward Balance between Employer Security and Employee Privacy. In: The 2011 International Conference on Security and Management (SAM 2011), Las Vegas, Nevada, USA, July 18-21 (2011)

Keywords Extraction via Multi-relational Network Construction

Kai Lei[*], Hanhong Tang, and YiFan Zeng

Shenzhen Key Lab for Cloud Computing Technology & Applications (SPCCTA),
School of Electronics and Computer Engineering, Peking University,
Shenzhen 518055, P.R. China
leik@pkusz.edu.cn, {segeon,evanzengcn}@gmail.com

Abstract. Keywords extraction can be regarded as a process of ranking the words in a given document (set) according to their importance to this document (set). Previous graph-based methods usually consider only one kind of relation between words, such as co-occurrence, ignoring the fact that words in a text interact with each other via multiple relations, which collaborate to decide the importance of words. Although some recently published methods use more than one relation type, they fail to consider the interactions between relations. Therefore, we propose a new approach for keywords extraction by constructing a multi-relational network from texts, which evaluates the various relations at the same time. Experiments shows that our approach is competitive compared with some typical methods.

Keywords: keywords extraction, multi-relational network, MultiRank.

1 Introduction

In the era of Internet, the amount of information grows exponentially on the web. On the one hand, people can get almost everything they want to know from the web. On the other hand, how to organize and retrieve the large amount of information becomes a compelling challenge. Automatic keywords extraction is an effective technique to help solve this problem. It extracts (generates) keywords from documents automatically, which in turn can be used to categorize and index documents. Also, the automatically extracted keywords can be used to generate summaries of the documents or used for user recommendation.

Keywords extraction (or keyphrase extraction) approaches can be roughly divided into two categories: supervised approaches and unsupervised approaches. The former category generally regards keywords extraction as a binary classification task, which needs large training datasets to train the classifiers. Training datasets are usually labeled manually, which is very time-consuming and cost-ineffective. Moreover, the learned classifier may have the problem of overfitting. Graph-based approaches are representative unsupervised keywords extraction approaches. They usually represent texts with graphs, and run ranking algorithms on these text graphs to obtain the importance of words within the texts. In this paper, we focus on unsupervised approaches.

---

[*] Corresponding author.

D. Nagamalai et al. (Eds.): *Adv. in Comput. Sci., Eng. & Inf. Technol.*, AISC 225, pp. 33–45.
DOI: 10.1007/978-3-319-00951-3_4

Previous graph-based methods, such as [1-3], usually consider only one kind of relation between words, such as co-occurrence, ignoring the fact that words in a text interact with each other via multiple relations, which collaborate to decide the importance of words. Although some recently published methods like [4],[14],[21], took more than one relation type into consideration, they failed to consider the interactions between relations. Motivated by that, we propose a new approach named MRN (multi-relational-network) model for keywords extraction by constructing a multi-relational network from texts, which considers the various relations at the same time to determine the importance of candidate keywords.

In MRN model, we represent a document as a multi-relational network, where words are nodes and various types of word relations are taken as various kinds of links between nodes. In Fig. 1(a), we give an illustration of multi-relational network in a text. There are five words (W1, W2, W3, W4 and W5) and three relations (R1, R2, and R3). Note that although word relations usually have no directions, we deliberately add bidirectional links to the words to reflect their interactions. In Fig. 1(b), we show the same network in a tensor format. In this figure, each two-dimensional slice represents an adjacency matrix in a single word relation type. A tensor can be represented as a multi-dimensional array of numerals. So the multi-relational network of the text can also be represented as a $5 \times 5 \times 3$ array, where (i,j,k) entry is nonzero if the ith word is connected to jth word via kth relation.

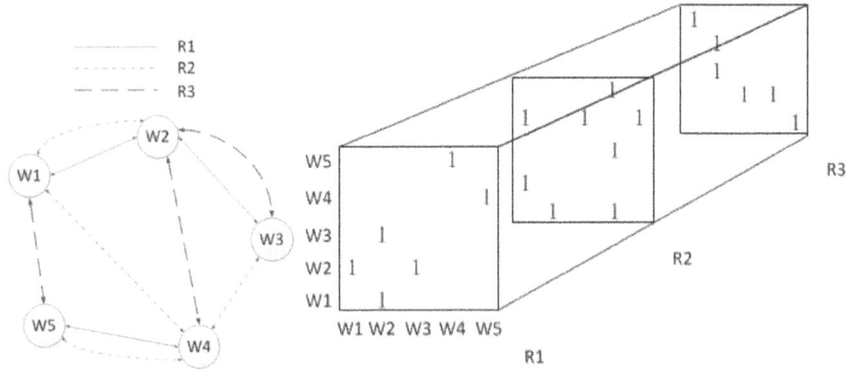

(a) Multi-relational network in a text

(b) A tensor representation of the same multi-relational network

Fig. 1.

In our study, we use three types of word relations: semantic similarity, co-occurrence, and topic similarity. It's worth mentioning that other relations can also be incorporated in our MRN model to construct a more sophisticated multi-relational network from documents.

Now the remaining question is how to decide the importance of words in such a multi-relational network. Ng et al. [7] proposed a framework named MultiRankto determine the importance of objects and relations simultaneously in such a multi-relational network. In MultiRank, the importance of an object depends on the importance of objects that connect to it via multiple relations and the importance of these relations; the importance of a relation depends on the importance of objects that to be linked. In our study, we use MultiRank to evaluate the interaction of words and their relations at the same time and finally obtain the importance of the words under such interaction. More detailed descriptions about MultiRank will be provided in Section 4.

We conduct experiments on the same dataset with [8]. Results show that our MRN model is competitive compared with typical methods.

The rest of this paper is organized as follows: Section 2 introduces the related work. Section 3 describes the MRN construction process in detail. We introduce the MultiRank algorithm used for words ranking in Section 4. Experiment results are demonstrated and analyzed in Section 5. Finally, we summarize and conclude this work in Section 6.

2 Related Work

For the task of automatic keywords extraction, an initial achievement was made by Turney[20], who treated keywords extraction as a supervised machine learning problem. Hulth[8] followed Turney's line of thinking, but he combined additional linguistic features with statistic features, such as term frequency, collection frequency, to discriminate keywords and non-keywords. As supervised machine learning algorithms generally need large training datasets with manually assigned labels, which is time-consuming and cost-ineffective, we focus on unsupervised algorithms in this paper.

Plas et al. [6] utilized lexical resources (EDR and WordNet) to automatically extract keywords from spoken texts. In their approach, words were first filtered with relative frequency ratio (RFR). For a given word, RFR is the ratio of word frequency in a document to the word frequency in a general corpus. Then words' concepts were grouped into clusters using their semantic similarities. Cluster level scores and concept level scores were calculated respectively, and were used for final ranking of candidate keywords.

Co-occurrence statistic information was used in [17] to extract keywords from a single document. First, frequent words were identified and each word was measured against the frequent words to get their co-occurrence distributions. If a word's co-occurrence distribution to frequent words is biased towards a small set of frequent words, the word was regarded as a keyword.

Wartena et al. [21]adopted a method similar to [17] by using words co-occurrence distribution to discover keywords. However, instead of only considering co-occurrence distributions within a single document, they compared co-occurrence distributions of individual texts to those of a corpus.

Wang et al.[3] represented a text with a weighted semantic graph using WordNet [15]. Then they applied PageRank [18] in the rough graph to do word sense disambiguation, prune the graph, and finally apply UW-PageRank (PageRank on undirected weighted graph) on the pruned graph to extract keywords.

TextRank[1]also represented a document as a graph. Unlike Wang et al.'s approach[3], TextRank used words co-occurrence relations as edges. Then PageRank was executed on this graph to rank the words. TextRank method considered only the word co-occurrence relations within a single document, assuming that documents were independent of each other. It didn't explore other types of relations within a single document or word relations across documents.

CollabRank [2] took advantage of information from other documents by first grouping documents into a few clusters using the clustering algorithm, then conducting the cluster level evaluation and document level evaluation consecutively to rank the words in each document. The encouraging performance of CollabRank indicates that the inter-document relations can benefit keywords extraction tasks.

Some researchers took topical relation into account when ranking candidate keywords. Topical PageRank [4] decomposed a traditional random walk into multiple ones specific to various topics to get topic specific ranking of words. Then candidate keyphrases' topic specific ranking was calculated simply by summing up the topical ranking score of composing words. After that, the overall raking scores of the candidate keyphrases were calculated based on the document's topic distribution and candidate keyphrases' topic specific ranking scores. The top ranked phrases were then chosen as keyphrases. This approach used two kinds of relations when ranking the candidate keyphrases. However, itopted to simply add the weighted ranking scores obtained from topical PageRank runs to get the overall ranking score of a candidate keyphrase, without considering the interactions between relations. In [14],Xin et al. modified Topical PageRank by introducing topic sensitive score propagation to rank individual words. Then a principled probabilistic phrase ranking method was used to rank phrases.

In this contribution, we propose a multi-relational network model for automatic keywords extraction, in which words links to each other via a variety of relations. The difference between our MRN model and the above mentioned approaches is that various word relations and their interactions are considered simultaneously when we rank candidate keywords.

3 Multi-relational Network Construction

Before starting to construct the multi-relational network from a document, we first perform part-of-speech (POS) tagging. In this paper, we use the Stanford Log-linear Part-Of-Speech Tagger [9]. As keywords are usually nouns, we only consider the words with tags in Table 1 during the following multi-relational network construction process.

3.1 Word Relations Choosing

In this paper, we use three word relations, semantic similarity derived from WordNet [15], word co-occurrence and topical similarity derived via LDA [11].The semantic similarity calculated from WordNet is corpus-independent, while the other two relations are corpus-specific. Compared with word co-occurrence, LDA based measure has been proved to be capable of capturing corpus-specific topical similarity between words[22].

In the following MRN construction process, we extract the three relations between each pair of candidate keywords, and add a bidirectional link for each relation type between them to reflect their interactions.

Table 1. POS tags used to filter candidate keywords

Tag	Description
NN	noun, common, singular or mass
NNP	noun, proper, singular
NNPS	noun, proper, plural
NNS	noun, common, plural

3.2 Semantic Similarity Calculation

WordNet [15], [16] is a lexical database for the English language. English nouns, verbs, adjectives, and adverbs are organized into sets of synonyms, each representing a lexicalized concept. WordNet is particularly well suited for similarity measures, since it organizes nouns and verbs into hierarchies of IS_A relations.

For each pair of candidate keywords, we use WordNet::Similarity [5] to calculate their semantic similarity, which is an open source package based on WordNet. As WordNet is a manually built database, some of the keywords may not appear in it. If we get a similarity score$V_{i,j}$ for a candidate keyword pair (W_i, W_j) successfully, we add a bidirectional semantic link with weight $V_{i,j}$ for this candidate keyword pair to the multi-relational network.

There are six measures of similarity and three measures of relatedness implemented in WordNet::Similarity package. We try different measures in our experiments and find that the simple Path measure produces the best results. More details about different measures' influence on our MRN model are provided in the experiment section. The Path measure simply counts the number of nodes on the shortest path between the concepts and takes the reciprocal of path length as their similarity. The smallest path length occurs when the two concepts are the same. Hence the maximum similarity is 1.

Unlike Wang et al. 's [3] strategy of word sense disambiguation, we adopt the same strategy on sense disambiguation with [6]. The most common sense is used as the candidate keyword's sense. We don't evaluate the influence of this approach in this study as it's not the focus of our study.

3.3 Word Co-occurrence Extraction

We use a sliding window to count the co-occurrence of candidate keywords within a document. For each pair of candidate keywords, if they co-appear in one window, their co-occurrence in that window will be counted only once, no matter how many times each word appears in that window. The following formula is used to calculate the co-occurrence score for each pair of candidate keywords:

$$C(i,j) = \ln\left(1 + \frac{p(i,j)}{p(i)\times p(j)}\right), \tag{1}$$

$$p(i) = \frac{n_i}{N}, \tag{2}$$

where $C(i,j)$is the co-occurrence score of word i and word j; $p(i)$is possibility that word i appears in a sliding window; $p(i,j)$is the probability that word i and word j co-appear in a sliding window; n_i is the number of times word i appears in a sliding window;$N$is the total number of sliding windows. We add one to the quotient of $p(i,j)$ and $p(i) \times p(j)$to make sure that $C(i,j)$ is non-negative.

When we obtained $C(i,j)$, we add a bidirectional link with weight $C(i,j)$ between the candidate keyword pair to the multi-relational network.

3.4 Topical Similarity Extraction

Topical similarity between words is a measure to indicate in what degree the words are talking about the same topics.

There are many methods in machine learning to infer latent topics of words and documents, such as Latent Semantic Analysis (LSA) [12], probabilistic LSA (pLSA) [10], and Latent Dirichlet Allocation (LDA) [11].LDA is probably the most popular topic model today. In LDA, a document is considered as a mixture of various topics which generate words with certain probabilities. It can learn latent topic models from a corpus and thereafter infer topic models of unseen documents. In this study, we use LDA to infer topic distributions of words. Then words' topical similarity is measured using their topic distributions.

Formally, given a document collection $C = \{D_1, D_2, D_3, ..., D_M\}$ with vocabulary $V = \{W_1, W_2, W_3, ..., W_N\}$, topic number K, we need to obtain the topic distributions of each word i, $p(Z_j|W_i)$, j=1,2,...K, where M is the number of documents in C, $N$is the vocabulary size, Z_j denotes topic j.

We use GibbsLDA++ [13], an open source implementation of LDA to infer topic model of words and documents. As GibbsLDA++ only calculates the probability $p(W|Z)$ and $p(Z|D)$, where $p(W|Z)$ is the conditional probability of word W for given topic Z and $p(Z|D)$is the conditiaonal probability of topic Z for given document D, we use Bayes' theorem to obtain the topic-word distribution.

$$p(Z_j|W_i) = \frac{p(W_i|Z_j)\times p(Z_j)}{p(W_i)} \tag{3}$$

$$p(W_i) = \sum_{j=1}^{K} p(W_i|Z_j) \times p(Z_j) \tag{4}$$

In above equations, $p(W_i|Z_j)$ is obtained from GibbsLDA++'s output and $p(Z_j)$is calculated using the following equation assuming that each document is produced with equal probability:

$$p(Z_j) = \frac{1}{M} \times \sum_{s=1}^{M} p(Z_j|D_s) . \tag{5}$$

After we get the topic distributions of each word, we use cosine similarity to compute the topic similarity of each pair of candidate keywords. Let *A* and *B* be the topic distribution vectors of word *i* and word *j*, then their topic similarity sim(*i*,*j*) is calculated using following formula:

$$\mathrm{sim}(i,j) = \frac{\sum_{s=1}^{K} A_s \times B_s}{\sqrt{\sum_{s=1}^{K}(A_s)^2} \times \sqrt{\sum_{s=1}^{K}(B_s)^2}} \tag{6}$$

When the topic similarity between word *i* and word *j* is obtained, we add a bidirectional link with weight $\mathrm{sim}(i,j)$ to the multi-relational network.

4 WordsRanking Using MultiRank Algorithm

So far, we have introduced the construction process of the multi-relational text network. Following these procedures, we can construct a MRN, in which the candidate keywords are nodes and the three types of relations between each pair of words are links. This kind of multi-relational network can also be represented in tensor format, just like we have illustrated in Fig. 1. In paper[7], the authors propose a framework named MultiRankto rank both objects and relations in this kind of multi-relational network simultaneously. Here we use MultiRank algorithm to rank the candidate keywords. Next, we will give an introduction of the MultiRank algorithm.

Let U be the real field. We call A = (a_{i_1,i_2,j_1}) wherea_{i_1,i_2,j_1} ∈U, $i_s = 1, \dots, M, s = 1,2$ and $j = 1, \dots, N$a real (2,1)th order $(M \times N)$-dimensional rectangular tensor. In this setting, we refer (i_1, i_2) to be the indices for objects (words in our case) and j_1 to be the index for relations.In this study, we have $|V|$(V is the vocabulary of a particular document waiting for keywords extraction) words $(M = |V|)$ and three relations $(N = 3)$.

Given the above setting, we can construct two transition probability tensors $\mathcal{O} = (o_{i_1 i_2, j_1})$ and $\mathcal{R} = (r_{i_1,i_2,j_1})$with respect to objects (words) and relations by normalizing the entries of$\mathcal{A}$as follows:

$$o_{i_1 i_2, j_1} = \frac{a_{i_1,i_2,j_1}}{\sum_{i=1}^{M} a_{i,i_2,j_1}}, i_1 = 1,2, \dots M, \tag{7}$$

$$r_{i_1,i_2,j_1} = \frac{a_{i_1,i_2,j_1}}{\sum_{j=1}^{N} a_{i_1,i_2,j}}, j_1 = 1,2, \dots N. \tag{8}$$

Then we consider the following probabilities:

$$\mathrm{p}[X_t = i_1] = \sum_{i_2=1}^{M} \sum_{j_1=1}^{N} o_{i_1 i_2, j_1} \times \ \mathrm{p}[X_{t-1} = i_2, Y_t = j_1], \tag{9}$$

$$\mathrm{p}[Y_t = j_1] = \sum_{i_1=1}^{M} \sum_{i_2=1}^{M} r_{i_1 i_2, j_1} \times \ \mathrm{p}[X_t = i_1, X_{t-1} = i_2] \ , \tag{10}$$

where $\mathrm{p}[X_{t-1} = i_2, Y_t = j_1]$ is the joint probability distribution of X_{t-1} and Y_t, and $\mathrm{p}[X_t = i_1, X_{t-1} = i_2]$ is the joint probability distribution of X_t and X_{t-1}.Suppose an equilibrium/stationary distribution of words and relations is reached, we need the ranking scores of words and relations given by

$$\bar{\mathbf{x}} = [\bar{x}_1, \bar{x}_2, \ldots, \bar{x}_M]^T \text{ and } \bar{\mathbf{y}} = [\bar{y}_1, \bar{y}_2, \ldots, \bar{y}_N]^T \tag{11}$$

respectively, with

$$\bar{x}_{i_1} = \lim_{t\to\infty} \mathrm{p}[X_t = i_1] \text{ and } \bar{y}_{j_1} = \lim_{t\to\infty} \mathrm{p}[Y_t = j_1] \tag{12}$$

for $1 \le i_1 \le$M and $1 \le j_1 \le$N.

Considering that (8) and (9) are coupled and they involve two joint probability distribution, $\bar{x}_{i_1}$ and $\bar{y}_{j_1}$are difficult to obtain. MultiRank employs a product form of individual probability distributions for joint probability distributions in (8) and (9), assuming that

$$\mathrm{p}[X_{t-1} = i_2, Y_t = j_1] = \mathrm{p}[X_{t-1} = i_2] \times \mathrm{p}[Y_t = j_1] \tag{13}$$

$$\mathrm{p}[X_t = i_1, X_{t-1} = i_2] = \mathrm{p}[X_t = i_1] \times \mathrm{p}[X_{t-1} = i_2]. \tag{14}$$

Under the above assumption, (8) and (9) becomes

$$\bar{x}_{i_1} = \sum_{i_2=1}^{M} \sum_{j_1=1}^{N} o_{i_1, i_2, j_1} \times \bar{x}_{i_2} \times \bar{y}_{j_1}, i_1 = 1,2, \ldots, M, \tag{15}$$

$$\bar{y}_{j_1} = \sum_{i_1=1}^{M} \sum_{i_2=1}^{M} r_{i_1, i_2, j_1} \times \bar{x}_{i_1} \times \bar{x}_{i_2}, j_1 = 1,2, \ldots, N, \tag{16}$$

when t goes to infinite.

Under the tensor operation for (14) and (15), the following tensor (multivariate polynomial) equations

$$\mathcal{O}\bar{\mathbf{x}}\bar{\mathbf{y}} = \bar{\mathbf{x}} \text{ and } \mathcal{R}\bar{\mathbf{x}}^2 = \bar{\mathbf{y}}, \tag{17}$$

are solved with

$$\sum_{i_1=1}^{M} \bar{x}_{i_1} = 1 \text{ and } \sum_{j_1=1}^{N} \bar{y}_{j_1} = 1 \tag{18}$$

to achieve the ranking value of words and relations.

In order to obtain $\bar{\mathbf{x}}$ and $\bar{\mathbf{y}}$, an efficient iterative algorithm is presented to solve (17), which is described in Algorithm 1. For more details about MultiRank, please refer to[7].

When the iterative algorithm stops, we get the words ranking scores and the relation ranking score. As we are only interested in the importance of words, we just sort the words according to their ranking scores and choose the top ones as keywords.

Algorithm 1. The MultiRank Algorithm

Input: Two tensors $\mathcal{O}$ and $\mathcal{R}$, two initial probability distributions $\mathbf{x}_0$ and $\mathbf{y}_0$ ($\sum_{i_1}^{M}[\mathbf{x}_0]_{i_1} = 1$ and $\sum_{j_1}^{N}[\mathbf{y}_0]_{j_1} = 1$) and the tolerance ϵ

Output: Two stationary probability distributions $\bar{\mathbf{x}}$ and $\bar{\mathbf{y}}$

Procedure:

1. Set $k = 1$;
2. Compute $\mathbf{x}_k = \mathcal{O}\mathbf{x}_{k-1}\mathbf{y}_{k-1}$;
3. Compute $\mathbf{y}_k = \mathcal{R}\mathbf{x}_k^2$;
4. If $\|\mathbf{x}_k - \mathbf{x}_{k-1}\| + \|\mathbf{y}_k - \mathbf{y}_{k-1}\| < \epsilon$, then stop, otherwise set $k = k + 1$ and goto step 2.

5 Experiments and Analysis

5.1 Dataset and Metrics

In this paper we use the same dataset with [8]. The dataset contains 2,000 abstracts from journal papers, which has 248,728 words and 19,254 manually annotated keyphrases. As we extract keywords instead of keyphrases, we simply split the annotated keyphrases into separate keywords. If one keyword appears in more than one keyphrase for an abstract, only one copy of that keyword will be kept for evaluation. After this preprocessing, each abstract has about 19 keywords on average.

As the dataset is not enough for LDA to learn useful topics, we use a subset of English Wikipedia snapshot[1]in 2008. The subset is a 5% sample of the English Wikipedia snapshot, and it contains 121,650 documents. After cleaning up the HTML tags, we keep only articles longer than 100 words. Finally, 107,290 documents are used to learn LDA models.

We use precision, recall and F-measure to evaluate the performance of our approach and the baseline approaches. The three metrics can be represented as follows:

$$p = \frac{C_{correct}}{C_{assigned}}, r = \frac{C_{correct}}{C_{standard}}, f = \frac{2 \times p \times r}{p + r}, \tag{19}$$

where $C_{correct}$ is number of correctly extracted keywords for a single document; $C_{assigned}$ is the number of extracted keywords for each document, which is set to 20 in our experiments, as the average number of standard keywords for the documents in our dataset is 19. $C_{standard}$ is the number of annotated keywords for the document. The overall precision, recall and f-measure are average score of all documents, which are represented as follows:

$$P = \frac{\sum_{i=1}^{n} p_i}{n}, R = \frac{\sum_{i=1}^{n} r_i}{n}, F = \frac{\sum_{i=1}^{n} f_i}{n}, \tag{20}$$

where n is the number of documents.

[1] http://en.wikipedia.org/wiki/Wikipedia_database

5.2 Influence of Parameters on Our MRN Model

There are three major parameters which influence the performance of our approach: the method chosen to measure words' semantic similarity, the co-occurrence window size and the LDA topic number. We use *Sim*, *Win* and *K* to denote them respectively in the following part. As our focus of this study is the design of our new approach, we just set $\alpha = 1, \beta = 0.01$ in LDA without further investigation.

In the next sections, we investigate the influence of these three parameters on our MRN model. Except the parameter under investigation, we set the parameters as follows: *Sim*= Path, *Win* = 2, *K* = 500, which are the settings when our approach achieves the best performance.

Semantic Similarity Measure. We try six different measuring methods provided by WordNet::Similarity[5]: Leacock&Chodorow (lch), Path, Wu&Palmer (wup), Jiang&Conrath (jco), Lin and Resnik. The results are shown in Table 2.

Table 2. Influence of different semantic similarity measures

Sim	Precision	Recall	F-measure
lch	0.462	0.495	0.478
Path	0.491	0.523	0.506
wup	0.433	0.458	0.445
jco	0.419	0.431	0.425
Lin	0.347	0.346	0.346
Resik	0.491	0.504	0.498

We can see that Path measure and Resnik measure perform better than other measures in our model, while Lin performs the worst. The reason why Path outperforms the other measures needs further investigation.

Window Size. In our experiments, the co-occurrence window size ranges from 2 to 15. The results are shown in Table 3.

Our algorithm achieves the best performance when co-occurrence window size is set to 2, which is consistent to the observation in [1]. It maybe result from that when window size becomes larger, words will be more densely connected in the text network and the importance of words tends to be more equally distributed, which makes our algorithm less effective to differentiate them.

Table 3. Infuence of co-occurrence window size

Win	Precision	Recall	F-measure
2	0.491	0.523	0.506
5	0.478	0.520	0.498
10	0.436	0.497	0.465
15	0.411	0.456	0.432

Table 4. Influence of topic number in LDA models

K	Precision	Recall	F-measure
100	0.435	0.480	0.456
300	0.486	0.520	0.502
500	0.491	0.523	0.506
1000	0.490	0.521	0.505

Topic number. The number of topics in our LDA models ranges from 100 to 1000. The results are listed in Table 4.

As we can see, the results don't change very much when the top number varies from 300 to 1000. However, when topic number is set to 100, the performance of our algorithm becomes comparatively low.

5.3 Comparing with Baselines

In this study, we compare our MRN model to three baselines: TFIDF [19], TextRank, and LDA based approach.

TFIDF is the most intuitive approach used for keywords extraction. Many previous researchers have used TFIDF approach as a baseline to evaluate their own keywords extraction approach. We also follow this convention.

As our MRN model is a graph-based method in nature, we compare our MRN model to the probably most representative graph-based method TextRank. Just like our MRN model, the baseline TextRank method also achieves the best results when window size is set to 2. The best results are compared in Table 5.

Table 5. Comparision with baselines

Method	Precision	Recall	F-measure
TFIDF	0.484	0.517	0.500
TextRank	0.445	0.484	0.464
LDA	0.486	0.521	0.503
MRN	**0.491**	**0.523**	**0.506**

In addition, a LDA based baseline approach is also used. In this approach, topic distribution of each document and topic distribution of each word within that document are obtained first. Then cosine similarity is measured against the word's topic distribution and the document's topic distribution to see in what degree the word and the document are talking about the same issue. In this LDA baseline, topic number is set to 500, too.

The comparing results are shown in Table 5. As we can see, the MRN approach outperforms the baselines on the three metrics, which indicates that the multi-relational network model is effective in reflecting the interactions between words and itcan successfully determine words' importance.

6 Conclusion and Future Work

In this paper, we present a new approach for keywords extraction by exploring the multiple relations between words. In our approach, a multi-relational network is constructed by taking words as nodes and three kinds of relations between words: semantic similarity, co-occurrence and topical similarity as links between nodes. Then MultiRank algorithm is used to evaluate the importance of words in this multi-relational network. Our MRN method is evaluated against typical keywords extraction methods in experiments. Results show that the MRN approach is competitive compared with some typical methods in keywords extraction.

Possibly, other relations between words can also be used to construct the multi-relational network and various relations may not have equal weight in deciding the importance of words. We plan to investigate the two issues in future work.

Acknowledgments. This work was supported by NFSC project (Grant No: 61103027), 973 project (No: 2011CB302305) and Shenzhen Gov Projects (JCYJ2012082917002 8558).

References

1. Mihalcea, R., Tarau, P.: Textrank: Bringing order into texts. In: Proceedings of EMNLP, Barcelona, Spain, vol. 4, pp. 404–411 (2004)
2. Wan, X., Xiao, J.: Collabrank: towards a collaborative approach to single-document keyphrase extraction. In: Proceedings of the 22nd International Conference on Computational Linguistics, vol. 1, pp. 969–976. Association for Computational Linguistics (2008)
3. Wang, J., Liu, J., Wang, C.: Keyword extraction based on pageRank. In: Zhou, Z.-H., Li, H., Yang, Q. (eds.) PAKDD 2007. LNCS (LNAI), vol. 4426, pp. 857–864. Springer, Heidelberg (2007)
4. Liu, Z., Huang, W., Zheng, Y., Sun, M.: Automatic keyphrase extraction via topic decomposition. In: Proceedings of the 2010 Conference on Empirical Methods in Natural Language Processing, pp. 366–376. Association for Computational Linguistics (2010)
5. Pedersen, T., Patwardhan, S., Michelizzi, J.: Wordnet: Similarity: measuring the relatedness of concepts. In: Demonstration Papers at HLT-NAACL 2004, pp. 38–41. Association for Computational Linguistics (2004)
6. Van Der Plas, L., Pallotta, V., Rajman, M., Ghorbel, H.: Automatic keyword extraction from spoken text. a comparison of two lexical resources: the edr and wordnet. arXiv preprint cs/0410062 (2004)
7. Ng, M.K.P., Li, X., Ye, Y.: Multirank: co-ranking for objects and relations in multi-relational data. In: Proceedings of the 17th ACM SIGKDD International Conference on Knowledge Discovery and Data Mining, pp. 1217–1225. ACM (2011)
8. Hulth, A.: Improved automatic keyword extraction given more linguistic knowledge. In: Proceedings of the 2003 Conference on Empirical Methods in Natural Language Processing, pp. 216–223. Association for Computational Linguistics (2003)

9. Toutanova, K., Klein, D., Manning, C.D., Singer, Y.: Feature-rich part-of-speech tagging with a cyclic dependency network. In: Proceedings of the 2003 Conference of the North American Chapter of the Association for Computational Linguistics on Human Language Technology, vol. 1, pp. 173–180. Association for Computational Linguistics (2003)
10. Hofmann, T.: Probabilistic latent semantic indexing. In: Proceedings of the 22nd Annual International ACM SIGIR Conference on Research and Development in Information Retrieval, pp. 50–57. ACM (1999)
11. Blei, D.M., Ng, A.Y., Jordan, M.I.: Latent dirichlet allocation. The Journal of Machine Learning Research 3, 993–1022 (2003)
12. Landauer, T.K., Foltz, P.W., Laham, D.: An introduction to latent semantic analysis. Discourse Processes 25(2-3), 259–284 (1998)
13. Phan, X.H., Nguyen, C.T.: Gibbslda++: A c/c++implementation of latent dirichlet allocation (lda) (2007)
14. Zhao, X., Jiang, J., He, J., Song, Y., Achananuparp, P., Lim, E.P., Li, X.: Topical keyphrase extraction from twitter (2011)
15. Miller, G.A.: Wordnet: a lexical database for english. Communications of the ACM 38(11), 39–41 (1995)
16. Stark, M.M., Riesenfeld, R.F.: Wordnet: An electronic lexical database. In: Proceedings of 11th Eurographics Workshop on Rendering. Citeseer (1998)
17. Matsuo, Y., Ishizuka, M.: Keyword extraction from a single document using word co-occurrence statistical information. International Journal on Artificial Intelligence Tools 13(01), 157–169 (2004)
18. Brin, S., Page, L.: The anatomy of a large-scale hypertextual web search engine. Computer Networks and ISDN Systems 30(1), 107–117 (1998)
19. Salton, G., Buckley, C.: Term-weighting approaches in automatic text retrieval. Information Processing and Management 24(5), 513–523 (1988)
20. Turney, P.D.: Learning algorithms for keyphrase extraction. Information Retrieval 2(4), 303–336 (2000)
21. Wartena, C., Brussee, R., Slakhorst, W.: Keyword extraction using word co-occurrence. In: DEXA, vol. 10, pp. 54–58 (2010)
22. Yin, W., Pei, Y., Huang, L.: Automatic multi-document summarization based on new sentence similarity measures. In: Anthony, P., Ishizuka, M., Lukose, D. (eds.) PRICAI 2012. LNCS, vol. 7458, pp. 832–837. Springer, Heidelberg (2012)

Accelerating Super-Resolution Reconstruction Using GPU by CUDA

Toygar Akgün and Murat Gevrekci

ASELSAN Microelectronics, Guidance and Electro-Optics Division,
Akyurt, Ankara / Turkey
{takgun,mgevrekci}@aselsan.com.tr

Abstract. This paper demonstrates a massively multi-threaded implementation of super-resolution image formation on the NVIDIA CUDA architecture. On the algorithm side maximum a-posteriori (MAP) reconstruction is adopted with sub-pixel translational motion estimation algorithm for spatial resolution enhancement. Resulting algorithm is implemented in CUDA using a low end $GT640$ GPU, and an overall speed up of $10-11$ times is achieved compared to ANSI C implementation running on a Core $i5$ CPU.

Keywords: Massive multi-threading, GPU, CUDA, super-resolution, multi-frame resolution enhancement.

1 Introduction

Graphic Processor Unit (GPU) had been used for general purpose programming since late 1990s by carefully leveraging OpenGL API. But neither OpenGL nor these older GPUs were designed with this goal in mind, leading to limited functionality and a steep learning curve. Starting early 2000s GPU evolved into a programmable, highly parallel, multi-threaded, many-core processor with tremendous computational float-point horsepower and very high memory bandwidth. In 2006 NVIDIA Corporation introduced the CUDA (Compute Unified Device Architecture) architecture with an accompanying programming model and API. CUDA and its accompanying API were *designed* to allow users with high computational needs to leverage GPU's compute power with minimal learning effort. Since then massive-multithreading on GPUs has been gaining traction and general purpose computing on GPUs is being utilized by researchers coming from vastly varying backgrounds [1].

This paper demonstrates a massively multi-threaded implementation of super-resolution reconstruction on the NVIDIA CUDA architecture. Super-resolution (SR) is a multi-frame video enhancement framework to obtain a high quality description from multiple degraded observations. SR reconstruction aims to compensate for image degradations i.e. aliasing, blurring, noise, interlacing, and low resolution. Sub-pixel shifts among consecutive frames are utilized to perform image enhancement. As discussed in [2], our multi-frame image enhancement system is composed of two main parts: registration and reconstruction. Work in

D. Nagamalai et al. (Eds.): *Adv. in Comput. Sci., Eng. & Inf. Technol.*, AISC 225, pp. 47–58.
DOI: 10.1007/978-3-319-00951-3_5

[2] is utilized for estimating the vertical and horizontal shifts among consecutive images. Reconstruction step uses a Bayesian framework to form a high-resolution image. This paper is structured as follows: We will first briefly go through the details of the super-resolution algorithm under discussion. Then we will discuss the CUDA mappings of the major processing blocks and present performance comparisons. Proposed methodology along with some visuals results are given in Section 2. Massively multi-threaded CUDA implementation of the algorithm is presented in Section 3 and performance boost of the proposed work is discussed in Section 4. The final section presents conclusions and future work.

2 Algorithm Details

Super-resolution system under discussion includes several building blocks. For a detailed discussion including references to prior art please refer to [2]. Consecutive images with various shifts are acquired using OpenCV image acquisition module. Pre-processing might be required to de-interlace in case interlacing is present. Registration step aims to align images on a common geometrical reference. Then a reconstruction step samples the aligned images on a sub-pixel grid to create finer details. Finally, contrast enhancement can be applied to emphasize local details as a post-processing step, which is planned as future work.

Representing the low-resolution degraded image as $\boldsymbol{y}_k$, and SR image (ground truth) as $\boldsymbol{x}$, image formation matrix can be represented as $\boldsymbol{H}_k$, which is the multiplication of down-sampling ($\boldsymbol{D}$), blurring ($\boldsymbol{B}_k$) and warping ($\boldsymbol{M}_k$) matrices.

$$\boldsymbol{y}_k = \boldsymbol{D}\boldsymbol{B}_k\boldsymbol{W}_k\boldsymbol{x} + \boldsymbol{n}_k = \boldsymbol{H}_k\boldsymbol{x} + \boldsymbol{n}_k, \tag{1}$$

$k = 1, ..., N$, where N is the number of images acquired and $\boldsymbol{n}_k$ is additive white Gaussian noise. Using Bayesian estimation, SR reconstruction can be written in the form of a cost function consisting of prior information and data fidelity terms. The optimization problem turns into following form using a discrete derivative operator ($\boldsymbol{L}$) as prior information

$$\boldsymbol{x}_{map} = \arg\min_{\boldsymbol{x}} \gamma^2 \parallel \boldsymbol{L}\boldsymbol{x} \parallel_2^2 + \sum_{k=1}^{N} \parallel \boldsymbol{y}_k - \boldsymbol{H}_k\boldsymbol{x} \parallel_2^2 . \tag{2}$$

Prior information provides smoothness to the SR estimate by penalizing high frequency components. Selecting the regularization parameter has critical importance to avoid over smoothing. During experiments we only kept data fidelity term in cost function, since analytic selection of regularization parameter (γ) is problematic and will violate the robustness we seek. The reduced cost function becomes:

$$C(\boldsymbol{x}) = \sum_i \parallel g_{r_i}(\boldsymbol{z}_i) - \boldsymbol{H}_i\boldsymbol{x} \parallel_2^2, \tag{3}$$

where $g_{r_i}(z)$ is the observation that passes through photometric conversion, also known as intensity mapping function, that compensates for the intensity fluctuations among infrared images. Time dependent intensity scaling or histogram equalization can be used for photometric mapping. This reduced cost function can be expanded using weighted least squares transform as

$$C(\boldsymbol{x}) = \frac{1}{2}\sum_i (g_{r_i}(\boldsymbol{z}_i) - \boldsymbol{H}_i\boldsymbol{x})^T \boldsymbol{W}_i (g_{r_i}(\boldsymbol{z}_i) - \boldsymbol{H}_i\boldsymbol{x}) . \tag{4}$$

Here $\boldsymbol{W}_i$ nothing but a certainty function of ith image that weights the IR intensities to suppress dark noise. Cost function in Equation 4 represents a weighted least squares optimization. Taking certainty matrix $\boldsymbol{W}_i$ as identity turns the problem into regular least squares. Then the super-resolved output can be solved using iterative gradient descent techniques

$$\boldsymbol{x}^{(k+1)} = \boldsymbol{x}^{(k)} + \gamma \sum_i \boldsymbol{H}_i^T \boldsymbol{W}_i \left(g_{r_i}(\boldsymbol{z}_i) - \boldsymbol{H}_i\boldsymbol{x}^{(k)} \right) . \tag{5}$$

Here γ is the step size and taken as a constant in our experiments. SR algorithm is visualized in Figure 1 for sake of clarity. Forward projection matrix $\boldsymbol{H}_k$, is the building block of the algorithm as depicted in Figure 1. Forward projection is composed of down-sampling ($\boldsymbol{D}$), blurring ($\boldsymbol{B}_k$) and warping ($\boldsymbol{M}_k$) operations applied sequentially. Note that all of these operations are suitable for parallel implementation. Likewise, back projection $\boldsymbol{H}_k^t$ consists of up-sampling with zero insertion ($\boldsymbol{U}$), blurring ($\boldsymbol{B}_k^t$) and back warping ($\boldsymbol{M}_k^t$) operations. Blurring operation is same in case a symmetric kernel is adopted such as Gaussian.

Selected parameter set is given in Table 1 for future references. Parameters are kept constant throughout the experiments to demonstrate the robustness of the system.

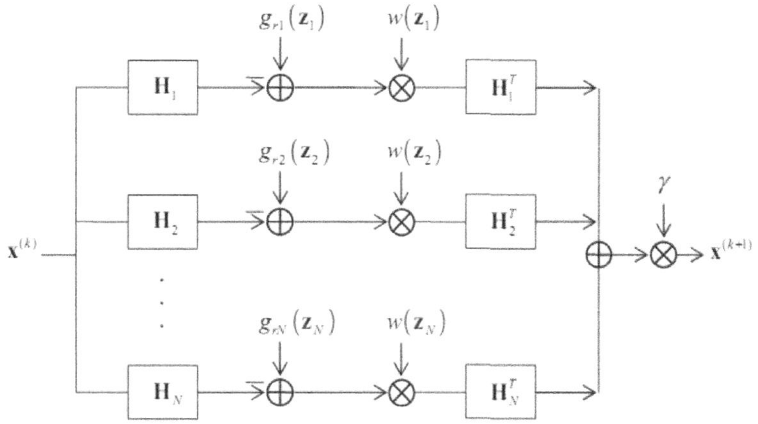

Fig. 1. SR system

Please note that vertical resolution enhancement factor should be selected twice as the horizontal resolution factor to compensate for the vertical decimation performed in preprocessing step in case frames are interlaced.

Table 1. Experiment setup

Number of low-res input frames	5
Vertical resolution enhancement factor	2
Horizontal resolution enhancement factor	2
Reconstruction iteration number	4
Gaussian kernel support	[5,5]
Gaussian sigma	1.0

Table 2. GPU specifications

CUDA cores	384
Graphics clock	900 MHz
Memory clock	900 MHz
Memory amount and type	2 GB DDR3
Memory interface width	128 bit
Memory bandwidth	28,5 GB/sec

Algorithm is suitable for parallel implementation as each frame and pixel inside the frames can be processed independently. In registration step translational shifts between input images and the reference image is computed independently. This independence paves the way for parallel computing in registration step. Reconstruction step is also parallel in nature since residual computation is independent for every input image. The scheme shown in Figure 1 illustrates how the reconstruction step is suitable for parallel implementation since we adopt an iterated back-projection method.

3 CUDA Implementation

The super-resolution algorithm described so far was implemented in CUDA to assess the potential performance boost. For the test results presented here a GT 640 NVIDIA GPU was used. The technical specifications of this card are as given in Table 2.

The competing platform is a Core i5 CPU clocked at 3,1 GHz with 8 GB RAM. Note that GT 640 is the smallest and weakest GPU from the Kepler architecture, which is the latest NVIDIA GPU architecture as of late 2012. As a result it has very low compute power and memory bandwidth compared to high end cards such as GTX 670, GTX 680 and GTX 690. In the following subsections we present the CUDA mappings of several major processing blocks. Note that there are several other CUDA kernels that will not be mentioned here, since they mostly handle simple data moving and updates.

3.1 Bilinear Scaling

Super-resolution reconstruction begins with an initial estimate image which is typically chosen as a bilinearly upscaled version of the original low resolution frame,

Table 3. Bilinear filtering code analysis

Thread numbers (x,y)	(32,32)
Shared memory bank conflict	N/A
Global memory BW efficiency	100%
Register usage	17
Occupancy	0,877
Total execution time per frame	161 microsec

where the vertical and horizontal scaling ratios are equal to the vertical and horizontal enhancement ratios. For the CUDA mapping of this bilinear upscaling block, the texture unit of the GPU hardware was used. Texture units are one of the available specialized hardware blocks on GPUs that handle simple pixel sampling operations. Such operations are quite common in graphics processing tasks. As a result, texture units are quite optimized and can provide substantial performance boosts. Extensive details of the texture hardware are beyond scope of this technical report, but we note one key property that was leveraged in this implementation. Texture units can bilinearly sample 2D arrays with very small computational load thanks to their specialized pixel sampling hardware. Initial implementation of the bilinear upscaling operation is presented below:

```
__global__ void bilinearResizeKernel(unsigned char *target,
                                     int width,
                                     int height,
                                     int pitch,
                                     float factor)
{
    const int x = blockIdx.x * blockDim.x + threadIdx.x;
    const int y = blockIdx.y * blockDim.y + threadIdx.y;

    if(x<width && y<height)
    {
        // Warped coordinates to sample from input
        float norm_x = ((float)x)/factor + 0.5f;
        float norm_y = ((float)y)/factor + 0.5f;

        // Read from texture and write to global memory
        target[y * pitch + x] =
                (unsigned char)(tex2D(texRef, norm_x, norm_y)*255.0f);
    }
}
```

As the code block presents, bilinear interpolation kernel is very simple. The heart of the kernel is the bilinear texture sampling operation (tex2D) at the very end. Since the output of the bilinear upscaling operation is also converted from 8 bit unsigned values to 32 bit float values for further processing, improvement is possible by slightly modifying the texture sampling operation to avoid an additional kernel launch. This faster implementation that combines bilinear interpolation with type conversion is presented in the code block below. Note that this very simple example demonstrates the difference between CPU and GPU programming in terms of surface access methodology.

```
__global__ void bilinearResizeToFloatKernel(unsigned char *target,
                                            float * target_f,
                                            int width,
                                            int height,
                                            int pitch,
                                            int pitch_f,
                                            float factor)
{
    const int x = blockIdx.x * blockDim.x + threadIdx.x;
    const int y = blockIdx.y * blockDim.y + threadIdx.y;

    if(x<width && y<height)
    {
        // warped coordinates to sample from input
        float normalized_x = ((float)x)/factor + 0.5f;
        float normalized_y = ((float)y)/factor + 0.5f;

        // Read from texture and write to global memory
        float temp = (tex2D(texRef, normalized_x, normalized_y)*255.0f);
        target_f[y * pitch_f + x] = temp;
        target[y * pitch + x] = (unsigned char)temp;
    }
}
```

Note that bilinear interpolation during texture fetch is only available for float type and there are some fine details regarding the setup and use of the texture units to get the best performance (including using surface writes to avoid additional CUDA surface copies). Thread mapping for the bilinear resampling kernel is very simple: Every thread handles a single output pixel. A slight speed-up (30 micro seconds faster) is possible by assigning four pixels to each thread, hence getting better memory bandwidth usage, but given texture units are cached and that bilinear sampling operation is called only once and constitutes only a small percentage of the overall execution time, the resulting speed-up would not be visible at the end. Execution configuration and performance figures for the second code block are given in Table 3.

Just to give an idea about how efficient this kernel is, the very same operation on a Core i5 CPU clocked at 3,1 GHz takes 5 ms, which means, for this specific block we have an average speed up of more than 30 times.

3.2 Image Warp

Image warp operation is modeled as a six parameter affine transform on the input image coordinates. Obviously, the transformed coordinates are not guaranteed to fall on a regular pixel grid. Off-grid pixel locations are simply obtained by bilinear interpolation that uses the four corner pixels of the grid block that includes the off-grid pixel location. As explained in the previous sub-section this is a perfect match for the texture hardware. The resulting CUDA implementation is presented in the code block below:

```
__global__ void warpAffineKernel4(float4 *target,
                                  float4 *grad,
                                  int width,
                                  int height,
                                  int pitch16)
{
   // Calculate normalized texture coordinates
   const int x = blockIdx.x * blockDim.x + threadIdx.x;
   const int y = blockIdx.y * blockDim.y + threadIdx.y;
   const int x4 = x * 4;

   float xf = (float)x4;
   float yf = (float)y;

   if(x<width && y<height)
   {
        float4 gradTemp = grad[y * pitch16 + x];
        // warped coordinates to sample from input
        float warped_x = c_H[0]*xf + c_H[1]*yf + c_H[2] + 0.5f;
        float warped_y = c_H[3]*xf + c_H[4]*yf + c_H[5] + 0.5f;
        float fx = tex2D(texRef, warped_x, warped_y);

        warped_x = c_H[0]*(xf+1.0f) + c_H[1]*(yf) + c_H[2] + 0.5f;
        warped_y = c_H[3]*(xf+1.0f) + c_H[4]*(yf) + c_H[5] + 0.5f;
        float fy = tex2D(texRef, warped_x, warped_y);

        warped_x = c_H[0]*(xf+2.0f) + c_H[1]*(yf) + c_H[2] + 0.5f;
        warped_y = c_H[3]*(xf+2.0f) + c_H[4]*(yf) + c_H[5] + 0.5f;
        float fz = tex2D(texRef, warped_x, warped_y);

        warped_x = c_H[0]*(xf+3.0f) + c_H[1]*(yf) + c_H[2] + 0.5f;
        warped_y = c_H[3]*(xf+3.0f) + c_H[4]*(yf) + c_H[5] + 0.5f;
        float fw = tex2D(texRef, warped_x, warped_y);

        // Read from texture and write to global memory
        target[y * pitch16 + x] = make_float4(fx,fy,fz,fw);
        grad[y * pitch16 + x] = make_float4(gradTemp.x-fx, gradTemp.y-fy,
                                            gradTemp.z-fz, gradTemp.w-fw);
    }
}
```

Note that this function is called at the inner most loop of the algorithm; hence it is required to be extremely optimized. As a result, the thread mapping is slightly modified compared to the bilinear interpolation kernel. Here each thread processes 4 pixels, which are modeled by the CUDA specific vector type float4.

Table 4. Affine warp code analysis

Thread numbers (x,y)	(32,32)
Shared memory bank conflict	N/A
Global memory BW efficiency	100%
Register usage	16
Occupancy	0,773
Total execution time per frame	245 microsec

To avoid an additional kernel launch, gradient surface update is also merged into this kernel by simply obtaining updated gradient values and conducting a global write. Execution configuration and performance figures for the previous code block are as given in Table 4.

3.3 Image Blur

Image blur block consists of 2D convolution with a separable kernel. The choice of a separable kernel is mostly due to its computational efficiency. For a 5×5 kernel, separable convolution requires 10 multiplications and 8 additions, whereas non-separable convolution would require 25 multiplications are 24 additions. Separable convolution is implemented as two consecutive kernel launches. The first kernel launch performs row-wise 1D convolution on the input image, and the second kernel launch performs column-wise 1D convolution on the output of the row-convolution kernel. Row and column convolution kernels are mostly adapted from NVIDIA's CUDA SDK with minor performance tunings. These kernels are presented in the following two code blocks.

As the code block below shows, the row convolution implementation uses shared memory. For the convolution operation shared memory usage provides substantial performance boost due to better memory bandwidth usage. To see this, simply consider what happens when we finish processing one pixel and move to the next pixel. The 5 pixel window that is used for filtering also shifts by one pixel and there is a 4 pixel overlap with the old filter window. Shared memory usage removes the need for dispatching a global read to obtain these pixels. At the beginning of the kernel each thread loads the main data to be processed by the thread block as well as the additional pixels that will be needed to obtain results at the boundary pixels. Note that row convolution operation uses a 5 tap blur filter and as a result boundary handling is required. This is simply due to the fact that the pixels at the thread block boundaries require 2 neighbors to their left and right sides. Finally, the filtering operation is performed and the results are written back to global memory. Note that every thread handles 4 pixels and loop unrolling (pragma unroll) is heavily used due to the fixed structures of the loops. Execution configuration and performance figures for row filtering are summarized in Table 5.

```
__global__ void convolutionRowsKernel(float *d_Dst,
                                      float *d_Src,
                                      int imageW,
                                      int imageH,
                                      int pitch)
{
    __shared__ float
    s_Data[BLOCKDIM_Y][(RESULT_STEPS + 2 * HALO_STEPS) * BLOCKDIM_X];

    //Offset to the left halo edge
    const int baseX = (blockIdx.x * RESULT_STEPS - HALO_STEPS) * BLOCKDIM_X + threadIdx.x;

    const int baseY = blockIdx.y * BLOCKDIM_Y + threadIdx.y;

    d_Src += baseY * pitch + baseX;
    d_Dst += baseY * pitch + baseX;

    //Load main data
    #pragma unroll
    for(int i = HALO_STEPS; i < HALO_STEPS + RESULT_STEPS; i++)
        s_Data[threadIdx.y][threadIdx.x + i * BLOCKDIM_X] = d_Src[i * BLOCKDIM_X];

    //Load left halo
    #pragma unroll
    for(int i = 0; i < ROWS_HALO_STEPS; i++)
        s_Data[threadIdx.y][threadIdx.x + i * BLOCKDIM_X] =
        (baseX >= -i * BLOCKDIM_X ) ? d_Src[i * BLOCKDIM_X] : 0;

    //Load right halo
    #pragma unroll
    for(int i = HALO_STEPS + RESULT_STEPS; i < HALO_STEPS + RESULT_STEPS + HALO_STEPS; i++)
        s_Data[threadIdx.y][threadIdx.x + i * BLOCKDIM_X] =
        (imageW - baseX > i * BLOCKDIM_X) ? d_Src[i * BLOCKDIM_X] : 0;

    //Compute and store results
    __syncthreads();
    #pragma unroll
    for(int i = HALO_STEPS; i < HALO_STEPS + RESULT_STEPS; i++){
        float sum = 0;

        #pragma unroll
        for(int j = -KERNEL_RADIUS; j <= KERNEL_RADIUS; j++)
            sum += c_Kernel[KERNEL_RADIUS - j]
                         * s_Data[threadIdx.y][threadIdx.x + i * BLOCKDIM_X + j];

        d_Dst[i * BLOCKDIM_X] = sum;
    }
}
```

As the following code block shows, the column convolution implementation uses shared memory, due the reasons discussed previously. At the beginning of the kernel each thread loads the main data to be processed by the thread block as well as the additional pixels that will be needed to obtain results at the boundary pixels. Note that just like row convolution, column convolution operation uses a 5 tap blur filter and as a result boundary handling is required. This is simply due to the fact that the pixels at the thread block boundaries require 2 neighbors to their top and bottom. Finally, the filtering operation is performed and the results are written back to global memory. Note that every thread handles 8 pixels and loop unrolling pragma unroll is heavily used due to the fixed structures of the loops. Execution configuration and performance figures column filtering are given in Table 6.

```
__global__ void convolutionColumnsKernel(float *d_Dst,
                                         float *d_Src,
                                         int imageW,
                                         int imageH,
                                         int pitch)
{
    __shared__ float s_Data[BLOCKDIM_X][(RESULT_STEPS + 2 * HALO_STEPS) * BLOCKDIM_Y + 1];

    //Offset to the upper halo edge
    const int baseX = blockIdx.x * BLOCKDIM_X + threadIdx.x;
    const int baseY = (blockIdx.y * RESULT_STEPS - HALO_STEPS) * BLOCKDIM_Y + threadIdx.y;

    d_Src += baseY * pitch + baseX;
    d_Dst += baseY * pitch + baseX;

    //Main data
    #pragma unroll
    for(int i = HALO_STEPS; i < HALO_STEPS + RESULT_STEPS; i++)
        s_Data[threadIdx.x][threadIdx.y + i * BLOCKDIM_Y] = d_Src[i * BLOCKDIM_Y * pitch];

    //Upper halo
    #pragma unroll
    for(int i = 0; i < HALO_STEPS; i++)
        s_Data[threadIdx.x][threadIdx.y + i * BLOCKDIM_Y] =
        (baseY >= -i * BLOCKDIM_Y) ? d_Src[i * BLOCKDIM_Y * pitch] : 0;

    //Lower halo
    #pragma unroll
    for(int i = HALO_STEPS + RESULT_STEPS; i < HALO_STEPS + RESULT_STEPS + HALO_STEPS; i++)
        s_Data[threadIdx.x][threadIdx.y + i * BLOCKDIM_Y] =
        (imageH - baseY > i * BLOCKDIM_Y) ? d_Src[i * BLOCKDIM_Y * pitch] : 0;

    //Compute and store results
    __syncthreads();
    #pragma unroll
    for(int i = HALO_STEPS; i < HALO_STEPS + RESULT_STEPS; i++){
        float sum = 0;
        #pragma unroll
        for(int j = -KERNEL_RADIUS; j <= KERNEL_RADIUS; j++)
            sum += c_Kernel[KERNEL_RADIUS - j]
                        * s_Data[threadIdx.x][threadIdx.y + i * BLOCKDIM_Y + j];

        d_Dst[i * BLOCKDIM_Y * pitch] = sum;
    }
}
```

Table 5. Row filtering code analysis

Thread numbers (x,y)	(32,4)
Shared memory bank conflict	0%
Global memory BW efficiency	100%
Register usage	22
Occupancy	0,957
Total execution time per frame	120 microsec

Table 6. Column filtering code analysis

Thread numbers (x,y)	(32,8)
Shared memory bank conflict	0%
Global memory BW efficiency	100%
Register usage	32
Occupancy	0,476
Total execution time per frame	140 microsec

4 Performance Boost and Comparison

In this section we present the overall performance boost obtained by the CUDA implementation. The results we compare against are obtained by a Core i5 processor clocked at 3,1 GHz with 8 GB RAM for images of dimension 384x256 (single channel, 8-bit).

- One iteration of the gradient computation on the CPU was timed to be around 18 ms.
- One iteration of the gradient computation on the GPU was timed to be around 2 ms.

Gradient computation is the main computational block of the super-resolution algorithm and consists of applying the forward imaging model, taking the difference between the resulting *constructed* observation and the corresponding true observation, and finally applying the backward imaging model on the error. Hence, any speed up obtained for this block has a direct effect on the overall execution time. Please note that the "QueryPerformance" functions provided by MS has a time resolution of 1 ms. We have reason to believe that actual GPU timing is somewhere between 1ms and 2 ms, but we will not discuss this here and simply assume a 9X speed up.

- The overall algorithm execution time per frame on CPU is between 410 - 440 ms.
- The overall algorithm execution time per frame on GPU is between 42 - 46 ms.

These numbers represent a worst case speed up factor of approximately 9X and a best case speed up factor of approximately 10X on a low end GT 640 GPU, which has about 3 times lower core clock rate and 4 times less memory compared to the competing CPU configuration. For GT 640, current performance profiling results show that the CUDA implementation is both memory bandwidth and compute limited. This suggests that a high end GPU is guaranteed to improve the overall performance. Just to give an idea, GTX 670 has 3,5 times more compute power and 6,5 times higher memory bandwidth compated to GT 640.

5 Conclusion and Future Work

In this paper we discussed a massively-multithreaded CUDA implementation of the super-resolution algorithm originally presented in [2]. Performance test results were presented comparing a low-end $GT640$ NVIDIA GPU to a Core i5 CPU. Current implementation does not support overlapping kernel execution with data read/write. As future work, the current implementation will be moved to a higher end GPU such as $GTX660Ti$, $GTX670$ or $GTX680$ and modified to overlap kernel executions with data copies. We expect these modifications to further enhance the overall performance and lower the total execution time per frame to below 10 ms per frame, allowing the final implementation to run at 60 frames per second for frames of size 384x256 with a resolution enhancement factor of 2X.

References

1. NVIDIA CUDA C Programming Guide, http://developer.download.nvidia.com/compute/DevZone/docs/html/C/doc/CUDA_C_Programming_Guide.pdf
2. Gevrekci, M., Gunturk, B.K.: Image Acquisition Modeling for Super-Resolution Reconstruction. In: IEEE Int. Conf. on Image Processing (ICIP), vol. 2, pp. 1058–1061 (September 2005)

A Tool for Comparing Outbreak Detection Algorithms

Yasin Şahin

Hacettepe University,
Computer Engineering Department
06800 Ankara, Turkey
yasin@cs.hacettepe.edu.tr

Abstract. Despite of the main objective of recent biosurveillance researches is bioterrorist attack threats, detection of natural outbreaks are also being tried to solve by governments all over the world. Such that, international foundations as WHO, OECD and EU publish public declaration about necessity of an international central surveillance system. Each data source and contagious disease carries its own patterns. Therefore, standardizing the process of outbreak detection cannot be applicable. Various methods have been analyzed and published on test results in biosurveillance researches. In general, these methods are the algorithms in literature of SPC and Machine Learning, although specific algorithms have been proposed like Early Aberration Reporting System (EARS) methods. Differences between published results show that, the characteristic of time series are tested with algorithm and the chosen parameters of this algorithm are also determine results. Our tool provides preprocessing of data; testing, analyzing and reporting on anomaly detection algorithms specialized at biosurveillance. These functionalities make it possible to use outputs for comparing algorithms and decision making.

Keywords: outbreak detection, data smoothing, anomaly detection, time series analyzing.

1 Introduction

Outbreak is the occurrence of monitoring disease symptoms on a particular area and time more than normally expected. The expected value is quite hard to be expressed by a formula because of the impact of changing geographical locations and climatic conditions on it. Even in the cases of outbreak definition is given as statistically significant increase, it is not possible to see a description or a formula of this increase. Outbreaks which caused great deaths in the history also brought economic burdens together. During the world war I when the world's population was around 2 billion, Spanish Flu that predicted spread of 500 million people, caused the death of 20 to 50 million people all over the world [1]Besides health problems, just SARS has costed 10 billions of dollars only for Hong Kong government as a recent experiment[1].

[1] http://en.wikipedia.org/wiki/1918_flu_pandemic

D. Nagamalai et al. (Eds.): *Adv. in Comput. Sci., Eng. & Inf. Technol.*, AISC 225, pp. 59–72.
DOI: 10.1007/978-3-319-00951-3_6

1.1 Surveillance Systems

Surveillance systems have a major mission to struggle against the outbreak. Nowadays, the surveillance systems are specialized as the early warning systems which are being used for outbreak detection of time series provided from diagnostic or prediagnostic data[2]. Not only surveillance systems should be limited in use of bio-terrorist attacks, but also should address the natural outbreaks in order to reduce the number of deaths and prevent the financial losses. An idealized surveillance system consists of three stages. First stage is data collection phase. Second stage is, detecting outbreak from dataset provided by first stage. Warning the authorities about the outbreak possibility is the last step of the system. Outbreak detection phase should have adequate performance and this expectation relies on data collection stage's organization quality. The data collected asynchronously causes unreliable results.

1.2 Biosurveillance Data

Modern surveillance systems are expected to identify a possible outbreak as quick as possible. For this purpose, prediagnostic data such as increases in drug sales, number of absenteeism or tendencies of searches on the internet by regions could be inputs for the detection of outbreak[3]. Common patient pattern may consist of three steps in general; internet searching, treatment with the available drugs and visiting hospital lastly. Prediagnostic data might make more sense for this estimation.

The datasets used in the outbreak detection contain patterns which lead to different statistics over time[4]. The occurrence of a surging on the surveillance between the weekdays and weekend can be an obvious sample to this. Considering the outpatient services does not work on weekends, visitors to the hospital earlier in the week and so the number of cases may increase cumulatively. Goldenberg observed that drug sales increases at weekends[5]. Therefore, the time series even recorded for the same purpose may lead to significant differences on what conditions are covered. Holidays and holiday returns could be a similar effect as the day of the week[6]. The complexity of surveillance data became more prominent when considering the foresight of seasonal effects and the changes in the air temperature[3].

1.3 Data Preprocessing Methods

Various algorithms are proposed in anomaly detection in time series. Outbreaks are also special cases that arise in time series. Traditional methods provide highly effective process monitoring independent identically distributed (iid) time series. To make an effective performance on surveillance data with traditional methods, the data should be made iid by cleaning from the explainable patterns such as day-of the week effect and seasonal effect[4]. Evaluation metrics of preprocessing methods are Mean Square Error(MSE), Root Mean Square Error(RMSE), Median Absolute Percent Error(MdAPE) and Median Absolute Deviation (MdAD) [3][3] [4].

Regression

Regression analysis is a statistical technique for estimating the relationships between dependent variable and independent variable(s). Number of regression methods has been recommended and eq. 1 is the notation all of these methods.

$$Y \approx f(\boldsymbol{X}, \boldsymbol{\beta}) \tag{1}$$

where Y is dependent variable, X is independent variable(s) and β is unknown parameters. Regression models are common methods in time series forecasting. It is also known, regression is most preferred forecasting technique for daily health series[2]. Regression models are common methods in time series forecasting. It is also known that regression is most preferred forecasting technique for daily health series. In the context of biosurveillance, linear regression handles the observations components such as DOW effect, seasonal effect, holiday effect etc. by dummy variables. Linear regression computes additive relation among variables. Sometimes predictors used have a multiplicative rather than additive effect[7]. This nonlinearity can be derived by using $\log(y_t)$ rather y_t.

Adaptive Regression

Adaptive regression model proposed for surveillance datasets by Burkom*et al.* [4] for the first time. Fricker *et al.* [8] applied Cusum to the prediction errors of adaptive regression and noted well-performing of that composed algorithm. Adaptive regression model predicts the day's observation count by sliding baseline consist of most recent observations. Although Burkom proposed 8 weeks sliding window, Fricker reported that the best window size could be varying from 15 to 90. See Dunfee and Hegler*et al.* [9] to study how to determine size of sliding window in detail.

$$y(i) = \beta_0 + \beta_1 \times (i - t + n + 1) + \varepsilon \tag{2}$$

where y(i) is observation count, β_0 is intercept, β_1 is slope and ε is prediction error.

7 Day Differencing

7 day differencing proposed by Muscotello *et al.* [10] is known as the simplest forecasting algorithm. It predicts observation of the same day of previous week as next forecast:$y_t = x_{t-7}$.Applying any detection algorithm after extracting residual from prediction may be a simple and useful algorithm.

Moving Average

Moving average smoothing generates a new smoothed dataset by using subsets belongs to raw dataset. Simple moving average predicts mean of sliding window for next observation: $y_t = \frac{1}{n}\sum_1^n x_{t-i}$. Each observation within window has same weight on this prediction. Actually some outbreak detection methods such as EARS methods involve simple moving average into method's algorithm.

Weighted moving average is not a familiar smoothing technique for biosurveillance researches. It provides weighting for more recent observations.

$$y_t = \frac{nx_{t-1}+(n-1)x_{t-2}+\cdots+2x_{t-n-1}+x_{t-n}}{n+(n-1)+\cdots+2+1} \tag{3}$$

Exponentially weighted moving average (EWMA) brings effect of any observation infinitely. But the effect of the observation decreases exponentially. One of the advantages of EWMA is that, it doesn't require historical data.

$$y_t = \alpha x_t + (1-\alpha)y_{t-1} \tag{4}$$

Holt

Charles H. Holt *et al.* [11] proposed Holt exponential smoothing method for the first time. Holt method adds linear trend component to computation in additional to EWMA smoothing method.

$$\begin{aligned} F_t &= \alpha x_{t-1} + (1-\alpha)(F_{t-1} + T_{t-1}) \\ T_t &= \beta(F_t - F_{t-1}) + (1-\beta)T_{t-1} \\ H_{t+m} &= F_t + mT_{t+m} \end{aligned} \tag{5}$$

where α is smoothing constant, β is linear trend smoothing constant, F_t is estimated level term at time *t* and T_t is estimated trend term at time *t*. H_{t+m} is overall forecast value of model for time *t+m*. *m* should be 1 if the model supposed to estimate next observation count. Although it is known as a simple and flexible smoothing technique, it performs well while extracting trends from datasets.

Level term initializes as $L_1 = x_1$ and T_1 can be computed by 3 separate ways: $T_1 = x_2 - x_1$ or $T_1 = \frac{(x_p - x_1)}{p-1}$ or $T_1 = 0$ where p is number of observation within cycle.

Holt& Winters

Peter Winters *et al.* [12] suggested handling seasonal components extended by Holt's method.

$$\begin{aligned} L_t &= \alpha(x_t - S_{t-s}) + (1-\alpha)(L_{t-1} + T_{t-1}) \\ T_t &= \beta(L_t - L_{t-1}) + (1-\beta)T_{t-1} \\ S_t &= \gamma(x_t - L_t) + (1-\gamma)S_{t-s} \\ HW_{t+m} &= L_t + mT_t + S_{t-s+m} \end{aligned} \tag{6}$$

where S_t is seasonal component of the model. Holt&Winters smoothing method cannot make any estimation at first seasonal cycle because of model's dependence on values provided by previous cycle. Holt&Winters needs also be initialized. Proposed initializing equations given as:

$$\begin{aligned} L_s &= \frac{1}{s}(x_1 + x_2 + \cdots + x_s) \\ \mathrm{T}_s &= [(x_{s+1} + x_{s+2} + \cdots + x_{s+s}) - (x_1 + x_2 + \cdots + x_s)]/m^2 \\ \mathrm{S}_i &= x_i - L_s \ \ (1 \le i \le s) \end{aligned} \tag{7}$$

Level and trend smoothing and their initializations are computed as additive model given above, but computations of seasonal component of model changes as eq.(8).

$$S_t = \gamma \frac{x_t}{F_t} + (1-\gamma)S_{t-s}$$
$$HW_{t+m} = (L_t + mT_t)S_{t-s+m} \tag{8}$$

Lastly, multiplicative model of Holt&Winters' seasonal initialization can be calculated as given eq.(9)[13].

$$\mathrm{S_i} = \frac{x_i}{L_s} \quad (1 \leq i \leq s) \tag{9}$$

Decomposition

Time series consists of four components in general. These are *Trend, Cyclical, Seasonal* and *Irregular* components. In many time series, occurrence of a pattern could be in two ways; additive and multiplicative. An additive model assumes that, the difference between the Monday and Saturday values is approximately the same each week for daily data. A component of a multiplicative model causes a pattern with equal ratio in time series under the same conditions. To give an example, if observation count of each Monday is 10 greater than Sunday of the same week, it means additive model exists, but if observation count each Monday is greater than Sunday of the same week 3/2 times, it means multiplicative model exists.

$$x_t = T_t + C_t + S_t + I_t \tag{10}$$
$$x_t = T_t \times C_t \times S_t \times I_t \tag{11}$$
$$\log(x_t) = \log(T_t) + \log(C_t) + \log(S_t) + \log(I_t) \tag{12}$$

Decomposition is a statistical technique to remove explainable pattern(s) from time series for smoothing. Decomposition has separate means for different study areas. While a statistician use it for curve-fitting, an electronics engineer may use it prevent to signals from noisy [14]. Subtraction of an additive component or division of a multiplicative component is adequate to decompose a pattern.

The day of week effect and seasonal effect are also smoothed by decomposition. Applying linear regression to dataset as given eq. (13) can provide the predictions values of the effects.

$$y_t = \beta_0 + \beta_{mon} I_{mon} + \cdots + \beta_{sun} I_{sun} + \beta_1 \sin\left(\frac{2\pi}{365.25}\right) i + \beta_2 \cos\left(\frac{2\pi}{365.25}\right) i \tag{13}$$

where β_0 is intercept, the I's are dummy indicators, where $I_{mon} = 1$ on Monday, otherwise $I_{mon} = 1$ to disable it. $\sin\left(\frac{2\pi}{365.25}\right) i$andcos$\left(\frac{2\pi}{365.25}\right) i$ components are estimator of the seasonal effects, where *i* is the distance to *last* 1 October at time *t*.

2 Outbreak Detection Methods

Control charts (Shewhart charts) were proposed by Walter Shewhart purpose of quality control at the beginnings of 1920s [15]. Statistical Process Control (SPC) has built on Shewhart charts in times. Control charts check two sided thresholds called Upper Control Limit (UCL) and Lower Control Limit (LCL) to monitor the process. Falling outside of this acceptable interval between limits generates an alarm in case of possibility of an anomaly existence. Surveillance systems only interested in UCL threshold. SPC algorithms such as Shewhart, EWMA and Cusum have been widely studied for adapting to biosurveillance. Methods extended by SPC such as EARS methods or methods does not belong to SPC also proposed with several studies such as Negative Binomial Cusum(NBC) [16], Historical Limits Method(HLM) [17], Hidden Markov Model(HMM) [18] etc.

Shewhart

In the biosurveillance studies, the Shewhart Control Charts, which are basic for the EARS System models, are discussed frequently.

$$S_t = \frac{x_t - \mu}{\sigma} \tag{14}$$

where μ and σ are estimated from observation counts of days that does not carry outbreak.It is checked that how much an observation moves away from the average with Shewhart method given eq. (14).Most important advantage of Shewhart for the biosurveillance systems is warning a quick response with a little historical data when the outbreak is sharp and suddenly raised. It is powerful when the shifts of process are not small.

Cumulative Sum

Counter to Shewhart, Cusum method is used to determine small shifts [19] encountered during process which is proposed by Page [20]. Calculating recent observations effect into score make Cusum method skillful algorithm in detecting small shifts. If the observations show seasonal waving, Cusum might be weak to make a right decision about process [21].

$$CS_t^+ = \text{maks}[0, \mathrm{x}_t - (\mu_0 + k) + CS_{t-1}] \tag{15}$$

In the context of biosurveillance, it needs to be prevented falling below zero in order not to have negative meaningless values obtained by Cusum because of increasing is only be dealt with [22]. k reference value could be middle of standard deviation [9]. See Fricker*et al.* [8] for choosing k in details.

EWMA

EWMA, which is also a smoothing method, is successful in the determination of the small changes during process.

$$E_t = \alpha x_t + (1 - \alpha)E_{t-1} \tag{16}$$

where,$\alpha$is EWMA constant can be interval of $0 < \alpha \leq 1$ equation.

$$\sigma_{ewma}^2 = \frac{\alpha}{2-\alpha}\sigma^2$$
$$E_t^+ = E_0 + k\sigma_{ewma} \tag{17}$$

Mean value of observations called *target* value is selected for initialization of EWMA score: $E_0 = \mu$. When the calculated EWMA score moves away from target value for k times, it exceeds the threshold and anomaly will have been detected. See Lucas *et al.*[28] which has study about determining k and threshold.

$$E_t = \text{maks}[\mu, \alpha x_t + (1-\alpha)E_{t-1}] \tag{18}$$

If EWMA algorithm is applied to processed data instead of raw data, eq.(18) could be changed as $E_t = \text{maks}[0, \alpha x_t + (1-\alpha)E_{t-1}]$.

EARS Methods

EARS methods adapted to detect sharp and sudden increments caused by bio-terrorism without long historical baseline. EARS has three methods called C1, C2 and C3. Although EARS methods are claimed as Cusum-based, in fact they are based on Shewhart.

EARS methods may be the most discussed methods within surveillance context. For this reason, a researcher, who proposes a new algorithm, considers comparing the algorithm performance with EARS methods firstly.

$$C_1(t) = \frac{x_t - \overline{X_t}}{s_t} \tag{19}$$

where $\bar{X}_t$ and s_t are mean and standard deviation of sliding baseline with fixed size 7. Minimum standard deviation can be applied to statistic to prevent it from zero division as 0.2. Alarm is raised when the statistic of C1 exceeds threshold. Although CDC, proposer of EARS methods, tests C1 and C2 with 3 as threshold, Fricker *et al.* [8]proposed adjusted thresholds determined by fixed ATFS to obtain comparable performance. The only difference between C2 and C1 is two days lag preference of C2. C2 calculates both of standard deviation and mean of baseline excluding last 2 days: $\overline{X_t} = \frac{1}{7}\sum_{i=t-3}^{t-9} x_i$ and $s_t = \sqrt{\frac{1}{6}\sum_{i=t-3}^{t-9}[x_i - \overline{X_\iota}]^2}$. Thus, manipulation of an outbreak caused by step-wise increments within time series could be prevented. C3, another method of EARS, is calculated by an equation uses C2 statistics at time t with most recent two statistics as given below.

$$C_3(t) = \sum_{i=t}^{t-2} max[0, C_2(i) - 1] \tag{20}$$

While threshold of C1 and C2 set as 3, C3 is being tested with 2 within EARS.

Negative Binomial Cusum

NBC is Cusum based method minimized false alert rates with right configuration. This ability can make it useful for designing a system purposes low false alert rate. Some of biosurveillance researches noted that NBC has better performance than EARS methods over outbreak detection [18]. NBC is described by two parameters; r and c. Anomaly is determined by parameter c.

$$c_0 = \frac{\mu}{\mu - \sigma^2}$$

$$r = \frac{\mu^2}{\mu - \sigma^2}$$

where mean and variance are being calculated based on sliding baseline. If there is an outbreak, in-control level c_0 will shift to out-of-control level c_1 . Decision interval of NBC can be monitored with the equation given below [16].

$$NBC_t = maks(0, NBC_{t-1} + x_t - k) \quad (21)$$
$$k = r.\ln\left[\frac{c_0(1+c_1)}{c_1(1+c_0)}\right] / \ln\left[\frac{1+c_0}{1+c_1}\right] .$$

Watkins et *al.* [18] proposed out of control level as value of a fixed interval of two baseline standard deviations greater than in control level c_0.

Historical Limits Method
HLM was designed to deal with seasonal effects [24].

$$\frac{x_0}{\mu} > 1 + \frac{2\sigma_x}{\mu} \quad (22)$$

where μ and σ_x are the mean and standard deviation of the historical data. If the size of sliding baseline set as 7 days for 3 years historical data, 45 daily points totally joins equation.15 observations for each year come from 7 preceding days, 7 subsequent days and current day. As seen in eq. (22), HLM needs a great historical data. This requirement makes it incomparable with EARS because of EARS' purposes. A study claims that HLM has better performance than EARS methods has also tested statistic with weekly datasets [24].

3 System Design of Tool

The tool has developed with one of the most popular programming language Java. All abilities provided by the tool could have been implemented with *Matlab* or *R* except availability on Internet and user interfaces. Author's experiment over Java and flexibility of Java made it one step ahead while language choosing phase. The tool was properly designed and implemented for requirements of Model View Controller (MVC) architecture. Thus, by the abstraction between layers, modifying business logic or user interface could be adapted quickly.

ZK ajax framework which is suitable for MVC preferred while developing. ZK has both of free community edition and enterprise edition but free version, Community Edition licensed with LGPL, was found adequate to the tool's development process. A small database was also modeled within tool development. Another free product has been chosen as Database Management System. PostgreSQL is a free DBMS application provides any functionalities tool needs. Communication between application and database was handled by Hibernate, which has been incorporated to standard library of Java Persistence API.

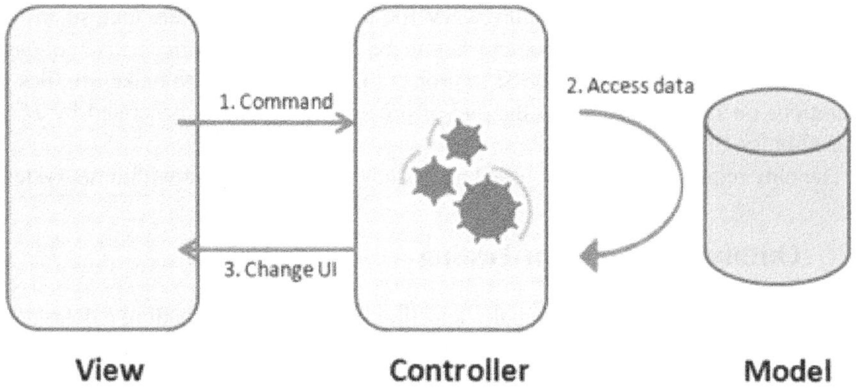

Fig. 1. MVC Architecture[2]

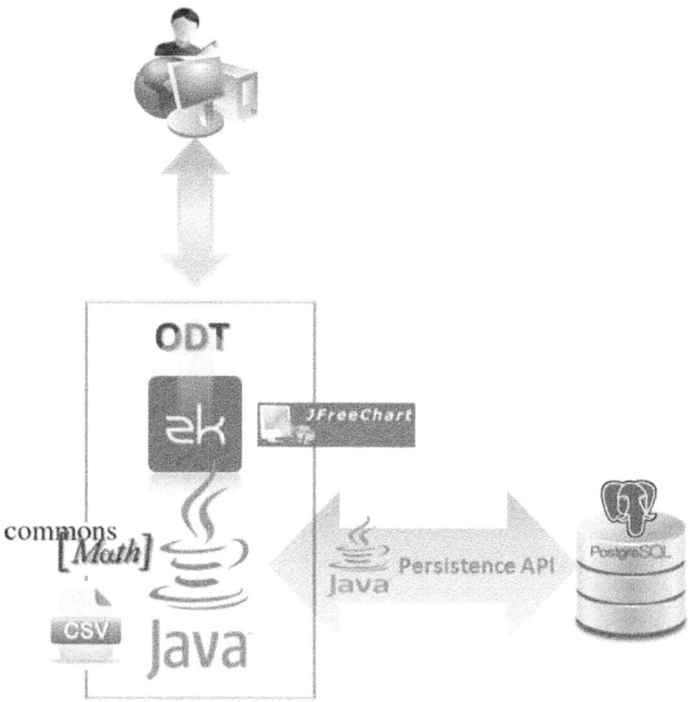

Fig. 2. Outbreak Detection Tool System Architecture

Apache Common Math was used to linear regression operations. Preprocessing methods such as decompositions, Simple Linear Regression, Multiple Linear Regression and Adaptive Regression exploits apache common math library. Comma-separated values (CSV) file format was chosen in order to users could freely and easily edit their

---

[2] `http://books.zkoss.org/wiki/ZK_Developer's_Reference/MVC`

files according to tool's file structure. CSV file format based on plain text, so any text editor is able to edit it. File structure has to be as *date, observation count, anomaly existence*. *OpenCsv* library imported to project to have an ability to make *csv* files I/O operations on Java language. At the end of process, result charts generated by *JFreeChart* library.

The only requirement of web-based tool is having a user profile within the system.

4 Outbreak Algorithm Testing

In this section, process of an anomaly detection beginning by uploading a dataset will be told. At the end of the process, comparison of the algorithms will be done.

Each step of the process can be saved by the user with this tool. Rest of the process can be done with separate ways to get different results. First step of the process after uploading is determining a smoothing process. One or more preprocessing method can be applied to a dataset or skipping this step may be more suitable for the researcher.

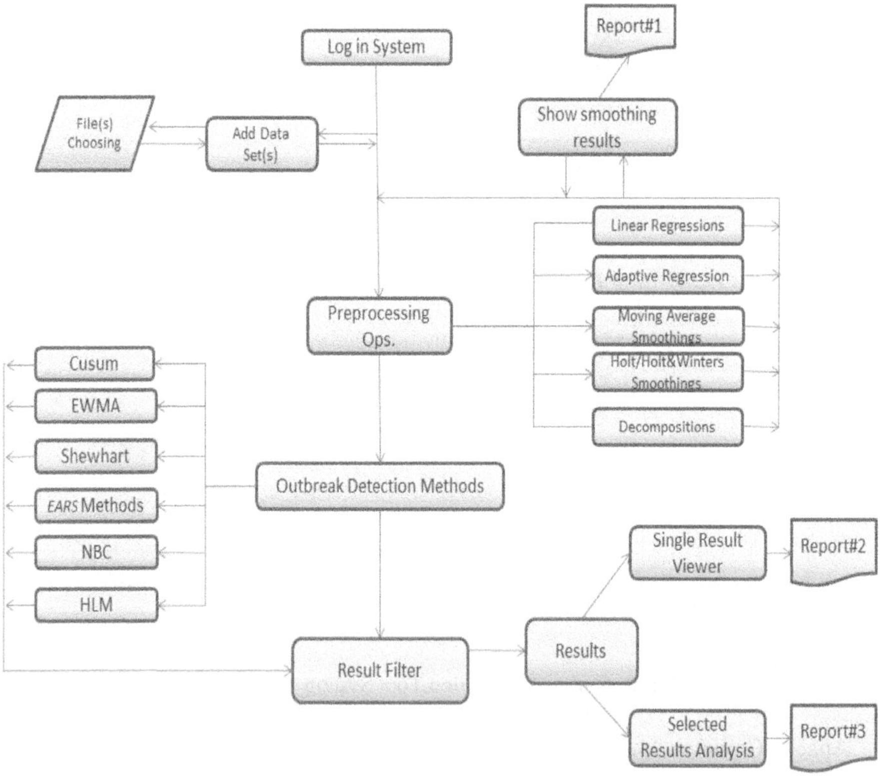

Fig. 3. ODSprocess flow chart

Report#1 is a chart for reporting processed dataset with its raw dataset.Report#2 shows daily scores of the algorithm with threshold. Report#3 provides XY line charts such as ROC, AMOC or whatever user wants to place within X,Y coordinates. AUC is calculated during XY line drawing to measure algorithm's performance.

After smoothing operations, detection methods would be applied to source. User can test and see the comparing results of a method with various parameters or various methods with similar parameters. To see summary of performance statistic of results, user needs to filter to make a comparing list. At this step, user can display a daily score chart or provides XY line charts such as ROC, AMOC or whatever user wants to place within X,Y coordinates. AUC is calculated during XY line drawing to measure algorithm's performance.

A sample analyzing process is comparison of Cusum and EWMA methods. Dataset we tested is available from CDC web site. Both of two algorithms were tested with ATFS values as 20,50,100 and 200.The threshold adjustment to achieve a particular ATFS estimated empirically.

Algorithm #1: Cusum method was tested with dataset that extracted residual from adaptive regression proposed as Burkom*et al.* [4]. Cusum scored reset after detection to prevent cumulative increasing. Reference value k was calculated from $\frac{1}{2}\sigma_{\varepsilon}\frac{(n+1)\times(n+2)}{n(n-1)}$ equation proposed by Fricker*et al.* [8].

Algorithm #2: EWMA method was applied to dataset that had been smoothed by DOW effect and seasonal effect decomposition. Sunday selected as reference day and both of seasonal components applied for multiplicative decomposition. EWMA was applied to processed data with smoothing factor as 0,3.

Parameters were not the best performers of the algorithms for applied dataset. The main idea of sampling is sampling the process within tool.

5 Results

Statistical evaluation metrics widely chosen for outbreak detection are *sensitivity, specificity* and *timeliness.* Starting and ending of the outbreaks are not same as anomalies which defined for other contexts. In general, definition of sensitivity is the proportion of conditional anomalies which signed as anomaly for it. But if the anomaly is an outbreak, then conditional anomaly is duration instead of observation. For example, consider an outbreak lasts 10 sequential days and it is detected only for 7th day of outbreak duration. Sensitivity would be measured 0,1 as usual way, but 1,0 for the context of biosurveillance. For this case, timeliness will be valuable metric to handle delay. Specificity is the proportion of conditional negatives which signed as negative for it. In other words, it is the probability that an outbreak test will correctly indicates that an individual does not have a particular condition.

Evaluation of the algorithms given previous section is based on sensitivity, specificity and timeliness metrics. The performance of each algorithm was also evaluated bycomparing the area under the receiver operating characteristic(ROC) curve (AUC).

According to AUC values, EWMA has better performance than Cusum. Areas under ROC curve of EWMA and Cusum are 0,021 and 0,009 respectively. Actually performance of Cusum could only approach to EWMA's for applied data source with chosen parameters when ATFS was leveled to 20.Both algorithms have similar timeliness and specificity but EWMA has better sensitivity than Cusum. It can be seen that, larger ATFS values make Cusum weak for detecting anomalies with adaptive regression.

Obtained performances of the algorithms were modeled without concern about finding best algorithms. Explaining skills of the tool we developed was purposed. These scores do not make better EWMA than Cusum or reverse. When we apply the algorithms to chosen dataset gives us these results. Defining parameters, determining comparison metrics or preprocessing methods are related to research's expectations, so the performance evaluation weighting depend on that expectations.

Using the skills of the tool, a user can processes modeling and get reports showed on fig. 3.

Table 1. Evaluation Metrics of Cusum and EWMA for different threshold adjusted to achieve given ATFS

ATFS (day)	Sensitivity (%)		Specificity (%)		Timeliness (day)	
	Cusum	*EWMA*	*Cusum*	*EWMA*	*Cusum*	*EWMA*
20	63,889	63,889	97,204	97,707	3,478	3,000
50	25,000	44,444	98,764	99,218	4,778	3,688
100	13,889	36,111	99,731	99,739	4,000	4,692
200	2,778	22,222	99,463	99,896	4,000	5,500

6 Conclusion

> *"The syndromic surveillance literature documents quite a number of efforts to develop and measure the performance of various individual detection algorithms. However, it contains very few comparisons between algorithms in order to assess the relative strengths and weaknesses of the algorithms. It is as if everyone is trying to develop a new hammer, but few are comparing among the hammers to determine which are to be preferred."*
>
> *Dunfee and Hegler et al.* [9]

Anomaly detection in time series is a common research area of network security (intrusion detection), credit card security (fraud detection) and epidemiology (disease outbreak detection). SPC methods and Machine Learning algorithms have been imported to outbreak detection recent years. Outbreak detection differs from other anomaly detections because of surveillance data features.

Determining the outbreak detection algorithms may play a key role for the surveillance systems. To give a right decision about algorithm, questions listed below must be answered smoothly:

- What is the outbreak's spreading type?
- Which type of outbreak spreading is more important to detect?
- What is the acceptable false alert rate?

- What is the least true alert rate?
- What is the longest acceptable out of control ARL(timeliness).

When we can give right answers to these questions, it is possible to determine the most convenient algorithm for data source. The tool we developed can give an idea about corresponding of expectations and results.

Adding new functionalities to search out patterns within time series and more algorithms from the literature of SPC and Machine Learning can be future works. Additional information such as provided from *GIS* can be added to process for detecting outbreak rapidly and robustly. It is also seen that, available surveillance datasets need to be increased to have field knowledge more than we do. An ideal surveillance system can be achieved by two ways. First one is defining the outbreak definition statistically for each disease and the second one is developing a system that is able to determine to the best algorithm related to expectations.

References

1. Siu, A., Wong, Y.C.R.: Economic Impact of SARS: The Case of Hong Kong. Asian Economic Papers 3(1), 62–83 (2004)
2. Lotze, T.H.: Anomaly Detection in Time Series: Theoretical and Practical Improvements for Disease Outbreak Detection, Ph.D. Thesis. University of Maryland (2009)
3. Shmueli, G., Burkom, H.S.: Statistical Challenges Facing Early Outbreak Detection in Biosurveillance. Technometrics (Special Issue on Anomaly Detection) (1), 39–51 (2010)
4. Burkom, H.S., Murphy, S.P., Shmeuli, G.: Automated Time Series Forecasting for Biosurveillance. Statist. Med. 26(22), 4202–4218 (2007)
5. Goldenberg, A., Shmueli, G., Caru, R.A., Fienberg, S.E.: Early statistical detection of anthrax outbreaks by tracking over-the-counter medication sales. Proceeding of the National Academy of Sciences 99(8), 5237–5240 (2002)
6. Fienberg, S.E., Shmueli, G.: Statistical issues and challenges associated with rapid detection of bio-terrorist attack. Statistics in Medicine 24(4), 513–529 (2005)
7. Kleinman, K., Lazarus, R., Platt, R.: A Generalized Linear Mixed Models Approach for Detecting Incident Clusters of Disease in Small Areas, with an Application to Biological Terrorism. Am. J. Epidemiol. 159(3), 217–224 (2004)
8. Fricker Jr., R.D., Hegler, B.L., Dunfee, D.A.: Comparing syndromic surveillance detection methods: EARS' versus a CUSUM-based methodology. Statistics in Medicine (17), 3407–3429 (2008)
9. Dunfee, D.A., Hegler, B.L.: Biological Terrorism Preparedness: Evaluating the Performance of the Early Aberration Reporting System (EARS) Syndromic Surveillance Algorithms (2007)
10. Muscatello, D.: An adjusted cumulative sum for count data with day-of-week effects: application to influenza-like illness. In: Syn. Surv. Conf., Boston (2004)
11. Holt, C.C.: Forecasting seasonals and trends by exponentially weighted moving averages. International Journal of Forecasting 20(1), 5–10 (2004)
12. Winters, P.R.: Forecasting Sales by Exponentially Weighted Moving Averages. Management Science 6(3), 324–342 (1960)
13. Makridakis, S.G., Wheelwright, S.C., Hyndman, R.J.: Forecasting: Methods and Applications. John Wiley and Sons, New York (1998)

14. Brown, R.G.: Smoothing, forecasting and prediction of discrete time series. Dover Publications, Mineola (2004)
15. Shewhart, W.A.: Quality Control Charts. Bell System Technical Journal 5(4), 593–603 (1926)
16. Hawkins, D.M., Olwell, D.H.: Cumulative Sum Charts and Charting for Quality Improvement. Springer, New York (1998)
17. Stroup, D.F., Williamson, G.D., Her, J.L.: Detection of aberrations in the occurrence of notifiable diseases surveillance data. Statistics in Medicine 8(3), 323–329 (1989)
18. Watkins, R.E., Eagleson, S., Veenendaal, B., Wright, G., Plan, A.J.: Applying cusum-based methods for the detection of outbreaks of Ross River virus disease in Western Australia. BMC Medical Informatics and Decision Making 8(37) (2008)
19. Gan, F.F.: An optimal design of CUSUM control charts for binomial counts. Journal of Applied Statistics 20(4), 445–460 (1993)
20. Page, E.S.: Continuous Inspection Schemes. Biometrika Trust 41(1/2), 100–115 (1954)
21. Kartal, M.: İstatistiksel Kalite Kontrolu, Erzurum: Şafak Yayınevi (1999)
22. Wagner, M.M., Moore, A.W., Aryel, R.M.: Handbook of biosurveillance. Academic Press Inc., Amsterdam (2006)
23. Lucas, J.M., Sanucci, M.S.: Exponentially Weighted Moving Average Control Schemes: Proporties and Enhancement. Technometrics 32(1) (1990)
24. Pelecanos, A.M., Ryan, P.A., Gatt, M.L.: Outbreak detection algorithms for seasonal disease data: a case study using ross river virus disease. BMC Medical Informatics and Decision Making 10(74) (2010)

Block Updates on Truncated ULV Decomposition

Jesse L. Barlow[1], Ebru Aydoğan[2], and Hasan Erbay[2]

[1] Department of Computer Science and Engineering, The Pennsylvania State University, 343G IST Building, University Park, PA 16802-6822 USA

[2] Computer Engineering Department, Kırıkkale University, Yahşihan, 71450, Kırıkkale Turkey

Abstract. A truncated ULV decomposition (TULV) of an $m \times n$ matrix X of rank k is a decomposition of the form $X = U_1 L V_1^T + E$, where U_1 and V_1 are left orthogonal matrices, L is a $k \times k$ non-singular lower triangular matrix and E is an error matrix. Only U_1,V_1, L, and $\|E\|_F$ are stored.

We propose algorithms for block updating the TULV based upon Block Classical Gram-Schmidt that in [4]. We also use a refinement algorithm that reduces $\|E\|_F$, detects rank degeneracy, corrects it and sharpens the approximation.

Keywords: Truncated ULVD, Block Classical Gram-Schmidt, Block Update.

1 Introduction

For a matrix $X \in \Re^{m \times n}$, $m \geq n$, the truncated ULV decomposition(TULV) is of the form

$$X = U_1 L V_1^T + E \tag{1}$$

where for some $k \leq n$, $U_1 \in \Re^{m \times k}$, $V_1 \in \Re^{n \times k}$ are *left orthogonal*, that is, $U_1^T U_1 = V_1^T V_1 = I_k$, $L \in \Re^{k \times k}$ is non-singular and lower triangular, and $E \in \Re^{m \times n}$ is an error matrix [2].

The matrices L and E satisfy

$$\|L^{-1}\|_2 \leq \epsilon^{-1}, \quad \|E\|_2 \leq \epsilon, \quad U_1^T E = 0 \tag{2}$$

where ϵ is some tolerence [2].

Using the TULV of X to produce the TULV of

$$\bar{X} = \begin{pmatrix} X \\ \mathbf{a}^T \end{pmatrix} \tag{3}$$

is called updating. In the literature there are plenty of updating algorithms such that given by Stewart [11] and Erbay, Barlow and Zhang [6] for computing the TULV of $\bar{X}$ in (3).

D. Nagamalai et al. (Eds.): *Adv. in Comput. Sci., Eng. & Inf. Technol.*, AISC 225, pp. 73–79.
DOI: 10.1007/978-3-319-00951-3_7

In this work, we develop the block matrix algorithms to support block update operation that is using the TULV of X produce the TULV of

$$\bar{X} = \begin{pmatrix} X \\ A^T \end{pmatrix} \tag{4}$$

where A is the new arrival matrix. Our algorithm uses the Block Classical Gram-Schmidt algorithm(BGCS2) [4] which is detailed in the section 2. In §3 we give a block update algorithm and show how the refinement algorithm in [1] may be used as a "clean up" procedure. In §4, we give numerical tests of our algorithm.

2 Computational Tools

As we mentioned in Section 1 our TULV block update algorithm based on the BGCS2 algorithm due to Barlow and Alicia [4]. The core of the BGCS2 algorithm is an orthogonal factorization routine **local_qr**, which for a matrix $\bar{X} \in \Re^{m \times p}$, $p \leq n \leq m$, produces

$$[\bar{Q}, \bar{R}] = \mathbf{local\_qr}(\bar{X}) \tag{5}$$

where $\bar{R} \in \Re^{p \times p}$ is an upper triangular matrix and $\bar{Q} \in \Re^{m \times p}$ is left orthogonal matrix, indeed, expected to be left orthogonal matrix. In [4] it is given that $\bar{Q}$ and $\bar{R}$ satisfy

$$\left\| I_p - \bar{Q}^T \bar{Q} \right\| \leq \epsilon_M L_1 (m, p) < 1 \tag{6}$$

$$\bar{X} + \triangle \bar{X} = \bar{Q}\bar{R}, \quad \left\| \triangle \bar{X} \right\| \leq \epsilon_M L_1 (m, p) \left\| \bar{X} \right\| \tag{7}$$

where $L_1 (m, p)$ is some modest function and ϵ_M is machine unit. **local_qr** routine can be coded using Householder or Givens QR-factorization. An error analysis on Householder QR-factorization given in [10, §19.3] yields $L_1 (m, p) = d_1 m p^{3/2}$ where d_1 is a constant. It is worthwhile to mention that Gram-Schmidt based algorithms play an important role in the modification of QR factorization when rows or columns are added or deleted [5],[9],[12],[3].

The BGCS2 algortihm consists of two application of the One Block CGS step outlined in Algorithm 1. The error bound of Algorithm 1 can be found in [4, §3.1].

Algorithm 1 One Block CGS Step

function $[\bar{Q}, \bar{R}, \bar{S}]$ = **block_CGS_step**(U, B)

$\bar{S} = U^T B;$
$\bar{X} = B - U\bar{S};$
$[\bar{Q}, \bar{R}]$ = **local_QR**$(\bar{X})$;
end block_CGS_step

The output $\bar{Q}$ is near left orthogonal and satisfies (6). Beside $\bar{R}$ is an upper triangular matrix, together with $\bar{Q}$ and $\bar{X}$ satisfies (7) [4].

The steps of BGCS2 algorithm are given in Algorithm 2. Detailed error analysis can be found in [4, §3].

Algorithm 2 Two Steps of Block CGS

function $[Q_B, R_B, S_B]$ = **block_CGS2_step**(U, B)

$[\bar{Q}_1, \bar{R}_1, \bar{S}_1]$ = **block_CGS2_step**(U, B);
$[Q_B, \bar{R}_2, \bar{S}_2]$ = **block_CGS2_step**$(U, \bar{Q}_1)$;
$S_B = \bar{S}_1 + \bar{S}_2\bar{R}_1$;
$R_B = \bar{R}_2\bar{R}_1$
end **block_CGS2_step**

The TULV Block Update algoritm also includes the refinement procedure given by Barlow and Erbay [2]. This algorithm assures us that the conditions in (2) are maintanied after the TULV Block Update.

3 Block Update Algorithm

Let $X \in \Re^{m\times n}$, $m \geq n$ be a matrix with the TULV given in (1). Let $\bar{X}$ be the matrix given in (4). Then $\bar{X}$ can be re-written as

$$\begin{aligned}\bar{X} &= \begin{pmatrix} X \\ A^T \end{pmatrix} \\ &= \begin{pmatrix} U_1 L V_1^T + E \\ A^T \end{pmatrix} \\ &= \begin{pmatrix} U_1 L V_1^T \\ A^T \end{pmatrix} + \begin{pmatrix} E \\ 0 \end{pmatrix}.\end{aligned}$$

Using **block_CGS2_step** [4] with the input arguments A and V we obtain

$$\left[V_{new}, L_{new}^T, S_{new}\right] = \textbf{block\_CGS2\_step}\,(A, V_1)$$

then

$$A^T = S_{new}^T V^T + L_{new} V_{new}^T$$

thus, $\bar{X}$ can be re-written as

$$\bar{X} = \begin{pmatrix} U_1 & 0 \\ 0 & I \end{pmatrix} \begin{pmatrix} L & 0 \\ S_{new}^T & L_{new} \end{pmatrix} \begin{pmatrix} V_1^T \\ V_{new}^T \end{pmatrix} + \begin{pmatrix} E \\ 0 \end{pmatrix}.$$

To get the final ULVD we apply refinement algorithm proposed by Barlow, Erbay and Slapničar [1] to the middle matrix $\begin{pmatrix} L & 0 \\ S_{new}^T & L_{new} \end{pmatrix}$.

The block TULV update algorithm is summarized in Algorithm (3).

Algorithm 3 TULV Block Update

function $[\bar{U}_1, \bar{L}, \bar{V}_1]$ = **TULV_block_update**(U_1, L, V_1, A)

$\bar{X} = \begin{pmatrix} X \\ A^T \end{pmatrix}$;

$[V_{new}, L_{new}^T, S_{new}]$ = **block_CGS2_step** (A, V_1);

$\tilde{U}_1 = \begin{pmatrix} U_1 & 0 \\ 0 & I \end{pmatrix}$; $\tilde{L} = \begin{pmatrix} L & 0 \\ S_{new}^T & L_{new} \end{pmatrix}$; $\tilde{V}_1 = (V_1 \; V_{new})$; $\tilde{E} = \begin{pmatrix} E \\ 0 \end{pmatrix}$;

$[\bar{U}_1, \bar{L}, \bar{V}_1]$ = **TULV_refinement**$(\bar{X}, \tilde{U}_1, \tilde{L}, \tilde{V}_1)$;

end TULV_block_update

Theorem 1. *Let $\bar{X} = \bar{U}_1 \bar{L} \bar{V}_1^T + \bar{E}$ be given. Then $\bar{U}_1^T \bar{E} = 0$.*

Proof. We know that $U_1^T E = 0$. Then

$$\begin{aligned}
\bar{U}_1^T \bar{E} &= \bar{U}_1^T \left(\bar{X} - U_1 \bar{L} \bar{V}_1^T\right) \\
&= \bar{U}_1^T \bar{X} - \bar{U}_1^T \bar{U}_1 \bar{L} \bar{V}_1^T \\
&= \bar{U}_1^T \bar{X} - \bar{L} \bar{V}_1^T \\
&= \begin{pmatrix} U_1^T & 0 \\ 0 & I \end{pmatrix} \begin{pmatrix} X \\ A^T \end{pmatrix} - \begin{pmatrix} L & 0 \\ S_{new}^T & L_{new} \end{pmatrix} \begin{pmatrix} V^T \\ V_{new}^T \end{pmatrix} \\
&= \begin{pmatrix} U_1^T X \\ A^T \end{pmatrix} - \begin{pmatrix} LV^T \\ S_{new}^T V^T + L_{new} V_{new}^T \end{pmatrix} \\
&= \begin{pmatrix} U_1^T X - LV^T \\ A^T - S_{new}^T V^T + L_{new} V_{new}^T \end{pmatrix} \\
&= 0
\end{aligned}$$

Let $\bar{U}_1^\dagger$ be the Moore-Pensore psuedoinverse of $\bar{U}_1$ [8, p.257–8]; that is the unique matrix satisfying the following equations:

$$\bar{U}_1 \bar{U}_1^\dagger \bar{U}_1 = \bar{U}_1 \tag{8}$$

$$\bar{U}_1^\dagger \bar{U}_1 \bar{U}_1^\dagger = \bar{U}_1^\dagger \tag{9}$$

$$(\bar{U}_1 \bar{U}_1^\dagger)^T = \bar{U}_1 \bar{U}_1^\dagger \tag{10}$$

$$(\bar{U}_1^\dagger U)^T = \bar{U}_1^\dagger \bar{U}_1 \tag{11}$$

Theorem 2. *Let $\bar{X} = \bar{U}_1 \bar{L} \bar{V}_1^T + \bar{E}$ be given. Then*

$$\bar{E} = \bar{P}\bar{X}, \quad \bar{P} = I - \bar{U}_1 \bar{U}_1^\dagger.$$

Proof. Let start with $\bar{P}\bar{X}$

$$\begin{aligned}\bar{P}\bar{X} &= \left(I - \bar{U}_1\bar{U}_1^{\dagger}\right)\bar{X} \\ &= \bar{X} - \bar{U}_1\bar{U}_1^{\dagger}\bar{X} \\ &= \bar{X} - \bar{U}_1\bar{U}_1^{\dagger}\left(\bar{U}_1\bar{L}\bar{V}_1^T + \bar{E}\right) \\ &= \bar{X} - \bar{U}_1\bar{U}_1^{\dagger}\bar{U}_1\bar{L}\bar{V}_1^T + \bar{U}_1\bar{U}_1^{\dagger}\bar{E}\end{aligned}$$

Usage of Moore-Pensore conditions (8) and (10) yield

$$\begin{aligned}\bar{P}\bar{X} &= \bar{X} - \bar{U}_1\bar{L}\bar{V}_1^T + \left(\bar{U}_1\bar{U}_1^{\dagger}\right)^T\bar{E} \\ &= \bar{E} - \left(\bar{U}_1^{\dagger}\right)^T\bar{U}_1^T\bar{E}\end{aligned}$$

The proof follows from Theorem 1.

4 Numerical Tests

This section presents some numerical results from our numerical experiments. The tests are run in Matlab Version R2009b on a personal computer.

We use block exponential window technique to test our algorithm. Initial decomposition of the matrix X is obtained by the method discussed in [7].

We tracked the numerical rank of the matrix after each TULV Block Update and used MATLAB's Singular Value Decomposition as a reference in checking the accuracy of the rank estimate.

We check the left orthogonality of the matrices V_1 and U_1 by computing

$$\left\|I - V_1^T V_1\right\|_2$$

$$\left\|I - U_1^T U_1\right\|_2$$

,respectively and plotted them on log 10 scale to have better view of errors.

We also computed the decomposition error

$$\|E\| = \left\|X - U_1 L V_1^T\right\|$$

and also

$$\left\|U_1^T E\right\|$$

after each TULV Block Update and plotted them on log 10 scale again.

Example 1. X, 150-by-8 random matrix, chosen from a uniform distribution on the interval $(0, 1)$. 145 randomly chosen rows of X multiplied by $\eta = 10^{-9}$ in order to vary the rank of the matrix, and $\epsilon \approx 10^{-8}$. The initial matrix $X(0)$ is of size 40-by-8. Later, $X(1)$ of size 4-by-8. After that at each step 2-by-8 matrices added one after another. The forgetting factor $\alpha = 0.9$.

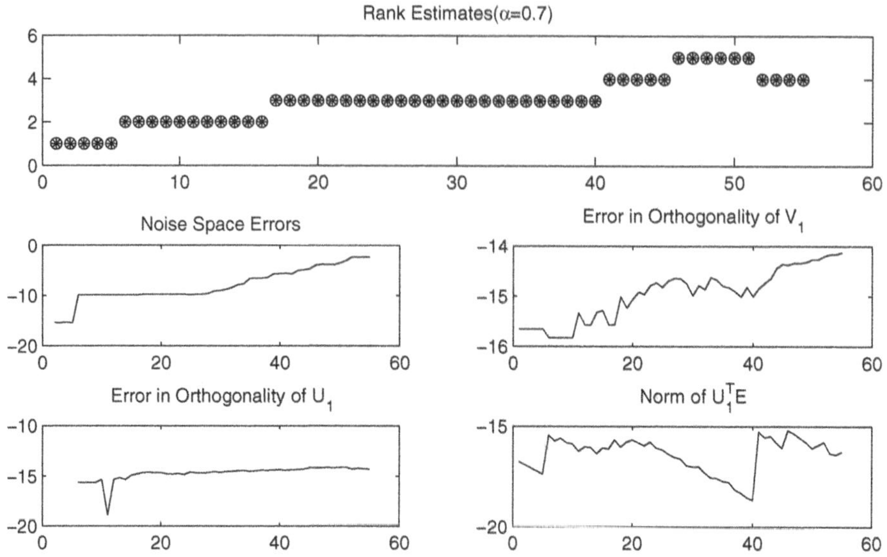

Fig. 1. Numerical results obtained by Example 1

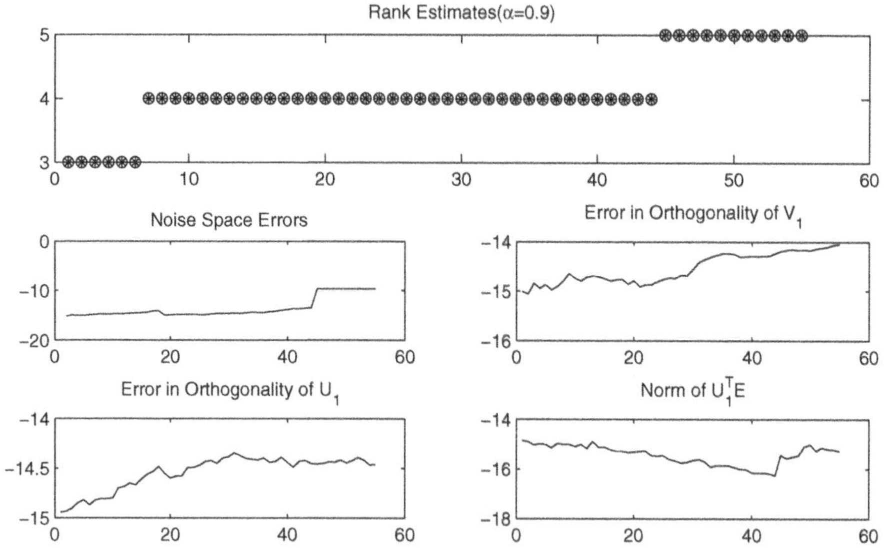

Fig. 2. Numerical results obtained by Example 2

Example 2. X, 150-by-8 random matrix, chosen from a uniform distribution on the interval $(0, 1)$. 145 randomly chosen rows of X multiplied by $\eta = 10^{-9}$ in order to vary the rank of the matrix, and $\epsilon \approx 10^{-8}$. The initial matrix $X(0)$ is of size 40-by-8. Later, $X(1)$ of size 4-by-8. After that at each step 2-by-8 matrices added one after another. The forgetting factor $\alpha = 0.7$.

The figures in Figure 1 and Figure 2 show the ability of TULV Block Update algorithm to track the numerical rank and also accuracy of the aproximated subspaces.

References

1. Barlow, J.L., Erbay, H., Slapničar, I.: An Alternative Algorithm for the Refinement of ULV Decompositions. SIAM Journal on Matrix Analysis and Applications 27(1), 198–211 (2005)
2. Barlow, J.L., Erbay, H.: Modifible Low-Rank Approximation to a Matrix. Numerical Linear Algebra with Applications 16, 833–860 (2009), doi:10.1002/nla.651
3. Barlow, J.L., Smoktunowicz, A., Erbay, H.: Improved Gram-Schmidt Type Downdating Methods. BIT Numerical Mathematics 45(2), 259–285 (2006)
4. Barlow, J.L., Smoktunowicz, A.: Reorthogonalized Block Classical Gram-Schmidt, Numerische Mathematik (to appear)
5. Daniel, J.W., Gragg, W.B., Kaufman, L., Stewart, G.W.: Reorthogonalization and Stable Algorithms for Updating the Gram-Schmidt QR Factorization. Mathematics of Computation 30, 772–795 (1976)
6. Erbay, H., Barlow, J.L., Zhang, Z.: A modified Gram–Schmidt-based downdating technique for ULV decompositions with applications to recursive TLS problems. Computational Statistics & Data Analysis 41(1), 195–209 (2002)
7. Erbay, H., Barlow, J.L.: An Alternative Algorithm for a Sliding Window ULV Decompostions. Computing 76(1-2), 55–66 (2006)
8. Golub, G.H., Van Loan, C.F.: Matrix Computations, 3rd edn. The John Hopkins Press (1996)
9. Mathias, R., Stewart, G.W.: A block QR algorithm and the singular value decomposition. Linear Algebra and Its Applications 182, 91–100 (1993)
10. Higham, N.J.: Accuracy and Stability of Numerical Analysis. Cambridge University Press, Cambridge (2002)
11. Stewart, G.W.: Updating A Rank-Revealing ULV Decomposition. SIAM Journal on Matrix Analysis and Applications 14, 494–499 (1993)
12. Yoo, K., Park, H.: Accurate Downdating of a modified Gram-Schmidt QR decomposition. BIT Numerical Mathematics 36, 166–181 (1996)

Complementary Problems for Subset-Sum and Change Making Problems

Asli Guler[1] and Urfat Nurıyev[2]

[1] Department of Mathematics, Faculty of Science and Letters, Yasar University, Izmir, Turkey
[2] Department of Mathematics, Faculty of Science, EgeUniversity, Izmir, Turkey
asli.guler@yasar.edu.tr, urfat.nuriyev@ege.edu.tr

Abstract. In this study, Change Making Problem (CMP) and Subset-Sum Problem (SSP), which can arise, in practice, in some classes of one dimensional cargo loading and cutting stock problems, are researched. These problems are often used in computer science, as well. CMP and SSP are NP-hard problems and these problems can be seen as types of the knapsack problem in some ways. The complementary problems for the change making problem and the subset-sum problem are defined in this study, and it is aimed to examine the CMP and SSP by means of the complementary problems.

Keywords: change making problem, subset-sum problem, complementary problem, greedy algorithm.

1 Introduction

It is known that most of the optimization problems are NP- complete problems; and there is little hope to find exact algorithms for these problems, which run in polynomial time , unless P = NP [3,4]. Therefore, designing approximate algorithms with guarantee value for NP-complete problems has become one of the attractive subjects recently [1, 7].

In this study, Change Making Problem (CMP) and Subset-Sum Problem (SSP), which can arise, in practice, in some classes of one dimensional cargo loading and cutting stock problems, are researched. These problems are often used in computer science, as well. CMP and SSP are NP-hard problems and these problems can be seen as types of the knapsack problem in some ways. The complementary problems for the change making problem and the subset-sum problem are defined in this study, and it is aimed to examine the CMP and SSP by means of the complementary problems.

Moreover, we will focus on approximation algorithms which run in polynomial time and find solutions that are guaranteed to be close to optimal solution. We will not seek for the optimal solution, and as a result, it becomes feasible to aim for a polynomial running time.

Let P be an optimization (minimization or maximization) problem with positive integral cost function c, and let A be an algorithm which, given an instance I of P returns a feasible solution $z^A(I)$, and let $z^*(I)$ be the optimal solution. Algorithm

D. Nagamalai et al. (Eds.): *Adv. in Comput. Sci., Eng. & Inf. Technol.*, AISC 225, pp. 81–88.
DOI: 10.1007/978-3-319-00951-3_8

A is an approximation algorithm with relative performance guarantee Δ if $\frac{z^A(I)}{z^*(I)} \geq \Delta$ holds for all problem instances I. Clearly, the given inequality makes sense for maximization problems and $0 < \Delta < 1$. If it is a minimization problem, then it must be $\frac{z^A(I)}{z^*(I)} \leq \Delta$, $\Delta > 1$ [2]. An algorithm with relative performance guarantee Δ will be called a Δ***-approximation algorithm*** and Δ is generally called the approximation ratio or the ***guarantee value*** of the algorithm.

The notations which are used in the study are given below:

$B = \{0,1\}$; $N = \{1,2,...,n\}$, set of items; $N_0 = \{0,1,2,...\}$; $Z^+ = \{1,2,...\}$,

$BI = \{x_j \in \mathbb{N}_0 \mid x_j \leq b_j,\ j \in N,\ b_j \in N\}$, set of variables with bounded integer,

For the notations that are given below, it is chosen one of the signs $i \in \{+,-\}$ as superscript; and it shows that the problem is a maximization or a minimization problem. For subscript, it is chosen one of B, BI, I ($ii \in \{B, BI, I\}$).

K_{ii}^{i} = Optimal solution value,

A_{ii}^{i} = Solution value found by greedy algorithm,

2 The Change Making Problem

The Change Making Problem (CMP) is to make up a specific amount of money with the minimum number of coins of given denominations. CMP is a kind of unbounded knapsack problem with $p_j = -1$, $(j = 1,...n)$ and equality in the capacity constraint instead of inequality. This problem can arise, in practice, in some classes of one dimensional cargo loading and cutting stock problems. It has shown that CMP is an NP-hard problem. The change making problem can be formally defined as follows:

Given n item types and a knapsack, with

w_j : weight of an item type j,

x_j : number of items of an item type j,

c: capacity of the knapsack

The problem is to determine the values of $x_j, (j = 1,...,n)$ so that the total weight equals c and total number of items is a minimum.

$$\text{minimize } Z = \sum_{j=1}^{n} x_j$$

$$\text{subject to } \sum_{j=1}^{n} x_j w_j = c$$

$$x_j \geq 0 \text{ and integer, } j \in N = \{1,...,n\}.$$

It is assumed, without loss of generality, that

w_j and c are integers

$w_j < c, \ j \in N$

$w_i \neq w_j, \ i \neq j$

A feasible solution to to the problem may not exist due to the capacity constraint.

One of the algorithms suggested for CMP is given below [12]; it is assumed that the items are sorted by

$$w_1 \geq w_2 \geq \ldots \geq w_n$$

in the following greedy algorithm.

Algorithm $A(A_I^-)$

A1) k = 1; $A_I^- = 0$; ĉ=c ;

A2) $x_k = \left\lfloor \frac{\hat{c}}{w_k} \right\rfloor$;

A3) $A_I^- = A_I^- + x_k$; $\hat{c} = \hat{c} - w_k x_k$;

A4) $k = k+1$;

A5) If $k \leq n$, then go to Step2;

A6) END.

After the algorithm is applied, if $\hat{c} = 0$ then it means that the solution found by the algorithm is a feasible solution.

Theorem: *Let* $\hat{c} = 0$ *and* $\left\lfloor \frac{c}{w_1} \right\rfloor = k$ *, then algorithm* $A(A_I^-)$ *has a guarantee value of* $\frac{k+1}{k} = \omega$

$$K_I^- \leq A_I^- \leq \omega K_I^-$$

2.1 Complementary Problem for CMP

In order to create the complementary problem of the change making problem, we compose $n_j = \left\lfloor \frac{c}{w_j} \right\rfloor$ for each item type j. Consequently, complementary problem is expressed below with the following statements.

$$W = \sum_{j \in N} n_j w_j, \ \ \bar{c} = W - c, \ \ y_j = n_j - x_{n-j+1},$$

$$\text{maximize} \quad Z = \sum_{j=1}^{n} y_j$$

$$\text{subject to} \quad \sum_{j=1}^{n} y_j w_j = \bar{c}$$

$$0 \le y_j \le n_j \ \text{ and integer }, j \in N.$$

It is seen that the complementary problem of change making problem is equivalent to the bounded integer maximization problem with $p_j = 1$ and equality in the capacity constraint instead of inequality [5,6].

One of the algorithms proposed for this problem is given below. It is assumed that the items are sorted according to

$$w_1 \le w_2 \le \ldots \le w_n$$

Algorithm $A(A_{BI}^{+})$

A1) $k = 1$ and $A_{BI}^{+} = 0$;

A2) $x_k = \left\lfloor \dfrac{\bar{c}}{w_k} \right\rfloor$;

A3) If $x_k > n_k$, then $x_k = n_k$ and $A_{BI}^{+} = A_{BI}^{+} + x_k$; $\bar{c} = \bar{c} - w_k x_k$;

otherwise; $A_{BI}^{+} = A_{BI}^{+} + x_k$; $\bar{c} = \bar{c} - w_k x_k$;

A4) $k = k + 1$;

A5) If $k > n$, then go to Step7;

A6) If $w_k \le \bar{c}$ then go to Step2;

A7) $x_k = x_{k+1} = \ldots = x_n = 0$;

A8) $A_{BI}^{+} = \max\left\{ A_{BI}^{+}, \max_{j \in N}\{n_j\} \right\}$

A9) If $A_{BI}^{+} = \max_{j \in N}\{n_j\}$, then let j* be the index that gives this result. In this case,

$x_{j*} = n_{j*}$ and $x_j = 0, \ \ j \ne j*$.

Otherwise, solution vector found before remains the same.

A10) END.

Theorem : *If* $\bar{c} = 0$, *algorithm* $A(A_{BI}^{+})$ *has a* guarantee *value of* $\dfrac{1}{2}$.

$$\frac{1}{2} K_{BI}^{+} \le A_{BI}^{+} \le K_{BI}^{+}$$

3 Subset-Sum Problem

The subset-sum problem (SSP) denotes the special case of knapsack problem where $p_j = w_j$ for all $j = 1,...,n$. It is also called the Value Independent Knapsack Problem or Stickstacking Problem. The most important applications of the problem are cargo loading, cutting stock and job scheduling [10].The SSP has been proven to be NP-compelete problem in [9].
The subset-sum problem is formally defined as follows:

Given a set of n items and a knapsack, with

w_j = weight of an item j,

c = capacity of the knapsack,

The problem is to select a subset of the items whose total weight is closest to, without exceeding, c.

$$\text{maximize} \quad Z = \sum_{j=1}^{n} x_j w_j$$

$$\text{subject to} \quad \sum_{j=1}^{n} x_j w_j \leq c$$

$$x_j = 0 \text{ or } 1, \; j \in N = \{1,...,n\}.$$

Here,

$$x_j = \begin{cases} 1 & \text{if item } j \text{ is selected;} \\ 0 & \text{otherwise.} \end{cases}$$

It is assumed, without loss of generality, that

w_j and c are integers

$$\sum_{j=1}^{n} w_j > c,$$

$$w_j < c, \; j \in N.$$

Many algorithms have been proposed for this problem [8]. By defining $p_j = w_j$ for all j, we can use an algorithm for the 0-1 knapsack problem. One of these algorithms is given below; it is assumed that the items are sorted by

$$w_1 \geq w_2 \geq ... \geq w_n$$

in the following greedy algorithm.

Algorithm $A(\, A_B^+ \,)$

A1) k = 1, $\hat{c} = c$;

A2) If $w_k > \hat{c}$ then $x_k = 0$ else

begin $x_k = 1$; $\hat{c} = \hat{c} - w_k$ end;

A3) k = k+1;

A5) If $k \le n$ then go to Step2;

A6) $z^g = c - \hat{c}$;

A7) END.

The guarantee value of this algorithm is $\frac{1}{2}$; and the time complexity is O(nlogn) because of the required sorting.

3.1 Complementary Problem for SSP

Minimization or maximization versions of some optimization problems can be considered in a similar way; one of them is known as the complementary problem of the other one. Complementary problem of SSP is expressed as below:
$W = \sum_{j \in N} w_j$, $\overline{c} = W - c$, $y_j = 1 - x_{n-j+1}$, and

$$\text{minimize } Z = \sum_{j=1}^{n} y_j w_j$$

$$\text{subject to } \sum_{j=1}^{n} y_j w_j \ge \overline{c}$$

$$y_j = 0 \text{ or } 1, \ j \in N = \{1, ..., n\}.$$

We can use the following algorithm to find an approximate solution for this problem. It is assumed that the items are sorted by

$$w_1 \le w_2 \le ... \le w_n$$

and $p_j = w_j$.

Algorithm $A(A_B^-)$

A1) $A_B^- = \sum_{j=1}^{n} p_j$, $R = \varnothing$;

A2) $\overline{k} = \min\left\{ l \middle| \sum_{j=1}^{l} w_j \ge c \right\}$;

A3) $L = \sum_{j=1}^{\overline{k}} p_j$;

A4) $A_B^- = \min\{A_B^-, L\}$, $N = N / \overline{k}$, $R = R \cup \{\overline{k}\}$;

A5) If $\sum_{j \in N} w_j \ge c$ then go to Step2;

A6) For $j \in N$ $\quad x_j = 1, \ j = 1, .., \overline{k} - 1$;

$$x_j = 0, \quad j = \bar{k}+1,...,n;$$

$$x_{\bar{k}} = 1;$$

For $j \in R / \{\bar{k}\}$ $x_j = 0$;

A7) END.

Theorem: *It holds for the minimization SSP that*

$$K_B^- \leq A_B^- \leq 2K_B^-$$

Some theorems for the problem and their proofs are given in [13,14].

4 Conclusions

In this study, maximization and minimization versions of subset-sum and change making problems and greedy algorithms for these problems are discussed. Guarantee values of the algorithms are calculated; and complementary problems are created in order to improve known guarantee values. Time complexities of the given algorithms are O(nlogn) because of the sortings.

We consider complementary problems for maximization versions of the problems, but they can be used for minimization problems to improve the guarantee values of their algorithms.

This study could be thought for other combinatorial optimization problems for improvement. Furthermore, the principle of complementary problem could be used with other different approaches as well as greedy algorithms.

References

1. Ausiello, G., Crescenzi, P., Kann, V., Marchetti-Spaccamela, A., Protasi, M.: Complexity and Approximation: Combinatorial Optimization Problems and their Approximability Properties, 524p. Springer, Berlin (1999)
2. Fisher, M.L.: Worst-case Analysis of Heuristic Algorithms. Management Science 26(1), 1–17 (1980)
3. Garey, M.R., Johnson, D.S.: Computers and Intractability: A Guide to the Theory of NP-Completeness, 338p. Freeman, San Francisco (1979)
4. Gens, G.V., Levner, E.V.: Efficient Approximate Algortihms for Combinatorial Problems, 66p. The USSR Academy of Sciences. CEMI Press, Moscow (1980)
5. Guler, A., Nuriyeva, F.: Algorithms with guarantee value for bounded knapsack problems. In: Proceedings of 24th Mini Euro Conference on Continuous Optimization and Information-Based Technologies in the Financial Sector (MEC EurOPT 2010), Izmir, Turkey, June 23-26, pp. 183–189 (2010)
6. Guler, A., Nuriyev, U., Berberler, M.E., Nuriyeva, F.: Algorithms with Guarantee Value for Knapsack Problems. Optimization 61(4), 477–488 (2012)
7. Güntzer, M.M., Jungnickel, D.: Approximate minimization algorithms for the 0/1 knapsack and subset-sum problem. Operations Research Letters 26, 55–66 (2000)

8. Hochbaum, D.S.: Approximation Algorithms for NP-Hard Problems, 596p. PWS Publishing, Boston (1997)
9. Horowitz, E., Sahni, S.: Computing Partitions with applications to the knapsack problem. Journal of ACM, 277–292 (1974)
10. Ibarra, O.H., Kim, C.E.: Fast approximation algorithms for the knapsack and sum of subset problems. Journal of ACM 22(4), 463–468 (1975)
11. Kellerer, H., Pferschy, U., Pisinger, D.: Knapsack Problems, 546p. Springer, Berlin (2004)
12. Martello, S., Toth, P.: Knapsack Problems, 296p. John Wiley& Sons, England (1990)
13. Nikitin, A.I., Nuriev, U.G.: On a method of the solution of the knapsack problem. Kibernetika 2, 108–110 (1983)
14. Nuriev, U.G.: On the solution of the knapsack problem with guaranteed estimate, Problems of Computing Mathematics and Theoretical Cybernetics. The Cybernetics Institute Academy of Sciences of Azarbaijan SSP, 66–70 (1986)

A New Method for Estimation of Missing Data Based on Sampling Methods for Data Mining

Rima Houari[1], Ahcène Bounceur[2], Tahar Kechadi[3], Tari Abdelkamel[1], and Reinhardt Euler[2]

[1] University of Abderrahmane Mira Bejaia
[2] Lab-STICC Laboratory - European University of Britanny - University of Brest
[3] University College Dublin, Ireland

Abstract. Today we collect large amounts of data and we receive more than we can handle, the accumulated data are often raw and far from being of good quality they contain Missing Values and noise.

The presence of Missing Values in data are major disadvantages for most Datamining algorithms. Intuitively, the pertinent information is embedded in many attributes and its extraction is only possible if the original data are cleaned and pre-treated.

In this paper we propose a new technique for preprocessing data that aims to estimate Missing Values, in order to obtain representative Samples of good qualities, and also to assure that the information extracted is more safe and reliable.

Keywords: Datamining, Copulas, Missing Value, Multidimensional Sampling, Sampling.

1 Introduction and Previous Work

During the process of knowledge extraction in databases, companies are trying to understand how to extract the values of all the data they collect. Consequently the presence of Missing Data devaluates the power of Data Mining algorithms causes a major problem on the way to achieve knowledge. Phase specific treatment of some data is often necessary to remove or complete them. Especially during the extraction of knowledge, these incomplete data are mostly removed. This sometimes leads to the elimination of more half of the base; the information extracted is no more representative and not reliable.

Many techniques for processing Missing Data have been developed [17][11][2]. According to [10] and [20], there are three possible strategies for dealing with Missing Data; the first technique use the deletion procedures [19] [16] [11]. These methods allow to have a complete database, therefore this method sacrifices a large amount of data [13], which is a major weakness in Data Mining. The following technique relies on the use of alternative procedures (substitution), that are intended to built to a comprehensive basis by finding an appropriate way

D. Nagamalai et al. (Eds.): *Adv. in Comput. Sci., Eng. & Inf. Technol.*, AISC 225, pp. 89–100.
DOI: 10.1007/978-3-319-00951-3_9

to replace Missing Values. Among these methods we can mention: the method of mean imputation [12], the regression method [13][16][24], and the method of imputation by the k-nearest neighbor [3] [5] [27][28]. Generally, these methods are not adapted to the characteristics of Data Mining when processing large databases or large percentage of missing values.

The latter technique is to estimate certain parameters of data distribution containing Missing Values, such as the method of maximum likelihood [4] [16] and the expectation maximization [1][29][14][24], these techniques are very costly in computation time, they require more specification of a data generation model. This task involves making a certain number of hypotheses, which is always difficult for they solutions because they are not always feasible.

The great advantage of the presented method is to create a complete database. However, it is not beneficial for Datamining unless the replaced data on the large databases with a great percentage of Missing Values are very close to the original data and they do not alter the relationship between the variables. For this reason, we propose a new approach based on the theory of Copulas which involves estimating Missing Values in a manner to better reflect the uncertainty of data when the most significant knowledge is to be extracted.

The paper is organized as follows: basic concepts are presented in Section 2, Section 3 contains a description of the proposed method, and experimental results are given in Section 4, Section 5 concludes the paper.

2 Basic Concepts

In this Section, we will introduce some basic concepts that will be used in the rest of the paper. These concepts include the inverse transform Sampling to generate random samples from a probability distribution given its cumulative distribution function (CDF). A range of family of Copulas will also be presented.

2.1 Inverse Transform Sampling

A classical approach for generating samples from a one dimensional CDF is the inverse transform sampling method. Given a continuous random variable X with a CDF $F(x) = P[X \leq x]$, the transform $U = F(X)$ results in a random variable U that is uniform in $[0, 1]$. Moreover, if U has a uniform distribution in $[0, 1]$ and X is defined as $X = F^{-1}(U)$, where $F^{-1}(u) = inf\{x, F(x) \geq u\}$ and inf denotes the greatest lower bound of a set of real numbers, then the CDF of X is $F(X)$. In order to generate an arbitrary large sample of X, we start with a uniformly distributed sample of U that can easily be generated using a standard pseudo-random generator. For each sampled value of U, we can calculate a value of X using the inverse CDF given by $X = F^{-1}(U)$. Figure 1 illustrates this method for the case of a Gaussian random variable.

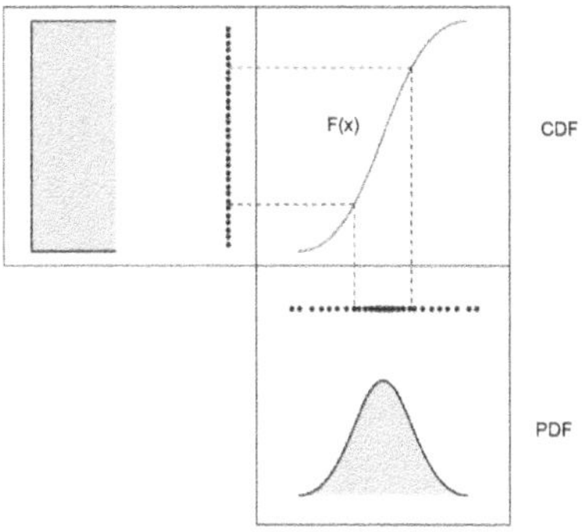

Fig. 1. The inverse method to generate a sample from a Gaussian distribution

2.2 Definition and Illustration of Copulas

Datamining seeks to identify the knowledge of massive data and the importance of the dependence structure between the variables is essential. By using Copulas, Datamining can take advantage of this theory to construct multivariate probability distribution without constraint to specific types of marginal distributions of attributes, in order to predict Missing Values in large databases.

2.3 Definition

Formally, a Copula [23] is a joint distribution function whose margins are uniform on $[0,1]$.

C $[0,1]^m \mapsto [0,1]$ is a Copula if $U_1 ... U_m$ are random variables which as uniformly distributed in $[0,1]$, such that [20]

$$C(u_1 ..., u_m) = p[U_1 \leq u_1, ..., U_m \leq U_m] \quad (1)$$

The most important theorem in the theory of Copulas is given by Sklar [22]

Let F be a distribution function with marginal distribution functions $F_1 ... F_m$. Then there exists a Copula such that $\forall (x_1 .... x_m) \in R^m$

$$F(x_1 ..., x_m) = C[F_1(x_1), ..., F_n(x_m)] \quad (2)$$

To illustrate the method of calculating a Copula from any sample, we consider the bivariate example in the Figure below

The above example illustrates that Gaussian Copulas can model the dependency between random variables that do not necessarily follow a Gaussian distribution. We consider two random variables X_1 and X_2 and we assume

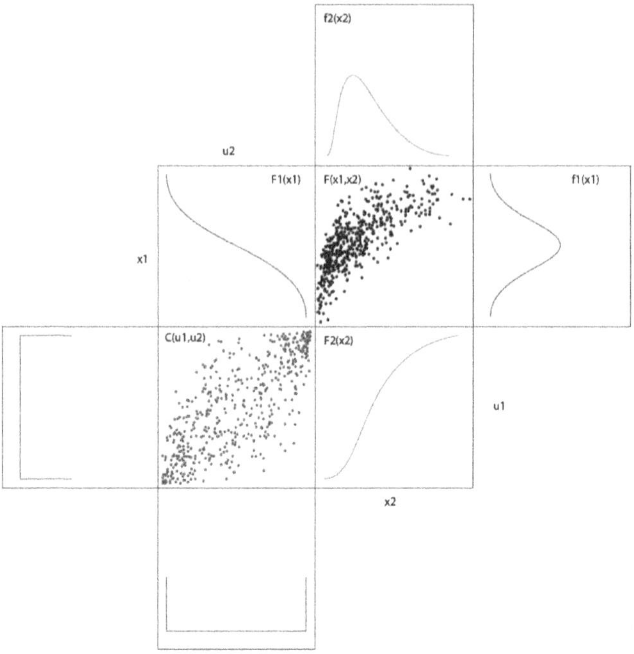

Fig. 2. Generation of a sample of a Gaussian copula from a sample of a bivariate distribution that has as marginal distributions Gamma and Gaussian

that $f_1(x_1)$ follows a Gaussian distribution and $f_2(x_2)$ a Gamma distribution, follow as shown in the top right corner of Figure 2. Using the transformation $U_i = F(x_i)$, we obtain the corresponding sample in the space (u_1, u_2). The result of this transformation is bivariate sample from the Gaussian Copula.

Empirical Copula

To avoid introducing any assumptions on the marginal CDF $F_i(x_i)$, we use the empirical CDF of $F_i(x_i)$, to transform the m samples of X into m samples of U. Empirical Copula is useful for examining the dependence structure of multivariate random vectors. Formally, the empirical Copula is given by the following equation:

$$c_{ij} = \frac{1}{m}(\sum_{k=1}^{m} I_{(v_{kj} \leq v_{ij})}) \tag{3}$$

$I(arg)$ is the indicator function, which is equal to 1 if arg is true and 0 otherwise. Here, m is used to keep the empirical CDF less than 1, where m is the number of observations, v_{kj} is the value of the kth row and j^{th} column; v_{ij} is the value of the ith row and j^{th} column.

2.4 Family of Copulas

There is a wide range of family of Copulas

Gaussian Copula

The difference between the Gaussian Copula and the joint normal CDF is that the Gaussian Copula allows having different marginal CDF types from the joint CDF type where as the joint normal CDF does not. Thus, the Gaussian Copula is indeed distinguished from the joint normal CDF. The Gaussian Copula is a link between a multivariate normal distribution and marginal distributions. It is defined as:

$$C(\Phi(x_1), \cdots, \Phi(x_m)) = \frac{1}{\mid \Sigma \mid^{\frac{1}{2}}} \exp(\frac{-1}{2} X^t(\Sigma^{-1} - I)X) \quad (4)$$

where $f_i(x_i)$ is standard Gaussian distribution, i.e., $X_i \sim N(0,1)$, and Σ is the correlation matrix. The resulting Copula $C(u_1, \cdots, u_n)$ is called Gaussian Copula. The density associated with $C(u_1, \cdots, u_n)$ is obtained by using Equation (6). Using $u_i = \Phi(x_i)$ we can write

$$C(u_1, \cdots, u_m) = \frac{1}{\mid \Sigma \mid^{\frac{1}{2}}} \exp[\frac{-1}{2} \xi^t(\Sigma^{-1} - I)\xi] \quad (5)$$

where $\xi = (\Phi^{-1}(u_1), \cdots, \Phi^{-1}(u_m))^T$.

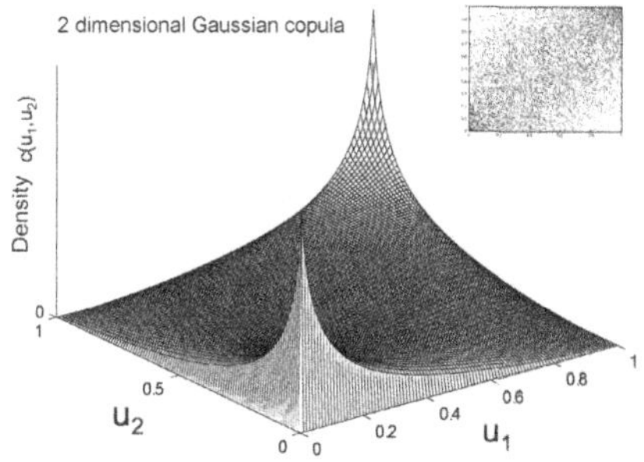

Fig. 3. 2-dimensional Gaussian Copula density resulting from a sample of a bivariate standard Gaussian distribution

Student Copula

The Copula t is extracted from the multivariate t distribution, which is given by the following:

$$\forall (u_1, \cdots, u_m) \in [0:1]^m \quad (6)$$

$$C(u_1, \cdots, u_m) = \frac{(f_{(v,\Sigma)}(t_v^{(-1)} u_1, \cdots, t_v^{(-1)} u_1))}{(\prod_{i=1}^{(n)} (f_{(v)} t_v^{(u_i)}))}$$

where $t_v^{(-1)}$ is the inverse of the t distribution centered and reduced to univariate degrees of freedom. $f_{(v,\Sigma)}$ is the probability density function of the Student distribution which is centered and reduced. Σ is the correlation matrix and $f_{(v)}$ is the density unirange of the Student distribution, centered and reduced ($\Sigma = 1$).

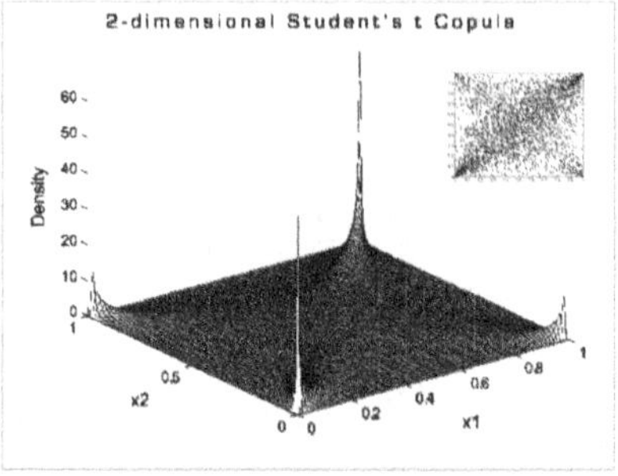

Fig. 4. 2-dimensional Student Copula density resulting from a sample of a bivariate standard Student distribution

Archimedean Copulas

Among various types of Copulas, one parameter Archimedean Copulas has attracted much attention owing to their convenient properties. For an Archimedean Copula, there exists a generator such that the following relationship holds:

$$C(u_1 \ldots, u_n) = \begin{cases} \varphi^{-1}(\varphi(u_1 + \ldots + u_n) & \text{if } \sum_{\varphi(u_i) \leq \varphi(0)}, \\ 0 & \text{else} \end{cases}$$

φ is called the generating function that checks the Copula: $\varphi(1) = 0$, $\varphi(\acute{u}) < 0$ and $\acute{u}\varphi(\acute{u}) > 0$, $0 \leq u < 1$. In the table belows (TABLE I) we have summarized some examples of Archimedean Copulas.

Table 1. Examples of Archimedean Copulas

Copules	$\varphi(u)$	C(u)
$\prod$	-ln u	$\prod_{i=1}^{d} u_i$
Gumbel	$(-lnu)^{\theta}, \theta \geq 1$	$exp\{-[\Sigma_{i=1}^{d}(-lnu)^{\theta}]^{1/\theta}\}$
Frank	$\frac{-lnexp((-\theta u)-1}{exp(-\theta)-1}$	$-\frac{1}{\theta}ln(1 + \frac{\prod_{i=1}^{d} exp((-\theta u_i)-1}{(exp(-\theta)-1)^{d-1}})$
Clayton	$u^{-\theta} - 1, \theta > 0$	$(\Sigma_{i=1}^{d} u^{-\theta} - d + 1)^{\frac{-1}{\theta}}$

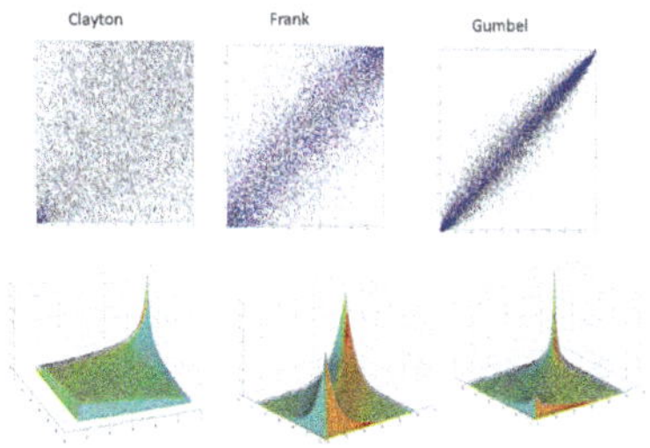

Fig. 5. 2-dimensional Archimedean Copulas : Clayton, Gumbel, Frank

3 Proposed Approach

To illustrate our approach, we give an overview in the following flowchart

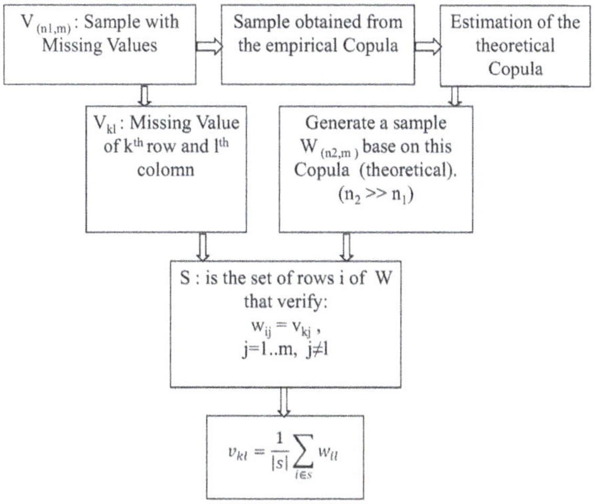

Fig. 6. General outline of the proposed approach

First, we calculate the empirical Copula from the sample containing Missing Values using Formula (3) to better observe the dependence between the variables of these data.

According to the marginal distributions from the observed and approved empirical Copula, we can determine the theoretical Copula adjusted by the family of Copulas presented in Section 2, in order to generate the theoretical sample of millions of points, having the same distribution as the empirical sample, by computing the inverse CDF of each empirical marginal of the sample.

To estimate a Missing Value, we will determine a subset of rows in the theoretical Sample whose its variables are the same as the ith row and k^{th} column of this Missing Value searched. At this time, we will calculate the average values found in order to acquire the Missing Values. Then, we calculate the standard deviation to estimate the accuracy of the mean and derive a confidence interval.

4 Results and Discussion

Data source
To evaluate the effectiveness of our solution, we have developed a large-scale experiment on a based UCI machine learning repository server. This is a version of database generator waveform of 21 column and 33367 rows.

In our case study, we noticed that most of the Copula obtained have the form of scatter plot shown in Figure 7, For this we choose a bivariate Copula among 21 and we will illustrate all the results from a this Copula.

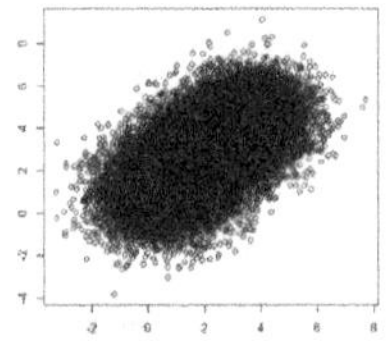

Fig. 7. The original sample with Missing Values for variables X^5, X^6

Generating an Empirical Sample
Given that the statistical model of the joint distribution is not known, we can calculate the empirical Copula of this sample to see if it follows a known Copula. Figure 8 shows the estimated Copula. This Copula is obtained by converting each point of the original sample by the cumulative distribution function of each marginal. The resulting Copula has an elliptical form just as a Gaussian Copula. To verify this formally, we used the fit test for Gaussian Copulas .

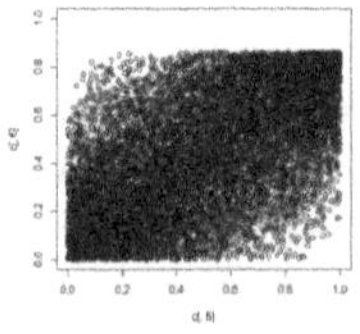

Fig. 8. The empirical sample from the empirical Copula of variables X^5, X^6

Generation of a Larger Theoretical Sample

Above, we have shown that we have a Gaussian Copula. To generate a large sample from this Copula and with the same parameters, we can calculate the inverse marginal $CDFs$ to obtain a sample of large size with the same statistical model as the empirical sample. To calculate the inverse $CDFs$, we observed the marginal distributions of variables X^5, X^6. These distributions are Gaussian. They were validated by fit test classic such as univariate Kolmogorov-Smirnov. Figure 9 shows the theoretical sample obtained with a million points.

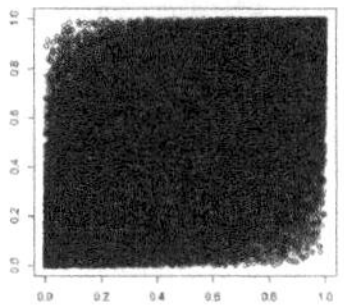

Fig. 9. The theoretical sample from the theoretical Gaussian Copula

The New Sample Obtained

To estimate a Missing Value, we have determined the subset of rows in the theoretical Sample (Figure 9) whose its variables are the same of the Missing Value searched. The average of these values has the value estimated.

We have repeated the same steps for all the Missing values and we have calculated standard deviation. We noticed that the new Sample obtained is Gaussian which is shown in the top left corner of Figure 10.

To test the performance of our method, we calculated the error rate for the percentage of Missing Values, and we also applied the technique of imputation of the mean and that of regression for the same database in order to better see the advantage of our approach.

Discussion

The comparison of the different graphs in Figure 11, shows the correspondance with the results obtained with the waveform database for the same evaluation criteria.

The increase in missing values by 5 % to 95 % caused a decrease in the minimum accuracy of 88 % for the mean imputation method, 14 % by the regression technique and a 1% by our method, for against the maximum error is 98 % for the imputation method, and 36 % by the regression technique and the average and 30 % our method.

Degradation of the error is always evident when increasing missing values. However our strategy (blue curve) based on Copula is much better than the mean imputation (red curve) and also superior to regression technique (green curve). The difference is most sensitive for all values.

Our approach should be a practical solution to estimate missing values for a very large database because it overcomes the Missing Values using Copulas

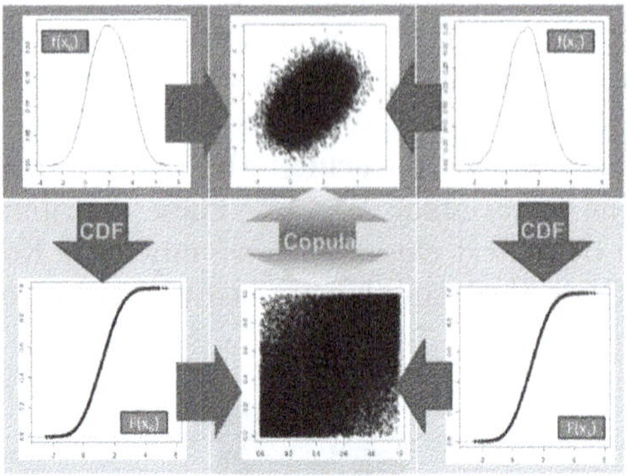

Fig. 10. The new sample obtained

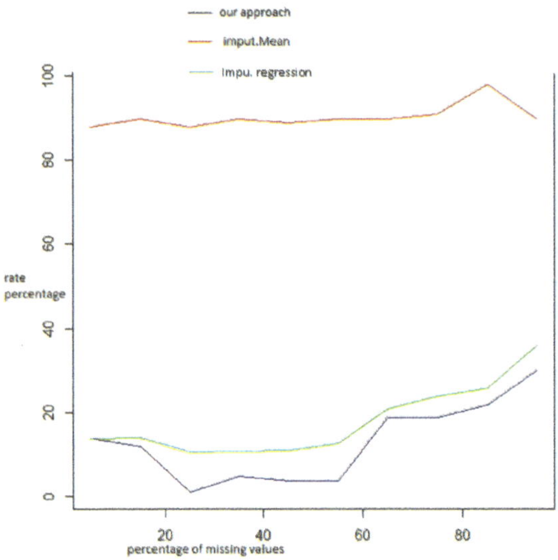

Fig. 11. Performance of substitution techniques according to missing values

with small errors even with a very large percentage of missing values which is contrary to the mean imputation if is much less conclusive and may be better on small databases which is not the case in the data Mining.

5 Conclusion and Future Work

Incomplete and noisy data, are a major obstacle in the pre-treatment process of KDD, those that lead to knowledge extraction from low quality and consequently the KDD process slows and the results that it provided are not reliable.

Within this context, we propose in this paper a new approach based Sampling techniques that essentially seeks to predict Missing Values, which constitute a major problem, because the information available is incomplete, less reliable and knowledge extracted from these data is not representative.

An experimental study on a large scale has provided very good results, those that show the effectiveness of our method. Future work will focus on the application at this same theory to eliminate redundant data to reduce the size of large amounts of data.

References

1. Dempster, A., Laird, N., Rubin, D.: Maximum likelihood from incomplete data via the EM algorithm. Journal of the Royal Statistical Society 39(1), 1–38 (1977)
2. Allison, P.D.: Multiple Imputation for Missing Data: A Cautionary Tale. Sociological Methods Research 28(3), 301–309 (2000)
3. Chen, J., Shao, J.: Nearest neighbor imputation for survey data. Journal of Official Statistics 16(2), 113–131 (2000)
4. DeSarbo, W.S., Green, P.E., Carroll, J.D.: Missing data in product-concept testing. Decision Sciences 17, 163–185 (1986)
5. Engels, J.M., Diehr, P.: Imputation of missing longitudinal data: a comparison of methods. Journal of Clinical Epidemiology 56(10), 968–976 (2003); Statistical Association 83, 1198–1202
6. Saporta, G.: Probabilités, analyse des données et statistique, Editions Technip, Paris (2006)
7. Frane, J.W.: Some simple procedures for handling missing data in multivariate analysis. Psychometrika 41, 409–415 (1976)
8. Han, J., Kamber, M.: Data Mining: Concepts and Techniques. Elsevier, New York (2006)
9. Joe, H.: Multivariate Models and Dependence Concepts. Monographs on Statistics and Applied Probability, vol. 73. Chapman and Hall, London (1997)
10. Kline, R.B.: Principles and Practice of Structural Equation Modelling. Guilford Press, New York (1989)
11. Kaufman, C.J.: The application of logical imputation to household measurement. Journal of the Market Research Society 30, 453–466 (1988)
12. Kim, J.O.: Curry, The treatment of missing data in multivariate analysis. Sociological Methods and Research 6, 215–241 (1977)
13. Little, R.J.A., Rubin, D.B.: Statistical Analysis with Missing Data, pp. 11–13. John Wiley and Sons, Inc., New York (2002)
14. Laird, N.M.: Missing data in longitudinal studies. Statistics in Medicine 7, 305–315 (1988)
15. Lee, S.Y., Chiu, Y.M.: Analysis of multivariate polychoric correlation models with incomplete data. British Journal of Mathematical and Statistical Psychology 43, 145–154 (1990)
16. Hu, M., Salvucci, S.M., Cohen, M.P.: Evaluation of some popular imputation algorithms. In: Section on Survey Research Methods, pp. 309–313. American Statistical Association (2000)
17. Cicognani, M.G., Berchtold, A.: Imputation des données manquantes:Comparaison de différentes approches. J. Statist, Plann. Inference, inria -00494698, version 1 (2010)

18. Malhotra, N.K.: Analyzing marketing research data with incomplete information on the dependent variable. Journal of Marketing Research 24, 74–84 (1987)
19. Deheuvels, La, P.: fonction de dépendance empirique et ses propriétés. Un test non paramétrique d'indépendance, Académie Royale de Belgique, Bulletin de la Classe des Sciences, 5me série
20. Song, Q., Shepperd, M.: A new imputation method for small software project data sets. Journal of Systems and Software 80(1), 51–62 (2007)
21. Nielsen, R.B.: An introduction to copulas, 2nd edn. springer (2005)
22. Ruschendorf, L.: On the distributional transform, Sklar's theorem, and the empirical copula process. J. Statist. Plann. Inference 139(11), 3921–3927 (2009)
23. Roth, P.L.: Missing data: A conceptual review for applied psychologists. Personnel Psychology 47, 537–560 (1994)
24. Ruud, P.A.: Extensions of estimation methods using the EM algorithm. Journal of Econometrics 49, 305–341 (1991)
25. Sinharay, S., Russell, H.S.: The use of multiple imputation for the analysis of missing data. Psychological Methods 6(4), 317–329 (2001)
26. Barnett, V., Lewis, T.: Outliers in statistical data. John Wiley and Sons (1994)
27. Zhang, S.: Parimputation From imputation and null-imputation to partially imputation. IEEE Intelligent Informatics Bulletin 9(1), 32–38 (2008)
28. Zhang, S., Zhang, J., Zhu, X., Qin, Y., Zhang, C.: Missing Value Imputation Based on Data Clustering. In: Gavrilova, M.L., Tan, C.J.K. (eds.) Transactions on Computational Science I. LNCS, vol. 4750, pp. 128–138. Springer, Heidelberg (2008)
29. Ghahramani, Z., Jordan, M.I.: Supervised learning from incomplete data via an EM approach. In: Cowan, J.D., Tesauro, G. (eds.) Advances in Neural Information Processing Systems 6, pp. 120–127. Morgan Kaufman (1994)

Multi-agent Based Intelligent System for Image Fusion

Ashok Kumar, Pavani Uday Kumar, Amruta Shelar, and Varala Naidu

University of Louisiana at Lafayette, LA, USA
ak@cacs.louisiana.edu

Abstract. The recent years have seen an increasing interest in developing algorithms for image fusion and several algorithms have been proposed in the literature. However, a process for assessing several fusion algorithms and coming up with the best solution for a given set of images has not been sufficiently explored so far. In this paper, a system is proposed that performs intelligent decision making in image fusion. The system uses the concepts of adaptive learning and inherent knowledge to present the best fusion solution for a given set of images. By automating the selection process, the system can analyze and exhibit intrinsic details of the images and adapt this knowledge to provide better solutions for varying types of images to provide better solutions for varying types of images.

Keywords: Image Fusion, Intelligent Systems, Learning, JADE.

1 Introduction and Related Research

Intelligent systems (IS) learn during their existence. They continually act, mentally and externally and reach their objectives more often than pure chance would indicate [1]. The development of such systems has revolutionized the areas of engineering and science. IS have also been known to be well suited to the complicated domain of biology and medicine as they are robust, able to extract weak trends and regularities from data, provide models for complex processes, cope with uncertainty and ambiguity, hold the potential to bring content-based retrieval to research literature, possess the ontological depth needed to diverse heterogeneous databases, and in general, aid in the effort to handle semantic complexity with grace [2]. IS have helped researches deal with the quintessential problems of time consuming and often exhaustive runs of their experiments. They have been used in the health sector [3], robotics [4], transportation and traffic management [5, 7]. Systems that use intelligent methods in manufacturing are called Intelligent Manufacturing Systems (IMS). IMS use software components with techniques such as expert systems, fuzzy logic, neural networks, and machine learning. Selection of the right software and choosing between adapting existing solutions and developing new tools is one of the main problems in IMS [6].

This work proposes and implements an IS for image fusion. The remainder of the paper focuses on the proposed architecture and implementation.

D. Nagamalai et al. (Eds.): *Adv. in Comput. Sci., Eng. & Inf. Technol.*, AISC 225, pp. 101–110.
DOI: 10.1007/978-3-319-00951-3_10

2 Proposed Architecture and Implementation

The main objective of system is to predict the best fusion method for a given set of images. The main tools used in the development of this system are MATLAB and JADE (Java Agent Development Toolkit). MATLAB is the backend of the system and it is used primarily to run the fusion algorithms. This work makes use of preexisting fusion algorithms and models them each as individual fusion agents. The proposed system uses a multi-agent approach, effective communication, self-learning and effective utilization of available resources. The architecture of the system is described next.

A. Architecture: The main architecture of the proposed system is shown in Fig. 1. The inputs to the system are the images that are to be fused. The images are then passed onto the Data Processing Unit (DPU). The DPU is mainly used to analyze the images and to recommend the most appropriate fusion algorithm. This fusion algorithm is then used to fuse the images. The output of the system is the fused image.

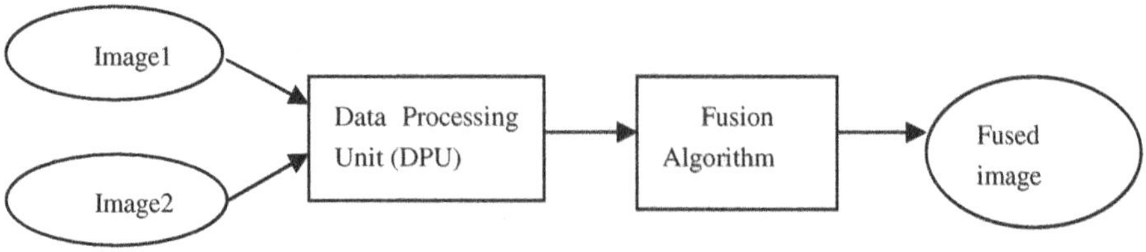

Fig. 1. Overview of the proposed system

The Data Processing Unit is the main part of the system. The duty of the DPU is to select the best fusion algorithm for the given set of images. The main components are the knowledge base and the fusion algorithms. The main aim of the knowledge base is to implement a learning algorithm to suggest most suitable fusion solutions. In order to provide the best possible fusion algorithm for the given set of images, the knowledge base uses information about the sensors used, the properties of the image, the area of application and the amount of resources available. For example, remote sensing images are fused using principal component analysis (PCA) or discrete wavelet transform (DWT) methods.

B. Fusion Algorithms: The Fusion Algorithm component is used to specify the pros and cons of the fusion methods we incorporate in the system. Each of these methods is designed as individual agents. Such agents then communicate with each other and find the best possible solution. The images are then fused using the selected algorithm. The advantage of using PCA is that it is well established and widely used. DWT performs well in several situations but it is not very good at edge-detection. Contrast and Laplacian pyramid methods of image fusion work well with both grayscale and color images. However, Laplacian pyramid method is better at edge detection. The fusion algorithms that have been used in this work include the traditional algorithms such as intensity hue saturation (IHS) and PCA; Wavelet based DWT, Shift-Invariant DWT; and Pyramid based contrast pyramid, and Laplacian Pyramid.

C. Quality Assessment: The fused image is tested to measure the performance of the algorithm. The measures used here are the Petrovic Metric, Blind Image Quality Index (BIQI), Visual Information Fidelity (VIF), and Mutual Information (MI). A threshold is established for each metric. If the fusion algorithm performs better in more than a threshold number of metrics then the Fused Image is concluded to be the best possible solution available. The brief architecture of the Knowledge Base is as shown in the Fig. 2. It can be seen the knowledge base is the most important component of the entire system. The knowledge base is responsible for determining the best solution. The more resources the knowledge base has for image categorization, the better will be the performance that can be expected from it.

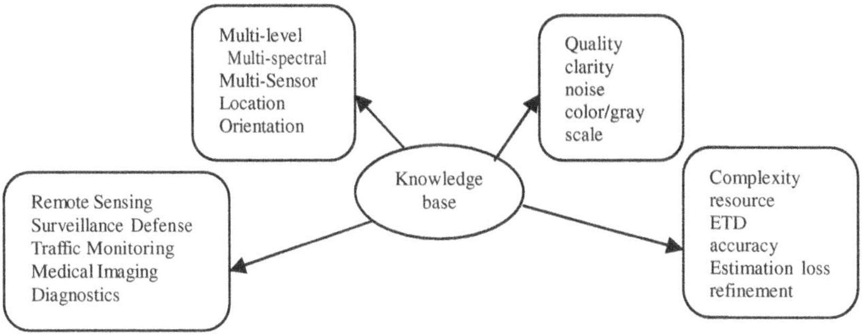

Fig. 2. Knowledge Base Architecture

D. Algorithm: The algorithm for the proposed system is described in six steps.

Step 1: Running a JADE program requires the user to initialize all agents and pass the runtime arguments during initialization. The Knowledge Base agent is the main agent that calls the individual agents. The individual agents are the *Sensor Agent*, *Image Agent*, *Application Agent* and *Resource Agent*. The Sensor Agent is the agent that generates the ranking of the algorithms based on the type of sensor used. For applications that do not have any particular sensor type, the location or orientation can be used as the decision criteria. The Image Agent bases its ranking on the characteristics of the image. The most important criteria considered are quality and if the image is color or grayscale. Any additional details such as clarity and noise are constant for majority of images and hence cannot be considered as important differentiating criteria. The Application Agent generates the ranking based on the type of application the images are intended to be fused. The initial rankings are generated on the data collected from previous runs of the images on different types of images. The Resources Agent ranks the algorithms on the basis of the number of resources needed. Images that are large in size, and have more natural characteristics require more number of system resources. The level of refinement needed i.e., the decomposition level and Estimated Time of Delivery (ETD) are also important user constraints that place demand on the resources available. The flow of the system is shown in Fig. 3.

Step 2: The knowledge base agent reads the initialized data and the images and passes the appropriate data to the individual agents using messages. The agents read the message and generate the ranking based on the message. The ranking is then passed to the knowledge base agent as a reply to the message received. The factors that influence the ranking of the algorithms are image characteristics and the accuracy of the ranking.

Step 3: The knowledge base agent waits till it receives the messages from the individual agents. The algorithms are then ranked based on the score provided by each agent. The final ranking is the average of the ranking of each algorithm. These algorithms are then sorted based on the final ranking. If more than one algorithm is ranked at the same position, the algorithm that has received the highest ranking in the most recent cycle is chosen.

Step 4: The algorithm with the best ranking is then used to generate the fused image. The fusion process is primarily the backend of the system. The fused image is developed using MATLAB. The fusion algorithms are basically MATLAB functions converted into jar files that take the images and decomposition level and the file to store the image as input. The MATLAB code fuses the two images and stores the fused image in the specified output file.

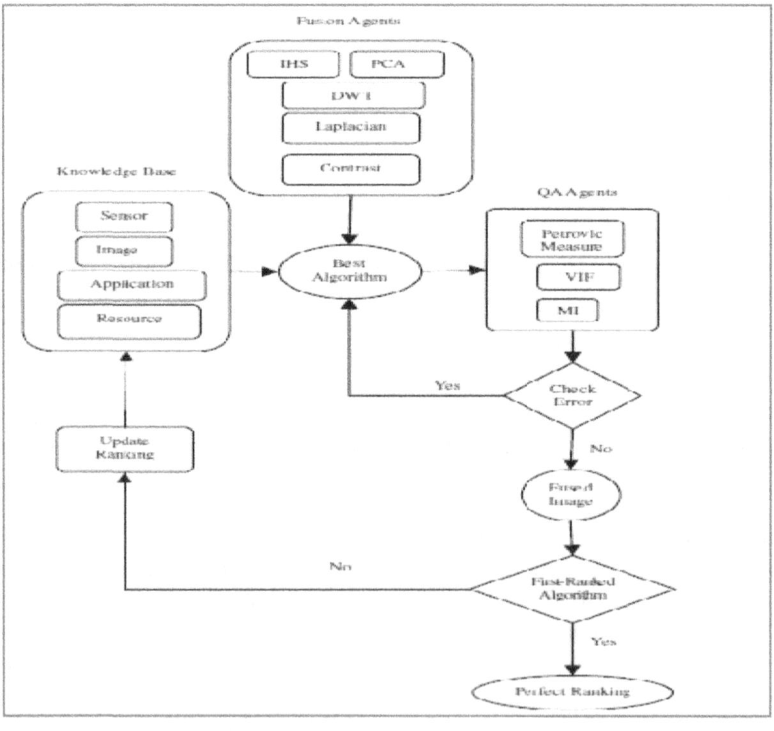

Fig. 3. Flow of the proposed system

Step 5: The fused image is evaluated to establish its validity and quality. The criteria used in the proposed system are Mutual Information (MI), Visual Information Fidelity (VIF) and Petrovic Measure. An image that satisfies 2 or more criteria is considered to be of acceptable quality. A brief description of the evaluation criteria has been given in section VI. The evaluation criteria can be image-specific as well as user-specific. Image-specific evaluation uses threshold values that are specific for the particular class of images being used. User-specific evaluation takes input from the user to judge the quality of the fused image. The input from the user can be either the number of criteria the user desires to be satisfied for the image to be accepted or the threshold values to be considered as benchmarks for evaluation.

Step 6: The image that satisfies the criteria in step 5 is the aim of the proposed system. If the system yields the correct output in the first iteration, the knowledge base is considered to be accurate. Else, the algorithm redirects to step 4 and chooses the next best algorithm. Steps 4, 5 and 6 are repeated in the cycle until the evaluation criteria are satisfied. At the end of each iteration, following the first iteration, the knowledge base is updated with the necessary changes to be made in the ranking of the algorithms. This is accomplished by sending the revised ranking of the algorithms to the individual agents i.e., Sensor Agent, Image Agent, Application Agent and Resource Agent. The agents then re-rank the algorithms based on the ranking provided. This accomplishes the task of Adaptive Learning by the Knowledge Base, which helps its agents in developing efficient ranking system of the algorithms by learning from every cycle of images analyzed.

3 Fusion Agents Implementing Known Algorithms

In this section a brief description of the fusion agents is given. Each of the agents takes the images to be fused as the input and generates the fused image as the output.

A. Intensity Hue Saturation (IHS) Agent: The agent that executes the Intensity Hue Saturation method is called the IHS Agent. This method aims to maintain as much of the original spectral information as possible, while maximizing the amount of spatial information from the high resolution data [8]. The intensity component contains the spatial information and the hue and saturation components contain the spectral information. The challenge in this method lies in keeping the intensity component correlated to the original images. A large number of IHS conversion algorithms exist. Calculation complexity and intensity retention are the two main characteristics taken into consideration when choosing any particular algorithm.

B. Principal Component Analysis (PCA) Agent: The Principal Component Analysis algorithm is run by the PCA Agent. Most of the sensors collect information in adjacent bands resulting in a lot of redundant information [10]. The PCA synthesizes the original bands creating new bands that recreate the original information. The first principal component (PC) collects the information that is common to all the bands used as input data in the PCA, i.e., the spatial information, while the spectral information that is specific to each band is picked up in the other principal components [9].

This makes PCA a very adequate technique when merging MS and PAN images. In this case, all the bands of the original MS image constitute the input data.

C. Discrete Wavelet Transforms (DWT) Agent: The most popular fusion method i.e., the DWT method is run by the DWT Agent. Most of the latest image fusion methods are based on the use of wavelets. The discrete wavelet transform (DWT) method is the most commonly used fusion technique as it is better at extracting the salient features in comparison to the Intensity-Hue- Saturation (IHS), Principal Component Analysis (PCA) or the Brovey Transform (BT) methods. Wavelet transforms are based on functions that are localized in both space and frequency and can be designed to have specific properties that are useful in the particular application of the transform. The detail information extracted from one image using wavelet transforms can be injected into another image by substitution, addition, selection, or by another model [11].

D. Pyramid Based Transform Agents: The image pyramid is a data structure designed to support efficient scaled convolution through reduced image representation. It consists of a sequence of copies of an original image in which both sample density and resolution are decreased in regular steps. The image is low pass filtered and subsampled by a factor of two resulting in the next pyramid level. Applying these steps continually yields the remaining pyramid levels.

4 Quality Assessment Agents

A brief description of the QA agents is provided next. Each agent analyzes the fused image and measures the quality of fusion.

A. Mutual Information (MI): One of the most widely used methods of assessment in image fusion is Mutual Information (MI). The lack of reference images led to the inception of MI. It is defined by the Kullback-Leibler measure which measures the degree of dependence between two variables A and B:

$$I_{A,B}(a,b)= \sum_{x,y} P_{AB}(a,b) \log [P_{AB}(a,b) / P_A(a)\, P_B(b)] \quad (1)$$

where $P_{AB}(a,b)$ is the joint probability and $P_A(a)P_B(b)$ is the probability associated with the case of complete independence. The Mutual Information (MI) between two images is given by the formula:

$$M_F^{AB} = I_{FA}(f,a) + I_{FB}(f,b) \quad (2)$$

B. Petrovic Measure: Xydeas and Petrovic in [14] derived a new metric that considers the edge information in each pixel to be important criteria in measuring the quality of image fusion. Given two images A, B and a fused image F, a Sobel edge operator is used to calculate the edge strength and orientation for each pixel. The relative strength and orientation values between A and F are calculated and used to determine the edge strength and orientation preservation values and. The edge information preservation value is the product of the above two values. For the given set of input images A and B the normalized weighted metric for a fusion process p is given as:

$$Q_P^{AB/F}(n,m)= \frac{\sum_{n=1}^{N}\sum_{m=1}^{M} Q^{AF}(n,m)W_A\,A(n,m)+Q^{BF}(n,m)W_B(n,m)}{\sum_{n=1}^{N}\sum_{m=1}^{M} W_A(n,m)+W_B(n,m)} \quad (3)$$

C. Visual Information Fidelity (VIF): The Visual Information Fidelity (VIF) [13] criterion is based on the HVS. This measure quantifies the information that could ideally be extracted by the brain from the reference image. Using the HVS model the test and reference images are defined as:

$$E = C+N \text{ (reference image)}$$
$$F = D+N' \text{ (test image)}$$

where E and F are the output visual signals of the HVS model and N and N' are the HVS noise. The *VIF* is then calculated as the mutual information between E and C over N elements. To tune the model to a specific reference image an additional criteria sn is used. Based on the above criteria the *VIF* based on the selection of a series of sub-bands is given as

$$\text{VIF}= \frac{\sum_{j\epsilon\, subbands} I\left(\overrightarrow{C}^{N,j};\overrightarrow{F}^{N,j}\,\middle|\, s^{N,j}\right)}{\sum_{j\epsilon\, subbands} I\left(\overrightarrow{C}^{N,j};\overrightarrow{E}^{N,j}\,\middle|\, s^{N,j}\right)} \quad (4)$$

If the VIF values is less than zero it indicates loss of information between the two images. A VIF value of unity indicates that the information is precise between the two images. However, if the VIF value is greater than 1 it indicates that the image is of superior quality than the reference image.

5 Results

The system has been analyzed for fusing two images from various sources and applications. The inputs to the system are the images to be fused and the type of sensors used. For each case, the ranking provided by each agent, the ranking developed by the knowledge base agent and the fusion output is shown.

Cases	Ranks given by agents of K.B and K.B.						Chosen technique	Evaluation using 3 criterias	
								criteria	result
Case 1. Multi-focus grayscale images have been chosen in first case.	Ranks given by various agents in K.B	IHS	PCA	DWT	LAP	CON	Chosen Technique is DWT	Evaluation results of fused images by DWT technique	
	sensor	5	4	1	2	3		VIF	0.7847
	application	5	2	3	1	4			

	resources	3	4	3	5	6		MI	6.0219
	image	2	2	0	1	1			
	Finalzed ranking developed by K.B with the help of its agents.	5	3	1	2	4		Petrovic measure	0.6457
Case 2. Here, we consider two multi-level brain scan images to be fused. Fusion of medical images is the most important area of application in the field of image fusion	Ranks given by various agents in K.B	IHS	PCA	DWT	LAP	CON	Chosen technique is PCA	Evaluation results of fused images by PCA technique	
	sensor	1	3	2	5	4		VIF	0.5065
	application	5	2	3	1	4			
	resources	3	4	3	5	6		MI	4.1657
	image	0	1	1	2	2			
	Finalzed ranking developed by K.B with the help of its agents.	3	1	4	2	5		Petrovic measure	0.5367
Case 3. Multi-spectral remote sensing images have been chosen to be fused. These images	Ranks given by various agents in K.B	IHS	PCA	DWT	LAP	CON	Chosen technique is DWT.But the evaluation results were not satisfactory so choose the next best algo-	Evaluation results of fused images by DWT technique	
	sensor	5	4	1	2	3		VIF	0.2930
	application	5	4	3	2	1		MI	1.9371

provide an interesting contrast to the set of images previously fused in terms of the sensor and application are to be fused	resources	**5**	**5**	**3**	**4**	**4**	rithm i.e. LAP.	Petrovic measure	0.4690
	image	**0**	**1**	**1**	**0**	**2**		Evaluation results of fused images by LAP technique	
	Finalzed ranking developed by K.B with the help of its agents.	5	4	1	2	3		VIF	0.2834
								MI	5.612
								Petrovis measure	0.7262
Case 4. The updated ranking system is put to test by running the system on a different set of images but, which fall under the same category. Here, we consider two multi-level thorax scan images to be fused	Ranks given by various agents in K.B	IHS	PCA	DWT	LAP	CON	Chosen technique is LAP	Evaluation results of fused images by LAP technique	
	sensor	**5**	**4**	**1**	**2**	**3**			
	application	**5**	**4**	**1**	**2**	**3**		VIF	0.7256
	resources	**5**	**5**	**3**	**4**	**4**		MI	3.6225
	imsge	**3**	**3**	**3**	**2**	**3**		Petrovic measure	0.5911
	Finalzed ranking developed by K.B with the help of its agents.	5	4	2	1	3			

6 Conclusion

This paper proposes a novel image fusion tool using JADE. The main aim of the tool is to offer the best image fusion solution using well-established fusion methods. The proposed system has been tested on various sets of images. The system has shown that it is effective, accurate and capable of adaptive learning.

Acknowledgment. The authors acknowledge support from Louisiana Board of Regents under research award number LEQSF (2009-12)-RD-A-22.

References

[1] http://www.intelligent-systems.com.ar/intsyst/glossary.htm#concept
[2] Lathery, R.: Intelligent systems in biology: why the excitement? IEEE Intelligent Systems (2001)
[3] Reddy, R.: Robotics and Intelligent Systems in Support of Society. IEEE Intelligent Systems 21(3), 24–31 (2006)
[4] Fukuda, T., Takagawa, I., Hasegawa, Y.: From intelligent robot to multi-agent robotic system. In: International Conference on Integration of Knowledge Intensive Mulch-Agent Systems, September 30-October 4, pp. 413–417 (2003)
[5] Klein, L.A.: Sensor Technologies and Data Requirements for ITS. Artech House Books, Boston (2001)
[6] Devedzic, V., Radovic, D.: A framework for building intelligent manufacturing systems. IEEE Transactions on Systems, Man, and Cybernetics, Part C:Applications and Reviews 29(3), 422–439 (1999)
[7] Patel, M., Ranganathan, N.: IDUTC: an intelligent decision-making system for urban traffic-control applications. IEEE Transactions on Vehicular Technology 50(3), 816–829 (2001)
[8] Joseph Carper, W., Lille, T.M., Liefer, R.W.: The Use of Intensity-Hue-Saturation Transformations for Merging SPOT Panchromatic and multicultural Image Data. Photogrammetricengineering and Remote Sensing 56(4), 459–467 (1990)
[9] Pohl, C., Van Genderen, J.L.: Multisensor image fusion in remote sensing: Concepts, methods and applications. Int. J. Remote Sens. 19, 823–854 (1998)
[10] Gonzalez-Audicana, M., Saleta, J.L., Catalan, R.G., Garcia, R.: Fusion of multispectral and panchromatic images using improved IHS and PCA mergers based on wavelet decomposition. IEEE Transactions on Geoscience andRemote Sensing 42(6), 1291–1299 (2004)
[11] Amolins, K., Yun, Z., Dare, P.: Applications of Wavelet Transforms in Image Fusion. In: Urban Remote Sensing Joint Event, April 11-13, pp. 1–7 (2007)
[12] Caire, G.: JADE tutorial JADE Programming for Beginners (June 30, 2009)
[13] Sheikh, H.R., Bovik, A.C.: Image Information and Visual Quality. IEEE Transaction on Image Processing 15(2), 430–444 (2006)
[14] Xydeas, C.S., Petrovic, V.: Objective image fusion performance measure. Electronics Letters 36(4), 308–309 (2000)

On the Nearest Neighbor Algorithms for the Traveling Salesman Problem

Gözde Kizilateş and Fidan Nuriyeva

Department of Mathematics, Faculty of Science, Ege University, Izmir, Turkey
{gozde.kizilates,nuriyevafidan}@gmail.com

Abstract. In this study, a modification of the nearest neighbor algorithm (NND) for the traveling salesman problem (TSP) is researched. NN and NND algorithms are applied to different instances starting with each of the vertices, then the performance of the algorithm according to each vertex is examined. NNDG algorithm which is a hybrid of NND algorithm and Greedy algorithm is proposed considering experimental results, and it is tested on different library instances.

Keywords: traveling salesman problem, hybrid algorithm, nearest neighbor algorithm, greedy algorithm.

1 Introduction

The Traveling Salesman Problem (TSP) is the most famous optimization problem in the set NP-hard [2, 6]. TSP aims to find a route for a salesman who starts from a home location, visits prescribed set of cities and returns to the original location in such a way that the total traveling distance will be minimum and each city will have been visited exactly once [3]. Many problems that are natural applications in computer science and engineering can be modeled using TSP [4, 8, 9, 14].

The algorithms for solving TSP can be divided into four classes: exact algorithms, heuristic algorithms, approximate algorithms and metaheuristics algorithms [7]. The exact algorithms are guaranteed to find the optimal solution. However, these algorithms are quite complex and very demanding of computer power [1]. The heuristic algorithms can obtain good solutions but it cannot be guarantee that the optimal solution will be found [5]. In general, the heuristic algorithms are subdivided into the following three classes: tour construction algorithms, tour improvement algorithms and hybrid algorithms [15]. The tour construction algorithms gradually build a tour by adding a new city at each step, the tour improvement algorithms improve upon a tour by performing various exchanges, and finally hybrid algorithms use both composing and improving heuristics at the same time [10-13]. The best results are obtained by using hybrid approaches [1, 15]. This paper presents a hybrid construction algorithm for solving TSP based on the nearest neighbor algorithm and the greedy algorithm.

D. Nagamalai et al. (Eds.): *Adv. in Comput. Sci., Eng. & Inf. Technol.*, AISC 225, pp. 111–118.
DOI: 10.1007/978-3-319-00951-3_11

2 The Nearest Neighbor Algorithm (NN)

The nearest neighbor (NN) algorithm for determining a traveling salesman tour is as follows. The salesman starts at a city, then visits the city nearest to the starting city. Afterwards, he visits the nearest unvisited city, and repeats this process until he has visited all the cities, in the end, he returns to the starting city.

The steps of the algorithm are as following [1]:

1. Select a random city.
2. Find the nearest unvisited city and go there.
3. Are there any unvisited cities left? If yes, go to step 2.
4. Return to the first city.

We can obtain the best result by running the algorithm over again for each vertex and repeat it for n times.

The next theorem, due to Rosenkrantz, Stearns and Lewis [16], shows that this approximation algorithm may have much worse behavior.

Theorem: For all m-city instances I of the traveling salesman problem with triangle inequality,

$$\text{NN(I)} \le \frac{1}{2}(\lceil \log_2 m \rceil + 1)\text{OPT(I)}$$

Furthermore, for arbitrarily large values of m, there exist m-city instances for which

$$\text{NN(I)} > \frac{1}{3}\left(\log_2(m+1)\frac{4}{3}\right)\text{OPT(I)}$$

The main import of this theorem can be stated quite succinctly: $R_{NN} = \infty$, it is not a promising guarantee. So that, $m \to \infty \Rightarrow NN(I) \to \infty$.

The performance of the Nearest Neighbor algorithm clearly leaves much to be desired. A different modification of this algorithm is proposd below. It is interesting to find similar guarantee values with the above theorem for this modification. The NN algorithm is not promising theoretically; however, it yields good results in practice.

3 The Nearest Neighbor Algorithm (NND) from Both End-Points

The algorithm starts with a vertex chosen randomly in the graph. Then, the algorithm continues with the nearest unvisited vertex to this vertex. We will have two end vertices. We add a vertex to the tour such that this vertex has not visited before and it is the nearest vertex to these two end vertices. We update the end vertices. The algorithm ends after visiting all vertices.

The steps of the algorithm are as following:

1. Choose an arbitrary vertex in the graph
2. Visit the nearest unvisited vertex to this vertex.
3. Visit the nearest unvisited vertex to these two vertices and update the end vertices
4. Is there any unvisited vertex left? If yes, go to step 3.
5. Go to the end vertex from the other end vertex.

4 Computational Tests for the NND Algorithm

NN and NND algorithms have been tested on well-known library problems. Table 1 and Figure 1 shows the experimental results for ulysses16 given in [17, 18].

Table 1. Comparison of NN and NND in terms of row-sum for ulysses16

Vertex index	Row-sum	NN	NND
1	111,039167	104,73494	84,427643
2	175,806969	85,46698	84,427643
3	168,175063	84,427643	84,427643
4	142,055212	84,427635	84,427643
5	144,262526	86,786415	86,786415
6	106,548756	88,588806	86,786415
7	99,585345	95,366249	86,786415
8	111,970472	104,076958	84,427643
9	147,71774	77,126892	84,269722
10	114,327834	84,652298	84,269722
11	334,310966	77,126884	77,126884
12	91,810072	96,719711	86,786415
13	91,23502	97,199722	86,786415
14	92,118889	92,334679	86,786415
15	104,784133	95,78952	86,786415
16	108,067343	84,433632	84,427643

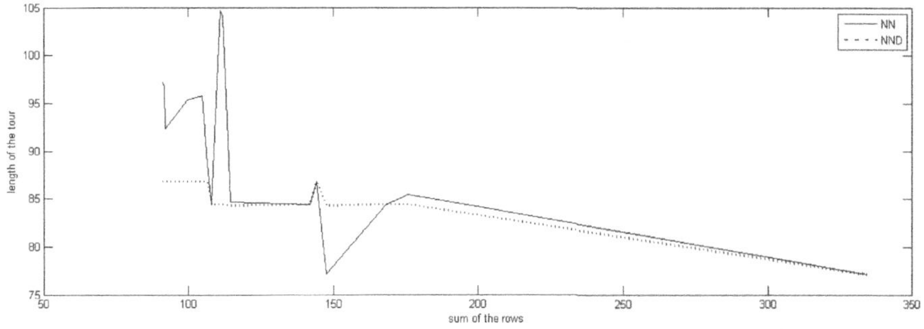

Fig. 1. Comparison of NN and NND in terms of row-sum for ulysses16

As it can be seen in Table 1 and Figure 1 above, the NND algorithm generally gives better results.

5 Greedy Algorithm

The Greedy heuristic gradually constructs a tour by repeatedly selecting the shortest edge and adding it to the tour as long as it doesn't create a cycle with less than *N edges, or increase the degree of any node by* more than 2. We must not add the same edge twice.

The steps of the algorithm are as following [1]:

1. Sort edges by increasing lengths.
2. Select the shortest edge and add it to our tour if it doesn't violate any of the above constraints.
3. Do we have *n* edges in our tour? If no, go to step 2.

6 An Hybrid of NND and Greedy Algorithms (NNDG)

After we create the adjacency matrix for the given problem, we compute the sum of the entries in each row, it is called row-sum here. We continue by finding the vertex which has the smallest row-sum. Then, we apply the NND algorithm proposed above for the vertex that has been found. After applying the NND algorithm, we continue the process by repeating the NND algorithm starting with the last vertex in the tour until we coincide with any vertex which has been added to the tour before. Then, we calculate $t = \lfloor n/k \rfloor$, where n is the number of vertices, and k is the number of implementations of NND. The first t edges are taken respectively from each solution found by NND and the edges which do not form subtours are added to our solution. The algorithm ends by applying Greedy algorithm to the other edges.

The steps of the algorithm are as following:

1. Find the vertex which has the smallest row-sum.
2. Apply NND by starting this vertex.
3. Continue to apply NND starting with the last vertex in the tour until you coincide with any vertex which has been added to the tour before.
4. Calculate $t = \lfloor n/k \rfloor$, where n is the number of vertices, k is the number of implementation of NND.
5. Take first t edges respectively from each solution found by NND and add the edges, which don't form subtours, to the solution
6. Apply Greedy algorithm to the other edges.

7 Computational Tests for NNDG Algorithms

We test the NNDG algorithm on different well-known library problems. Table 2 and Figure 2 shows the experimental results for ulysses22 given in [17, 18].

In these tests, the distance from each vertex to other vertices is computed by summing all entries in the row of the adjacency matrix. The relation between these distances and the results of the algorithm is examined in Table 2 and Figure 2.

Table 2. The results for ulysses22 obtained by choosing a starting vertex with respect to the row-sum

Row-sum	Vertex index	Tour length
123,2764	13	83.018
123,4718	12	82.861
126,0229	14	82.875
133,8718	21	87.166
133,9167	20	87.166
136,9844	7	87.166
138,2799	16	86.899
138,9798	19	87.166
140,9329	1	86.899
142,1713	8	86.899
147,0848	15	87.166
150,0691	6	87.166
151,096	10	87.166
163,0422	22	87.166
175,6999	18	87.166
175,9675	4	87.166
188,9961	17	87.166
206,9365	9	87.166
207,4082	5	87.166
213,945	3	79.671
223,5139	2	87.166
478,9774	11	87.166

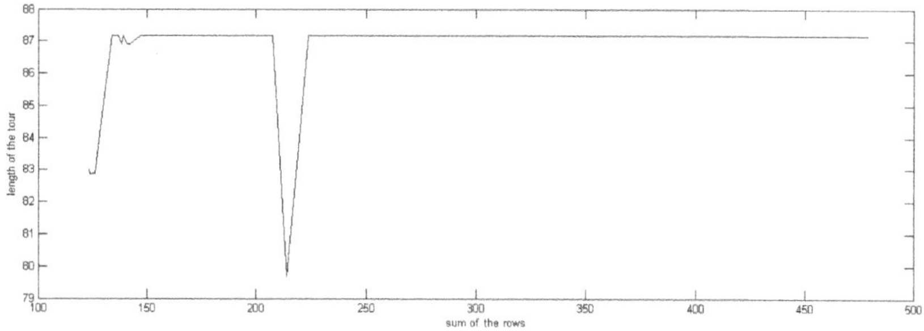

Fig. 2. The graph of the results for ulysses22 obtained by choosing a starting vertex with respect to the row-sum

According to Table 2 and Figure 2, it is seen that better results are obtained when starting from the vertices which have smaller sums of distances.

In Table 3, the experimental results obtained for 11 different values of *t* parameter on 5 problems in [17, 18] are given.

Table 3. The results depending on different values of t

graph	ulyses16	ulysses22	eil51	berlin52	eil76
optimum	74.108	75.665	429.983	7544.365	545.387
k	3	3	3	3	6
$t=\lfloor n/k \rfloor - 5$	77.126	79.671	521.961	8542.874	620.270
$t=\lfloor n/k \rfloor - 4$	77.126	87.166	521.961	8542.874	620.270
$t=\lfloor n/k \rfloor - 3$	77.126	87.166	521.961	8542.874	620.270
$t=\lfloor n/k \rfloor - 2$	77.126	87.166	513.562	8542.874	620.270
$t=\lfloor n/k \rfloor - 1$	77.126	85.501	513.562	8542.874	620.270
$t=\lfloor n/k \rfloor$	77.126	83.018	513.562	8542.874	620.270
$t=\lfloor n/k \rfloor + 1$	77.126	83.018	513.562	8542.874	620.270
$t=\lfloor n/k \rfloor + 2$	77.126	83.018	513.562	8542.874	620.270
$t=\lfloor n/k \rfloor + 3$	77.126	83.018	513.562	8542.874	620.270
$t=\lfloor n/k \rfloor + 4$	77.126	87.166	530.621	8542.874	620.270
$t=\lfloor n/k \rfloor + 5$	77.126	87.166	530.621	8542.874	620.270

According to Table 3, the best results are obtained for $t=\lfloor n/k \rfloor$.

In Table 4, the computational results which are obtained by the implementation of the NNDG algorithm for 5 different origin vertices, are given for 27 problems in [17, 18].

Table 4. Computational results for different problems and different origin vertices

Problem	Optimum	The vertex which has smallest row-sum	$\lfloor n/4 \rfloor$th vertex according to row-sum	$\lfloor n/2 \rfloor$th vertex according to row-sum	$\lfloor 3n/4 \rfloor$th vertex according to row-sum	The vertex which has biggest row-sum
ulysses16	74.108	77.126	77.126	77.126	77.126	77.126
ulysses22	75.665	83.018	87.166	87.166	87.166	87.166
bayg29	9074.148	10810.481	10810.481	9963.589	10810.481	10810.481
att48	33523.708	40068.035	40068.035	40068.03	40068.035	41182.40
eil51	429.983	513.562	527.034	524.255	539.811	530.370
berlin52	7544.365	8542.874	8542.874	8542.874	8542.874	8542.874
st70	678.597	797.815	778.911	778.911	797.815	797.815
eil76	545.387	620.270	519.751	620.270	605.206	620.270
pr76	108159.43	134284.812	139582.203	139582.203	139582.203	139582.203
rd100	7910.396	9900.226	9591.957	9474.017	9474.017	9516.281
kroA100	21236.951	27008.218	27008.218	27008.218	26261.531	26264.730
kroB100	22141	26383.210	26383.210	26383.210	26383.210	26383.210
kroC100	20750.762	23401.554	24761.794	25193.234	25759.095	25388.687
kroD100	21294.290	25752.627	27006.322	26193.382	23729.011	26600.332
kroE100	22068	26288.539	26288.539	26288.539	26288.539	26288.539
eil101	642.309	771.399	771.399	771.399	771.399	771.399
lin105	14382.995	17362.769	17067.777	17362.769	17362.769	17362.769
pr107	44303	46872.058	46872.058	46872.058	46872.058	46617.173
gr120	1666.508	2002.396	2002.396	1993.938	2002.396	1877.776
pr124	59030	67344.921	65940.471	67344.921	67344.921	67344.921
ch130	6110.860	7255.568	7255.568	7255.568	7017.948	7255.568
pr136	96772	118856.343	118856.343	118856.343	115966.906	118856.343
pr144	58537	62543.738	62543.738	62543.738	62543.738	62543.738
ch150	6528	7028.432	7028.432	7028.432	7028.432	7028.432
kroA150	26524	33104.113	33104.113	33104.113	33104.113	33104.113
kroB150	26130	32176.169	32176.169	32176.169	32176.169	32017.155
pr152	73682	84780.023	84780.023	84780.023	84780.023	84780.023

As it is seen from the table above, the results are also better when we start from the vertices which have smaller distances.

8 Conclusion

In this study, a new modification (NND) of the NN algorithm has been proposed for solving TSP. Computational experiments have been conducted on known library problems for this algorithm. Moreover, we have presented a new hybrid algorithm

(NNDG) based on NND (which is presented for improving the results) and Greedy algorithms. Computational experiments are conducted on known library problems for this algorithm, as well. These experiments have shown that the proposed algorithm (NNDG) is efficient. We are planning to improve the proposed hybrid algorithm in future works.

References

1. Appligate, D.L., Bixby, R.E., Chavatal, V., Cook, W.J.: The Travelling Salesman Problem, A Computational Study. Princeton Univesity Press, Princeton (2006)
2. Davendra, D.: Travelling Salesman Problem, Theory and Applications. InTech (2010)
3. Gutin, G., Punnen, A. (eds.): The Traveling Salesman Problem and Its Variations. Combinatorial Optimization, vol. 12. Kluwer, Dordrecht (2002)
4. Hubert, L.J., Baker, F.B.: Applications of Combinatorial Programming to Data Analysis: The Traveling Salesman and Related Problems. Psychometrika 43(1), 81–91 (1978)
5. Johnson, D., Papadimitriou, C.: Performance guarantees for heuristics. In: Lawler, et al. (eds.), ch. 5, pp. 145–180 (1985)
6. Johnson, D.S., McGeoch, L.A.: The Traveling Salesman Problem: A Case Study, Local Search in Combinatorial Optimization, pp. 215–310. John Wiley & Sons (1997)
7. Lawler, E.L., Lenstra, J.K., Rinnoy Kan, A.H.G., Shmoys, D.B.: The Travelling Salesman Problem: A Guided Tour of Combinatorial Optimization. John Wiley & Sons (1986)
8. Lenstra, J., Kan, A.R.: Some simple applications of the travelling salesman problem. Operational Research Quarterly 26(4), 717–733 (1975)
9. Lenstra, J.K.: Clustering a Data Array and the Traveling-Salesman Problem. Operations Research 22(2), 413–414 (1974)
10. Lin, S., Kernighan, B.: An effective heuristic algorithm for the traveling-salesman problem. Operations Research 21(2), 498–516 (1973)
11. Nuriyeva, F.: New heuristic algorithms for travelling salesman problem. In: 25th Conference of European Chapter on Combinatorial Optimization (ECCO XXV), Antalya, Turkey, April 26-28 (2012)
12. Nuriyeva, F., Kizilates, G., Berberler, M.E.: Experimental Analysis of New Heuristics for the TSP. In: IV International Conference "Problems of Cybernetics and Informatics, Baku, Azerbaijan, September 12-14 (2012)
13. Nuriyeva, F., Kızılateş, G.: A New Hyperheuristic Algorithm for Solving Traveling Salesman Problem. In: 2nd World Conference on Soft Computing, Baku, Azerbaijan, December 3-5, pp. 528–531 (2012)
14. Punnen, A.: The Traveling Salesman Problem: Applications, Formulations and Variations. In: Gutin, Punnen (eds.), ch. 1, pp. 1–28 (2002)
15. Reinelt, G.: The Traveling Salesman: Computational Solutions for TSP Applications. Springer, Germany (1994)
16. Rosenkrantz, D.J., Stearns, R.E., Lewis, P.M.: An Analysis of SeveralHeuristicsfortheTravellingSalesman Problem. SIAM J. Comput. 6, 563–581 (6.1)
17. http://comopt.ifi.uni-heidelberg.de/software/TSPLIB95/
18. http://www.iwr.uni-heidelberg.de/groups/comopt/software/TSPLIB95/tsp/

Path Guided Abstraction Refinement for Safety Program Verification

Nassima Aleb and Samir Kechid

USTHB-FEI,
BP 32 EL ALIA Bab Ezzouar, 16111, Algiers, Algeria
{naleb,skechid}@usthb.dz

Abstract. This paper presents a new compositional approach for safety verification of C programs. A program is represented by a sequence of assignments and a sequence of guarded blocks. Abstraction consists to abstract the program in a set of blocks relevant to the erroneous location (EL). As in the CEGAR paradigm, the abstracted model is used to prove or disprove the property. This checking is performed for each block backwardly, using Weakest Preconditions to generate a formula which Satisfiability is checked. If the abstraction is too coarse to allow deciding on the Satisfiability of the formula, then a path-guided refinement is performed. Our technique allows handling programs containing function calls and pointers. All aspects described in this paper are illustrated by clarifying examples.

Keywords: Program Verification, Abstraction, Refinement, Weakest Precondition, Compositional analysis, Interprocedural analysis, Backward Analysis.

1 Introduction

The state explosion problem remains a major obstacle in applying formal methods for the verification of large programs. Abstraction is the most important technique for handling this problem. In fact, it is essential for verifying designs of industrial complexity. In order for these methods to be used more widely in software verification, automatic techniques are needed for generating abstractions. CEGAR[6]: Counterexample Guided Abstraction Refinement is a well-known paradigm in software verification. It is an automatic abstraction technique which is based on an analysis of the structure of formulas appearing in the program. In general, the technique computes an upper approximation of the original program. Thus, when a specification is true in the abstract model, it will also be true in the concrete design. However, if the specification is false in the abstract model, the counterexample may be the result of some behavior in the approximation which is not present in the original model. When this happens, it is necessary to refine the abstraction so that the behavior which caused the erroneous counterexample is eliminated. Refinement technique uses information obtained from erroneous counterexamples. The scheme of CEGAR was first implemented for verifying software by the Slam project [3], and applied successfully to

D. Nagamalai et al. (Eds.): *Adv. in Comput. Sci., Eng. & Inf. Technol.*, AISC 225, pp. 119–130.
DOI: 10.1007/978-3-319-00951-3_12

find bugs in device drivers. The BLAST model checker [4] pioneered the use of Craig interpolants in CEGAR-style verification. In this paper, we present another approach for program abstraction refinement. Our methodology has two key features:

1. It is a backward analysis: so it starts from the erroneous location and try to infer the preconditions necessary to reach this location;
2. It is a path-guided analysis.

So, having a program and a safety property expressed as a Reachability of some "erroneous" location, we start investigations by the location and try to decide incrementally if there is some execution leading to it. To attain this objective, we first consider an initial abstraction of the program, if we can prove or disprove the reachability of the erroneous location then the process terminates , else, the abstraction must be refined and the process is reiterated. The refinement is performed backwardly. The use of backward analysis is motivated by the following points:

1. It is a goal-directed analysis.
2. Backward analysis is more scalable than a forward symbolic execution because it doesn't explore unnecessary program paths (i.e., paths not leading to location of interest) and unnecessary regions of code (i.e., assignments that don't influence the truth value of the considered predicate).
3. Backwards investigation takes away all paths and predicates which are not necessary for our analysis
4. It takes advantage of the locality property. In fact, the backward mode allows using, at each region of the program, the most recent variables values. So as a very simple example : suppose that a variable v has been assigned several values along the program and suppose that in the location 1000 v is set to 0 and in the location 1010 we must track some predicate containing v so, in a backward analysis it is sufficient to remount up of 10 locations to find that v has the value 0, inversely in a forward analysis all the assignments to the variable v from the location 1 to 1000 must be considered.

Our approach is compositional. Functions are modeled independently of the caller program; each call to a function is replaced by expressions summarizing the effect of this call on the caller program. Pointers are organized in equivalence classes, each class having one representing element. Every modification performed on one element of a given class is expressed on its representing element.

Organization: The rest of the paper is organized as follow: Section 2 introduces the programs representation. Section 3 describes the overall BIMC technique: Initial access path computing; weakest precondition with respect to (w.r.t.) blocks; path extension and splitting. Section 4 is devoted to elucidating examples. Functions are described in section 5 and pointers in the sixth one. The section 7 is dedicated to experimental results. In section 8 some related works are exposed. Finally, we conclude with a conclusion highlighting contributions of this paper, and exposing some future perspectives for this work.

2 Path Guided Abstraction Refinement Framework

The algorithm describing the overall methodology is given in figure 2. For a program Pg and a safety property expressed as a reachability of a given erroneous location EL in Pg, the aim is to check whether there is some execution of Pg leading to EL. In this section, we introduce an overview of our methodology; details concerning each step are described in the following sections. Our approach consists of the following steps:

1. We, first, compute an initial abstraction H_0, in the following manner: i=0; $Pi=H_0(Pg,EL)=a_1 a_2 .....a_{n-1}a_n$: a string where n is the number of blocks coming before EL. A block is a conditional or a loop statement. Let us designate the block number k of the program Pg by: BPg[k]. Each component a_k is either the character '1', '0' , 'x' or '*' regarding the position of the location with respect to the block Bpg[k]: in its 'then' branch, 'else' branch, after it, or in the opposite branch.
2. We generate a formula F from the abstraction. F is introduced to the feasibility analysis procedure to check if it represents a feasible path. This means that there is some execution such that all the guards are satisfiable. The feasibility analysis procedure uses the concept of weakest precondition [8].
3. Three cases are possible concerning the Satisfiability of F:
   - F is satisfiable: The process terminates, the path expressed by Pi is presented as a counterexample.
   - F is not satisfiable: *Pi* is sufficient to prove that E*L* is not reachable. The process is stopped. The program is safe.
   - We cannot decide: The abstraction is to coarse, it must be refined.
4. P_{i+1}= Refine(P_i). Goto (2).

In the subsequent, we describe each step of our approach. An example program is introduced in the figure 4. This program is used in each step to illustrate it.

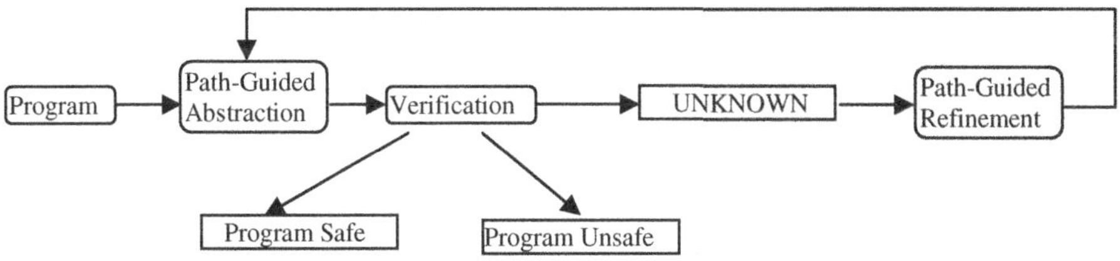

Fig. 1. Path-guided Abstraction Refinement framework

Algorithm 1.

Inputs: A program Pg ; An erroneous location EL :
Outputs: An execution E_i leading to EL.
Compute the intial abstraction : P0=H_0(Pg,EL); i=0
If **Exist_Satisfiable_Formula**(Pi,F)
Then Exit: The assignments set making F True is an execution leading to EL.
Else If **ALL_Blocks_Refined**()
then Exit: EL is not reachable, Pg is safe.
Else P_{i+1} = **Refine**(P_i,EL)
Goto 4
END.

Fig. 2. Path-Guided Abstraction Refinement Algorithm

2.1 Initial Abstraction Computing

The construction of the initial abstraction H_0(Pg,EL) is straightforward . It is directly related to the structure of the program, and the position of the erroneous location within this structure.

Algorithm 2. H_0 (Pg,EL)

Inputs: A program Pg, an erroneous location EL.
Outputs: A string P0
Initialization: P_0 = '',k=1;
While(True)
Do Begin
If End of Pg
Then exit
Else if EL occurs before BPg[k]
Then Exit
Else **If**(L∈ BPg[k])
Then Case 1: E=BPg[k] is a conditional statement
If L∈ Then(E)
Then P_0=P_0.'1'
Else P_0=P_0.'0' ; k++;
while BPg[k] in Then(E) P_0=P_0.'*';k++
Case 2: BPg[k] is a loop P_0=P_0.'1'
Else P_0=P_0.'x'
k=k+1;
End.

Fig. 3. Algorithm computing the Initial Abstraction H_0 (Pg,EL)

```
1 : int x,y,z,t,m;
6: scanf("%d",&x);
7: scanf("%d",&y) ;
[1]   if(x<>y)
8: { z=0 ;
 [2] if(x>y)
9:      m=x ;
         else
10:     m=y;
    }else
  {  [3] if(x<0)
11: m=-x ;else
12:     m=x;
     }
13: z=1;
    [4] if(m==z)
14:  t=1 ;
    else
15: t=0;
16: m=z+t;
```

- Variable declarations are numbered in sequence (from location 1 to 5). Each declaration is considered as an assignment of an unknown value having the type float.
- Input statements assign an unknown constant to the variables. They are considered as assignments of symbolic constants. An input of a variable a is modeled v=$v.
- Integers before statements represent locations. For example: 11:m=-x indicates that the assignation of –x to the variable m has the location 11 in the program.
- Guarded blocks are numbered in the format [i] where i is the rank of the block in the program. For example: [3] if (x<0) indicates that the guarded block having the condition (x<0) is the third block in the program. We designate the condition (x<0) by C[3].

Examples of Initial abstraction computing:

EL= 10: $P_0=H_0(Pg,10)=10$

EL=12: $P_0=H_0(Pg,12)=0*0$

EL=14: $P_0=H_0(Pg,14)=xxx1$

Fig. 4. An Example Program

2.2 Property Checking

2.2.1 Formula Computing

For an abstraction Hi, we first generate a formula Fi from the string Pi=Hi(Pg, EL), then Fi is checked with respect to the path expressed by Pi. Figures 5 exposes the algorithm computing Fi.

Algorithm 3. Formula Computing

Inputs: Pi: a string corresponding to Hi(Pg,EL)
Outputs: A Formula Fi
Initialization: Fi=True
For k=1 to length(Pi)
Do If (Pi[k]='1')
        Then Fi=Fi∧C[k]
        Else **If** (Pi[k]='0')
                Then Fi=Fi∧¬C[k]
End.

Fig. 5. Algorithm computing the Formula Fi

Example: For the precedent program, and the location 14, we have P0=H0(Pg,14)= 'xxx1' so, F0= True∧C[4], so, F0=(m=4). To verify if the formula F0 is satisfiable, we use the well-known concept of weakest precondition. So, let's recall it first.

2.2.2 Satisfiability of the Formula

As we said before, we perform backward symbolic executions. So, we use the concept of weakest precondition. For a statement S and a predicate C, let WP(S,C) denotes the weakest precondition of C with respect to (w.r.t.) the statement S. WP(S,C) is defined as the weakest predicate whose truth before S entails the truth of C after S terminates. Let v=exp be an assignment, where v is a variable and exp is an expression. Let C be a predicate, by definition WP(v=exp,C) is C with all occurrences of v replaced with exp, denoted C[exp/v]. For example: WP(v=v+2, v>8) = (v+2)>8 = (v>6). In the subsequent, we denote WP(Si,C) the weakest precondition of the predicate C w.r.t. the statement having the location Si. We use the concept of weakest precondition to evaluate F0. So, let : Then(k) and Else(k) be respectively: the THEN branch and the ELSE branch of the block BPg[k] representing an IF statement; and Body(k) , the body of the corresponding loop. We call the two first intervals: simple intervals: the sequence of statement from the beginning of the branch to its end; and the last one iterative interval. So, in the subsequent, we extrapolate the definition of weakest precondition to be applied to simple locations intervals. Weakest precondition for iterative intervals will be exposed afterward. We define the weakest precondition of a predicate C w.r.t. an interval [Si,Sj[, denoted by WP([Si,Sj[,C), as the weakest predicate whose truth before Si entails the truth of C after Sj-1 terminates. The idea is to compute successively the weakest preconditions of C with respect to each location within [Si,Sj[starting by the end until we attain Si, or we obtain a constant meaning that there are no variables occurring in C. For each location Sk∈[Si,Sj[, the result obtained from computing WP(Sk,Ck) is given as predicate to compute its weakest precondition w.r.t. Sk-1 and so on. So:

$$\text{WPI}([\text{Si},\text{Sj}[,\text{C})\text{v}=\begin{cases}\text{C} & \text{If no variable occurs in C}\\ \text{WP}(\text{Si},\text{C}) & \text{If Sj=Si}\\ \text{WPI}([\text{Si},\text{Sj-1}[,\text{WP}(\text{Sj-1},\text{C})) & \text{Otherwise}\end{cases}$$

Execution Path

We define the execution path P as the succession of locations intervals (i.e. branches) targeted in each guarded block. Let P[k] the portion of the path associated to BPg[k]. The path P is defined as the union :P=P[1]∪P[2]..∪P[m] where :

$$\text{P[k]}=\begin{cases}\text{Then(k)} & \text{if BPg[k] is an if statement and P0[k]='1'}\\ \text{Else (k)} & \text{if BPg[k] is an if statement and P0[k]='0'}\\ (\text{Body(k)})^{n} & \text{if BPg[k] is a loop statement and P0[k]='1'}\\ & \text{n is the iterations number.}\\ (\text{Then(k)},\text{Else(k)}) & \text{if "P0[k]='x'}\\ \emptyset \ \text{Otherwise} & \end{cases}$$

(Then(k),Else(k)) is a notation regrouping the two branches of BPg [k], it is used to represent the fact that we do not take a precise branch but we consider the overall block with its two branches in the same time; and we try to summarize the effect of the block on the considered condition without specifying a given branch. Ø corresponds to the case where P0[k] is '*'. In fact, in this case, the corresponding block is in the opposite branch, so it is not executed. $(Body(k))^n$ represents the union of the interval Body(k) n times. Let's call P_k the prefix having the length k of the path P.

1- Case 1: BPg[k] is not a loop :

$$WP(Pk,C[k]) = \begin{cases} WP([0,Begin(k)[) & \text{If } k=1 \\ C[k-1] \wedge WP(P_{(k-1)}, WP(Then(k),C[k])) \vee \\ \neg C[k-1] \wedge WP(P_{(k-1)}, WP(Else(k),C[k])) \\ \quad \text{if } P[k] \text{ is of the form(Then(k),Else(k))} \\ WP(P_{(k-1)}, WP(P[k],C[k])) & \text{Otherwise} \end{cases}$$

2- Case 2: BPg[k] is a loop: we note $(C[k])_j$ the value of C[k] in the path P in the iteration j:

$$(C[k])_j = \begin{cases} WP(Pk,C[k]) & \text{If } j=1 \\ WP(P_{(k-1)} \cup body[k]^{j-1}, C[k]) & \text{If } j>1 \text{ and } (C[k])_{j-1}=True \end{cases}$$

The loop iteration number is the least integer n such that $(C[k])_{n+1}$=False

2.3 Refinement

In the abstraction phase, we considered only required paths, and we leaved away other branches. So, if the abstraction allows proving or refuting the property then the process terminates, else it is necessary to refine the abstraction. So, we consider blocks which have not been considered before. Hence, let's suppose we have the abstraction Pi=Hi(Pg,EL) and we must refine it. So, we start with the beginning of the abstraction, each 'x' character in Pi represents two possible executions. In the abstraction process, we considered the block corresponding to the 'x' character as a unique entity without distinguishing between the two branches into this block. The refinement step, must explore inside the block, so, it consists to consider the two branches. So, we generate two strings, in the first string the first 'x' is replaced by '1' and in the second, it is replaced by '0'.

Example: Refine(1x01*1xx1) = (1101*1xx1,1001*1xx1)

For an abstraction Hi a formula Fi is generated in the same manner than as described in the algorithm of the figure5.

The process is reiterated from the point 2-2 until the property is: proved, refuted or no refinement is possible i.e. all the blocks have been entirely explored.

3 Illustrating Example

Let's consider the program in the figure 1 and the erroneous location 14.

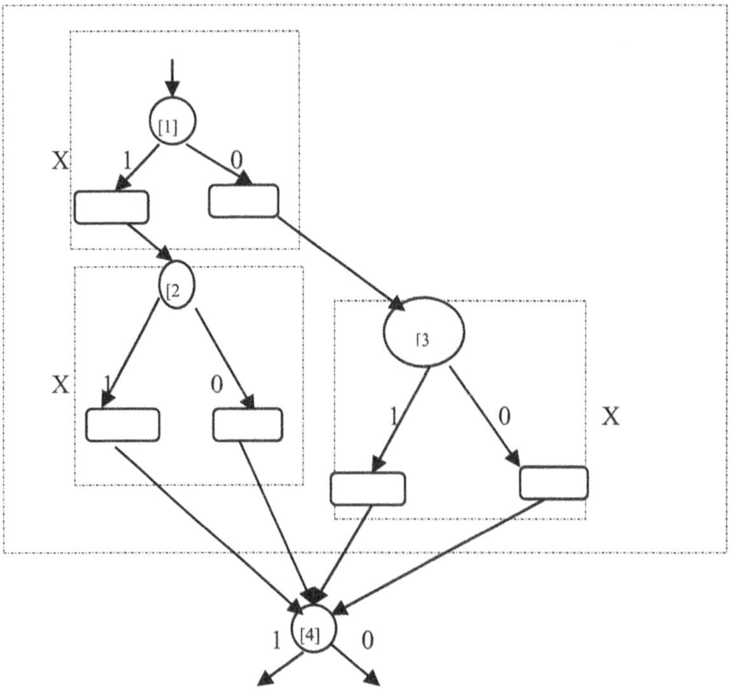

We have: P0=H0(Pg,14)=xxx1; F0=(m=z); WP($P_{03,}$,C[4])=WP([1,14[,m=z): WP(13,m=z)=(m=1). WP(12,m=1)=? We can not decide ➔ refine(P0)
P1=1xx1 ;

WP(Then (1),Else(1)),m=1)=(x<>y)∧WP([8,11[,m=1)∨(x=y)∧WP([11,13[,m=1) WP([8,11[,m=1): WP(10,m=1) ➔F1=(m=1) ➔Refine(P1)

P2=11x1: WP((Then(2),Else(2)) ,m=1) =(x>y)∧(x=1)∨(x<=y)∧(y=1)= C1.
WP(Then((3),Else(3)),C1)=WP([8,9[,C1) = C1.
WP([1,8[,C1) =($x>$y)∧($x=1)∨($x<=$y)∧($y=1)

Formula :F2=True ∧($x<>$y)∧(($x>$y)∧($x=1)∨($x<=$y)∧($y=1))

Satisfiability of F2: F2 is true if and only if : $x=1 and $y<1 or $y=1 and $x<=1. So F2 is satisfiable.

Conclusion: The location 14 is reachable. The program is unsafe.

4 Program with Functions

Functions are modeled independently of the caller, in the same way as ordinary programs. We have two cases regarding the position of the location L we verify w.r.t. a function *F*:

- The location L is in the body of the function: Let's call *L0* the location of the function call. To reach *L*, *L0* must be reached first. So the problem is transformed in two parts: First, check the reachability of *L0* in the caller program. If *L0* is reachable, then check the reachability of *L* in the function. The formula *F* is expressed with effective parameters instead of formal ones.
- The location *L* is outside the function, we perform function summaries: we try to 'capture' the effect of calling *F* on the caller. Thus, since the consequences of a function call are its returned values and the modifications performed on global variables, we resolve these two points in the following manner:

1. The following local variables: *#Ret* and *#F* having the same type as the function *F* are added in *F*. For each global variable *v* modified in *F*, a local variable *#v* is declared.
2. Each return statement: *return (exp)* within *F* is modeled by *#Ret=exp*. Since several return statements can exist in a function, then, *#Ret* can have several expressions.
3. At the end of *F*, the statement: if *(#F == #Ret)* {*#F=#Ret*} is added. This line of code has no effect on variables, but is used for verification purposes.
4. Then we check, by the process described previously, if this statement is reachable, this will imply the computing of weakest preconditions of *(#F == #Ret)* which gives as result the expression of *#F* with respect of all possible expressions of *#Ret* i.e. all the values computed by the function *F*.
5. The same idea is performed to quantify the influence of *F* on global variables: At the end of *F*, for each global variable modified in *F*, the statement if *(#v==v)* *{#v=v}* is added, and we check if this location is reachable.
6. In the feasibility analysis process, in each predicate, the call of *F* is replaced with the formula obtained in (4) expressed with effective parameters instead of formal ones. So is for predicate containing global variables used in *F* (see the example below)

5 Pointers and Aliasing

We regroup variables references in sets representing equivalence classes. Each class has one representing element. Every modification of any element of a class is expressed on its representing element meaning that all the elements of the considered class are modified in the same manner. This method allows us to resolve the problem of aliased variables in a very natural way. So, we perform these actions:

1. For every variable *x*, the first assignment having the reference of *x* as a right hand side i.e. having the form *Si:v=&x* , has as effect to create the class corresponding to *x* references, containing the elements *(&x,0)* and *(v,Si)*. meaning that *&x* is a reference of *x* since the location 0 and *v* is a reference of *x* since the location *Si*. *&x* is the representing of the class.
2. To each declaration of the form *type *v*, the class corresponding to *v*: Cv={(*v*,0)} with *v* its representing element is created.
3. For every assignment of the form *Si:a=b* such that *b* is an element of a given class *C*, the couple *(a,Si)* is added to the class *C*.

4. Each assignment of the form **c=d* such that *c* is an element of a class having *e* as representing element, has as effect to assign the value *d* to **e*. So, it is modeled by: **e=d*. In this manner, all the operations done on variables referenced by different names are expressed on the class representing element.
5. In the feasibility analysis phase, the weakest precondition of a predicate with a dereference **a* is computed in the same manner than precedent cases except that the representing element of equivalence class is used instead of the variable *a*.

Example

Let's consider the following portion of program and generate its variable table:

Program	Representation	
i=0;	1: i=0	
a=&i;	2: a=&i	/* Creation of C&I insert(a,2) */
b=&i;	3: b=&i	/* Insert (b,3) in C&i */
*b=1;	4: i=1	/* Since b∈ C&i, see (4) */
a=&j;	5: a=&j	/* Creation of C&j, insert (a,5) */
c=a;	6: c=a	/* Insert (c,6) in C&j */
*a=2;	7: j=2	/* Since a∈ C&j */

The created classes are: C&i={(&i,0) , (a,2) , (b,3)} and C&j={(&j,0) , (a,5) , (c,6)}. So, we have in these two sets all aliasing information. For example, for the variable *a*, we have: from the statement 2 to 4 the variable *a* is a reference to *i*, while from the instruction 5 to the end it is a reference to *j*. We have also, for example, the information that from 3 to 5 *a* and *b* are aliased and from 6 to the end *a* and *c* are aliased. The computing of weakest preconditions is performed on the representing element, which allows us to express it as in precedent cases. We have for example: WP([1,5[,*a=1) = WP(4,*&i=1)= WP(4, i=1)=(1=1)=True.

6 Related Work

Software model checking has been an active area of recent research. The overall focus is usually on reducing the size of the resulting verification models, by use of appropriate abstractions, in order to manage verification complexity. In terms of general abstraction techniques, predicate abstraction has emerged to be a popular technique for extracting verification models from software. It allows abstracting out data, by keeping track of predicates which capture relationships between data variables. A lot of work has been performed in this direction. The majority of existing approaches to software verification are based on the construction and analysis of an abstract reachability tree(ART). Usually, Each transition of the ART represents a single block of the program this approach is called Single Block encoding SBE. The model checkers SLAM [1,2,3] and BLAST[19] are typical examples of SBE approach, both based on counterexample-guided-abstraction-refinement (CEGAR) [6]. The tool SatAbs [7] is also based on CEGAR, but it performs a fully symbolic search in the abstract space. [5] is an example of Large Block Encoding : LBE. In this approach, transitions represent larger portions of the program. In [5], it is shown that the LBE approach outperforms the SBE one. A different approach to software model checking is bounded model checking (BMC) [6], with the most prominent example CBMC. Programs are unrolled up to a given depth, and a formula is constructed which is satisfiable iff one of the considered program executions reaches a certain error location. The BMC approaches are targeted towards discovering bugs, and cannot be used to prove program safety. Yogi[9,10] simultaneously searches for both a test to establish that the program violates the property, and an abstraction to establish that the program satisfies the property. If the abstraction has a path that leads to the violation of the property, Yogi attempts to focus test case generation along that path. If such a test case cannot be generated, Yogi uses information from the unsatisfiable constraint from the test case generator to refine the abstraction.

Static analysis is attractive because it is complete but has scalability limitations this is why we adopt exclusively the backward analysis which is goal-oriented and consequently it contributes to lighten scalability issues.

7 Conclusion

We have presented a new approach for software model checking problem. Our work presents several contributions:

1. The incremental construction of refinements P, starting by P0, and extending P only if needed, allows examining parts of program only if necessary. This constitutes a sort of abstraction since we move forward all areas of program that are not required for our objective. Moreover, the extensions are done in a way that does not require a great supplementary effort. In fact, since we extend P by the beginning, by adding only one element each time, we just need to continue the computing of the weakest preconditions of each element w.r.t. the newly added one.
2. Feasibility analysis procedure presents several advantages :
    - The definitions of weakest precondition w.r.t. intervals and control structures are novel; they have never been defined before. Furthermore, these definitions are done in a natural way.
    - The computing of weakest precondition in a reverse order, and by using in each location, L, the result found in the location L+1, allows using at each location, the more recent value of variables.
    - Weakest preconditions are computed only on the considered path in the needed location, and not over the entire program as in other approaches.
    - A key feature of our method is the computing of successive weakest preconditions of all the elements of P until arriving to the location at the entry of P. In fact, we have a great benefit in accumulating the checking of the Satisfiability in a unique point instead of verifying each condition separately which requires a great number of theorem prover calls.

3. A substantial characteristic of our approach is modularity: each program can be analyzed by our method given only the expressions of functions that it calls (one for each function and one for each global variable modified). Furthermore, the computing of these expressions can be determined independently from the rest of the program, and is expressed as an instance of the initial problem itself (reachability analysis).
4. Pointers are represented and manipulated in a simple and natural way. A lot of aliasing and points-to information can be deduced from the variable table without significant effort.

References

1. Ball, T., Rajamani, S.K.: The Slam project: Debugging system software via static analysis. In: Proc. POPL, pp. 1–3. ACT, New York (2002)
2. Ball, T., Bounimova, E., Kumar, R., Levin,V.: Slam2: Static Driver Verification with Under 4% False Alarms. In: FMCAD (2010)
3. Ball, T., Majumdar, R., Rajamani, S.: Automatic predicate abstraction of C programs. In: SIGPLAN Conference on Programming Language Design and Implementation, pp. 203–213 (2001)
4. Beyer, D., Chlipala, A.J., Henzinger, T.A., Jhala, R., Majumdar, R.: The BLAST Query Language for Software Verification. In: Giacobazzi, R. (ed.) SAS 2004. LNCS, vol. 3148, pp. 2–18. Springer, Heidelberg (2004)
5. Beyer, D., Cimatti, A., Griggio, A., Keremoglu, M.E., Sebastiani, R.: Software Model Checking via Large-Block Encoding
6. Clarke, E.M., Grumberg, O., Jha, S., Lu, Y., Veith, H.: Counterexample-guided abstraction refinement. In: Emerson, E.A., Sistla, A.P. (eds.) CAV 2000. LNCS, vol. 1855, pp. 154–169. Springer, Heidelberg (2000)
7. Clarke, E., Kroning, D., Sharygina, N., Yorav, K.: SATABS: SAT-based predicate abstraction for ANSI-C. In: Halbwachs, N., Zuck, L.D. (eds.) TACAS 2005. LNCS, vol. 3440, pp. 570–574. Springer, Heidelberg (2005)
8. Dijkstra, E.: A discipline of programming. Prentice-Hall (1976)
9. Gulavani, B., Henzinger, T.A., Kannan, Y., Nori, A., Rajamani, S.K.: Synergy: a new algorithm for property checking. In: Proc. FSE. ACT, New York (2006)
10. Nori, A.V., Rajamani, S.K.: An Empirical Study of Optimizations in Yogi. In: ICSE 2010: International Conference on Software Engineering (May 2010)

Integration Islamic Banking System Based on Service Oriented Architecture and Enterprise Service Bus

Ako A. Jaafar and Dayang N.A. Jawawi

Faculty of Computing,
Universiti Technologi Malaysia
Johor, Malaysia
akodyar@yahoo.com, dayang@utm.my

Abstract. Integration is one of the most important parts for the complex system like Islamic Banking System (IBS). Most of the Islamic Banking System have becomes in two different parts of financial and deposit part that made IBS to becomes more complex for integration than other type of banking system. Despite to the current technologies ability to integrate different application together, it also makes integrating IBS more complicated due to the poor reusability and loosely coupling in the present technologies and approaches like traditional Enterprise Application Integration (EAI) or Point to Point Web Services (P2PWS). This paper present the concept of Service Oriented Architecture (SOA) based application integration, by proposing an application integration framework for IBS using Enterprise Service Bus (ESB) and Business Process Execution Language (BPEL). The outcome of this paper demonstrates that applying ESB/BPEL in IB that increase the reusability and loosely coupled of IB services.

Keywords: IBS, SOA, Application Integration, Web Services, ESB/BPEL.

1 Introduction

Today, software and systems becomes to the key of business success. The efficiency of a system has had a major role to determine the success of a business. In fact, banks industry as a distribute system environment depends on the speed and the agility of IT implementation in order to provide banking services and products. The Islamic banking (IB) is a financial system that based on the Islamic law and diverted by the Islamic economics. In addition, the Islamic law prohibits the payment and collection of interest or usury [1]. The rise of Islamic banks and newer ways of doing business has significantly influenced the needs for a better assessment of the business value realized from the IT investments. Therefore, from the last several years, service oriented architecture (SOA) with integration middleware like Enterprise service bus (ESB) have became a sophisticated and refined architectural model and an alternative to integrate these kinds of systems.

IB has to follow the interpretations of the holy Qur'an in their products and services by the local Islamic scholars as well as the secular laws of their country. Therefore, it

D. Nagamalai et al. (Eds.): *Adv. in Comput. Sci., Eng. & Inf. Technol.*, AISC 225, pp. 131–140.
DOI: 10.1007/978-3-319-00951-3_13

is more complex than traditional banking [2]. In fact, the process of IB is different from the tradition banking, the reason for that is because; IB acts as a trader in their financing and services or a broker between costumer and supplier. In the other hand, traditional banks act as lender as demonstrated in Fig. 1 and Fig. 2.

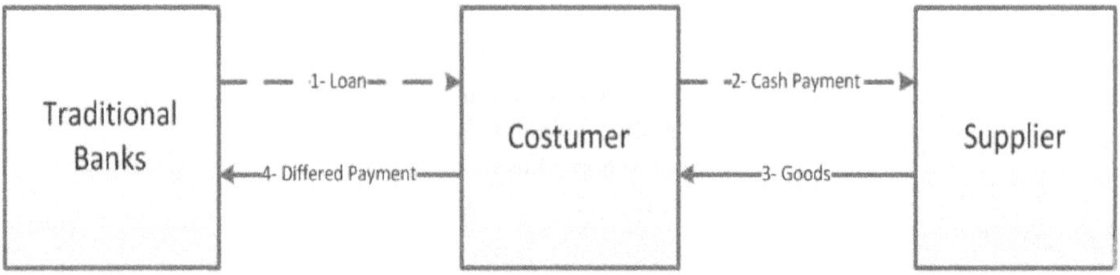

Fig. 1. Traditional banking process

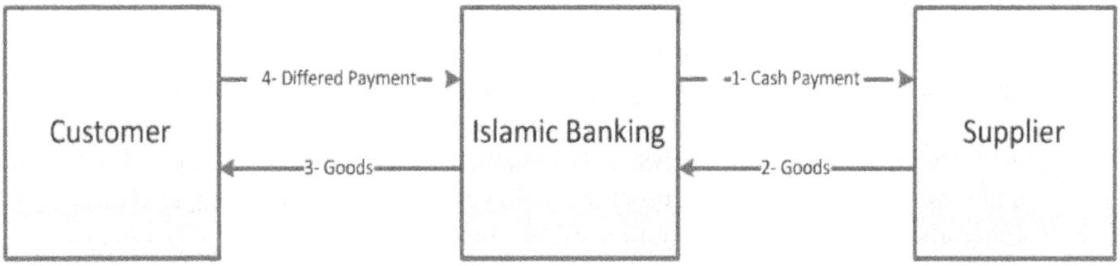

Fig. 2. Islamic banking process

In the Islamic banks, there are two types of services that are basic banking services and Islamic banking services. The basic banking services are consisting of basic operation in banks such as withdrawal, and the other type is consisting of all the other Islamic services in banks [3]. In addition, all products must be defined upon Islamic principles, like *Wadiah Yad Dhamanah, Mudharabah, Qard, and Murabaha.*

The majority of banking industries are still have a legacy system that developed based on using different architecture and technology, some of those systems are works on different platforms that usually have not been designed for integration or they are not expandable [4]. However, Bank's industry cannot afford to writing and developing a new system from the begging and replace it with the legacy system, because this operation required a huge effort, high cost of volume as well as time consuming.

On the other hand, current IB consists of two different parts, financing and deposit part [3]. Another challenge in IB is the integration with other banks. The aims of this scenario is the integration of IB with other banks that mostly to transfer money between them by using deposit cheque, direct transfer, swift and clearing, but there is not such a scenario in traditional banks to integrate with IB to shear Islamic Finance (IF) process such as Murabaha product. Furthermore, IB interests to integrate with suppliers to provide product for their customer. Therefore, the first important strategic is the integration, because current business solution innovations require integration of various business units, applications, business systems and data enterprise. The Integrated information systems have a great advantage to improve its competitiveness with unified and efficient access to the information. Hence, it is easy to see the importance of integration [5].

SOA based application integration is one of the successful solution that transforms IT systems and applications into highly reusable and flexible services. The core capability of SOA integration is to provide a framework for enabling services, events, data, and connecting them to better support business requirements.

SOA-based Application is enabled using various technology platforms such as RPC, Message Queues, CORBA, and XML Web Services. in fact, each of these technologies has a set of capabilities that make it suitable and compatible option of realization for a specific demands and cases [6]. SOA utilize services such as building blocks with several different directions to organize and architect the application within an enterprise. SOA shifts IT from an application-centric to service-centric.

Web services are popular approach to implement SOA, each services can be accessed through the Internet independent of platforms and programming languages [7], but the web services alone are insufficient for SOA integration projects. The Web Services are more providing a simple mechanism to define web services and call web services. In addition, the current Web Services standards lack specifications for management of the enterprise qualities of service required [5]. The developing applications that are supporting web service interfaces will not be sufficient to provide complete and coordinated business processes. Therefore, another approach was needed to compose and organize each these web services together in order to form processes definition [8]. Using XML based orchestration business process that execution language (BPEL) enables task sharing across enterprises using a combination of Web services [9]. ESB can be utilized to implement SOA where it is a software infrastructure to simplify the integration and flexible reuse of business components using a service-oriented architecture. An ESB makes it easy to dynamically connect, mediate, and control services and their interactions [5]. Using ESB as middleware to resolve integrating issue and BPEL to support system integration and collaboration that can improve IBS integration by improving IB service reusability, loosely coupled, and flexibility. Hence, this paper provides SOA based application integration framework to integrate IBS using ESB/BPEL in order to achieve reusable IB services and flexible integration to decrease IBS complexity.

The layout of this paper is structured as. Section 2 is proposes of integration framework for IBS as well as IBS service. Section 3 describes the implementation of the proposed framework using ESB/BPEL. Moreover, a brief introduction of related work has been discussed in Section 4 where followed by conclusion of this paper in Section 5.

2 The SOA Based Application Framework

In order to solve the observed problems in IBS we propose SOA based application integration framework to be applied in IBS. The implementation steps were series as following:

1. Analyze Islamic Banking demands and requirements for integration based on SOA.
2. Define deposit and finance services that be integrated.
3. Propose SOA based application integration framework model for IBS using ESB/BPEL approach.
4. Orchestrate service to come up business processes.

The following subsection describe step 1 and step 2

2.1 IBS Services

The IB services can be divided into two types finance service and deposit service. Finance services exported from finance part of IB such as *Murabaha*, *Musharakah*, and *Mudharabah* whereas deposit services exported from deposit part like Islamic current account and Islamic deposit account. Fig. 3 shows the IBS services.

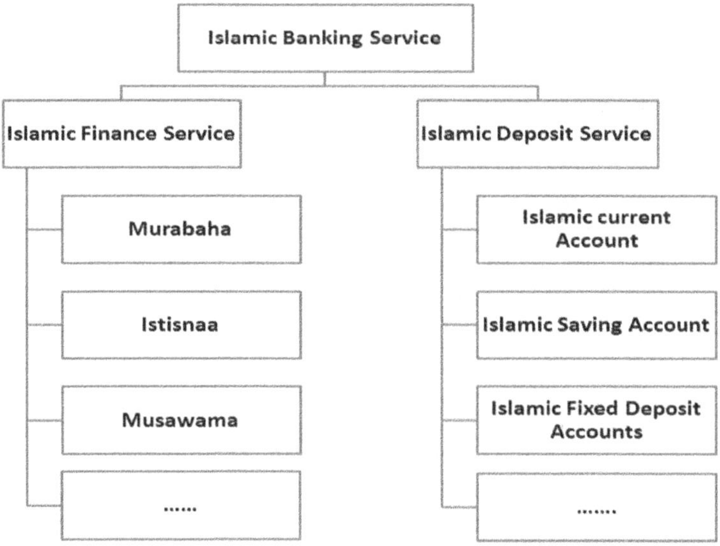

Fig. 3. IBS Services

2.2 Proposed Integration Framework Model for IBS

SOA framework model of application integration used to reach the requirements of IB integration and solve the weaknesses of current IB integration because this framework represents multilayer architecture style which can achieve reusability and loosely coupled between layers. As well as the main integration approach in integration layer is ESB to increase service reusability and loosely coupled, whilst BPLE is used in business process layer to orchestrate service and increase process reusability.

As shows in fig. 4 ESB is used in the integration layer. In the ESB system, service represents integrated objects of application without considering its underlying implementations; it must be standard based, as long as the services meet the SOA standards. Furthermore all finance services, deposit service and third party services connected to the ESB, which can be composite and orchestrate using BPLE in the business process layer.

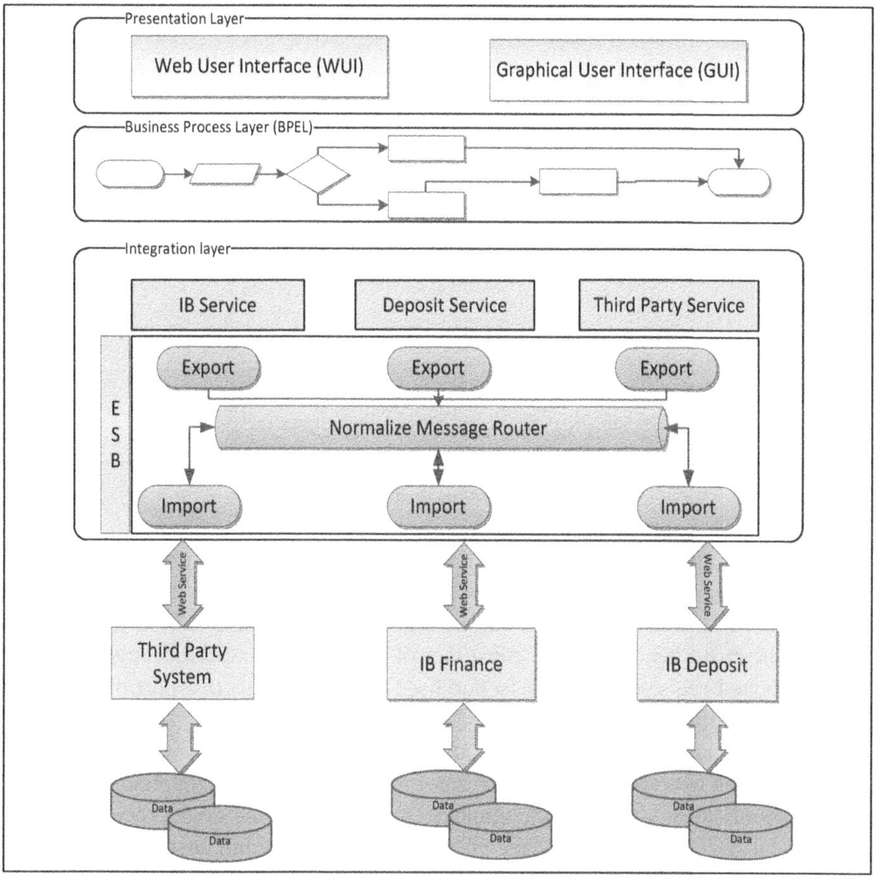

Fig. 4. SOA based Application Integration framework model for IBS

3 Integration Implementation

One example of IB principles is Murabaha, Islamic banks utilize a sale-based transaction Murabaha (cost plus profit) instead of a term loan for financing the purchase of assets by their customer, especially for working capital requirements. Murabaha is a particular kind of sale where the transaction is done on a Murabaha basis. The seller discloses the cost to the buyer and adds a certain profit to it to arrive at the final selling price. Murabaha steps has been illustrated in fig. 5.

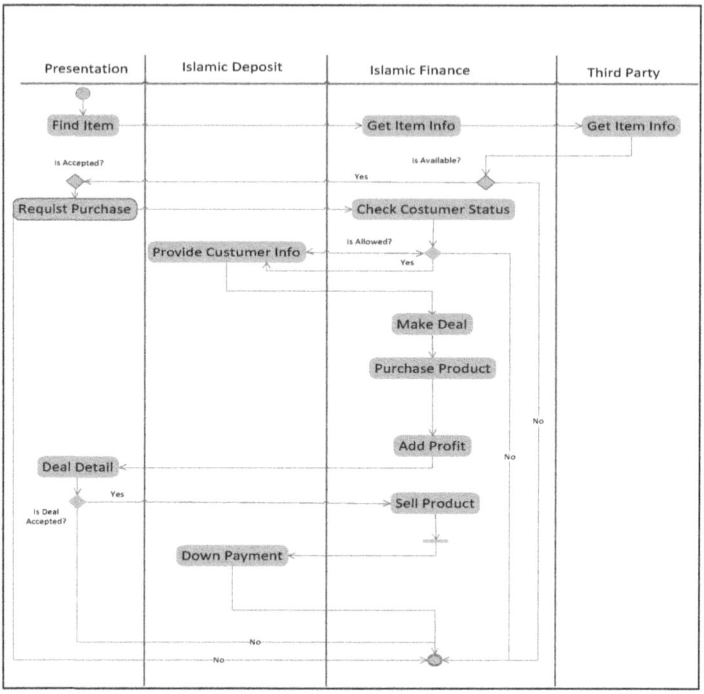

Fig. 5. Murabaha Steps

Fig. 5 shows two parties that banks deal with, that are the customer and Suppliers or third party. Bank provides IB finance service; at the same time third party provide goods for the IB for sale. Example can be sighted for a car shop which provides cars for sale. The customer who is the consumer of third party service, through IB system finds goods and requests the bank to purchase it. Also, IB deposit part provides customer information.

Fig. 6 illustrates Murabaha solution in IBS which is the most used in IB, the business module of Murabaha will be create from three module, the finance module, deposit model, third party model. Finance module provides deal process and deal management. Deposit model provide customer information and customer transaction, whereas third party module provides vehicle information. All modules are pluggable which lead to improve reusability of services. To create `NewDeal` the Murabaha process will be invoked to check customer account and ascertain if the transaction is allowed or not. A new Murabaha will be opened if the customer is allowed. The dealstatus will be New, also banks and branches will be checked. As long as deal is opened the deal should be initialized. The user selects enter customer account to return back the deals that belong to this customer and select the appropriate deal from the list of deals provided by `getDeals` process. After that the information of vehicle will be returned through invoking `getVehicle` process. Amount of profit, pay off, and deal amount will return after entering down payment and profit rate, `getDeal`

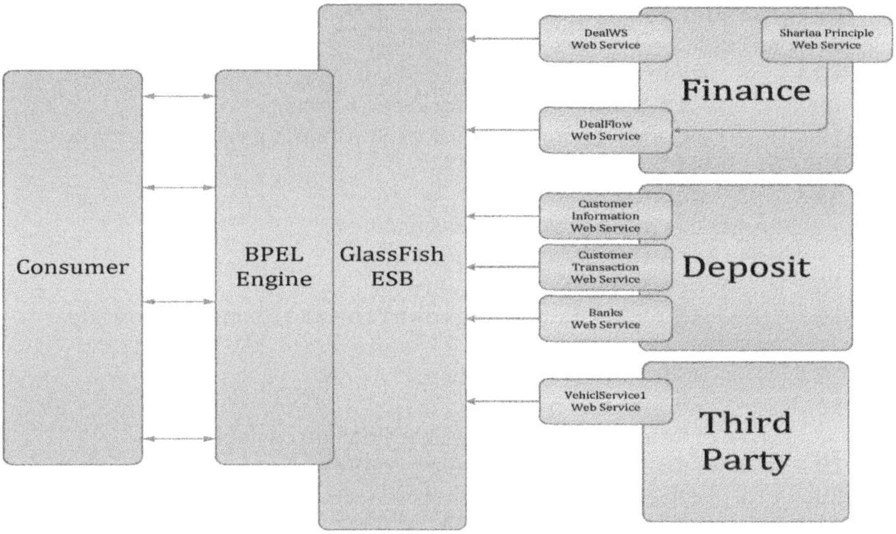

Fig. 6. Murabaha solution

process calculates the rates, all initialized data of the deal will be sent to `initialDeal` process. The status of deal will change to Agreement because this step is like initial agreement between the customer and the banks.

In the next step of Murabaha process, after the initial agreement or (commitment to buy); bank receive the Goods (Vehicle) by buying it from supplier and the vehicle price will be sent to supplier account number. Initially, `AccountEntries` process will be invoked to return back account entries of goods bought from the supplier, after that user confirms the account entries and buys the goods from the supplier by invoking `DealStep` process. The customer then signs a contract with the bank and selling the car to the customer; also, the down payment will be transferred from customers account to down payment account. Just like the Receive Goods step for sign contract account entries also will be returned after that `DealStep` process will be invoked. The last three steps (Initial, Receive Goods, and Sign Contract) will be checked by the Shariaa Principle before executing the process. The following BPEL code illustrates the Sell and Buy transaction in Murabaha process sequence.

```
1. <if name="BankChoice"> <condition>1 =
   $InsertTransactionIn.parameters/ns4:transaction/ns4
   :Transaction/ns4:_BankID</condition>
2. <sequence name="Sequence2">
3. <invoke name="InvokeCustomerInformation".../>
4. </sequence>
5. <else>
6. <sequence name="Sequence1">
7. <invoke
   name="InvokeCorrespondingBanksCustomer"..../>
```

```
8. </sequence>
9. </else>
10.      </if>
11.      <if name="IfCustomerAllowed">
12.      <condition>$CustomerStatusOut</condition>
13.      <if name="IfBuy">
14.      <condition>'Buy' =
   $InsertTransactionIn.parameters
   /ns4:ActivityType</condition>
15.      <sequence name="Buy">
16.      <invoke name="InvokeTransactionBuyProvider"
   ../>
17.      <if name="IfBuyTran">
18.      <condition>-1 =
   $InsertTransactionProviderOut.parameters
19.      /ns1:insertTransactionResult</condition>
20.      </sequence>
21.      <else>
22.      <sequence name="Sell-CheckCredit">
23.      <if name="IfSell">
24.      <condition>'Sell' =
   $InsertTransactionIn.parameters/ns4:ActivityType</c
   ondition><invoke name="InvokeCheckCredit" .../>
25.      <if name= "IfCheckCredit">
   <condition>$CheckCreditOut.
   parameters/ns1:checkCreditResult</condition>
26.      <invoke name="InvokeTransactionSellProvider"
   .../>
27.      <if name="IfSellTran">
28.      <condition>-1 =
   $InsertTransactionProviderOut.parameters/ns1:insert
   TransactionResult</condition>
29.      </if></sequence>
30.      <else><sequence name="SellFail">
```

This code illustrates Buy and Sell transaction for *Murabaha* process, line 11 and 25 shows invoking deposit part to check customer status and credit. Line 13 and 23 illustrates invoking finance part to accomplish transaction.

4 Related Work

The aim of emerging SOA was to design and architecture that allow software vendors to establish different software systems in the form of services that could be published easily and accessed by business customers and partners [4]. So in recent years SOA became most popular architecture method, which provides new technical platform [10].

Reference [4] Propose SOA and Web Service based integration to solve point to point integration in the current scenario of the banking system. They are dividing integration in the banking system into two parts which are 1) Using web services to integrate front end systems together by exposing internal functionalities as a services to receive callers (consumers) requests in the order process them and finally forward back providers (responses). 2) Using an integration bus to integrate front-end systems with back-end system that will act as a single access point for all coming requests from front-end system. In the second part, they build Web Service layer to acts as integration bus for wrapping back-end legacy systems with new systems. Reference [12] address the problem of current banking system in Taiwan to integrate and implement front end banking systems, which they use traditional EAI middleware, they proposes a new approach to integrate banking system based on SOA and Web Sphere ESB Message Broker and compose business processes by service chorography approach.

Bank application integration based on SOA and EAI proposed by [13] which describe how to integrate SOA based banking application using EAI. These works does not address the BPEL power with ESB in order to provide more flexible integration. Also they did not address the IBS problem, IBS is another type in the banking industry; also, we mentioned before IBS is more complex than conventional banking system. IBS like another complex enterprise application need to be agile, flexible, scalable, and reusable. Reference [3] Address the problems of current IB scenario which is a limitation of available IB system in the market. In other words, none of them is totally holistic IBS (that covers both deposits and financing) so they develop Holistic Islamic Banking System (HiCORE), based on the SOA and parameter-based semantic approach but redevelop and redesign a new IBS need huge cost and it is time consuming.

5 Conclusion

The variety of application in IBS that follows Islamic rules made IBS more complex than other banking system, as well as obtain IBS from two different parts to add more complexity into IBS. Therefore, this research were proposed SOA based application integration to integrate IBS. This paper used ESB/BPEL to integrate IB sub systems together such as deposit part with finance part that transform IBS service into highly distributed and loosely-coupled along with IB service and process that can be easily managed, swapped, removed, or reused as compare to other integration approaches such as P2PWS or traditional EAI. It also helped IB to expose their services and finances to other banks.

However, this research demonstrates the advantages of ESB/BPEL for integration but the evaluation of ESB/BPEL as compare to other approaches has not been carried out to demonstrate the improvement level of reusability and loosely coupled of services. in addition, the ESB/BPEL may have the disadvantages in terms of performance. Therefore, the next step of this paper will be the reusability and performance evaluation of ESB/BPEL as compare to other approaches.

Acknowledgment. Special thanks to the Universiti Technologi Malaysia (UTM) and MOSTI for funding this research. Also, to our software engineering laboratory members for their continuous support.

References

1. Al-Roubaie, A., Barhrain, Alvi, S.: Islamic Banking and Financ. Critical Concepts In Economic, vol. 1. Routledg (2009)
2. Hunt, R.: Islamic Banking: Core Vendors Fill Growing Demand for Shari'ah-Compliant Banking: TowerGroup (2007)
3. Halimah, B.Z., Sembok, T.M., Azlina, A.S., Sharu, A.M.N., Azuraliza, A. B., Zulaiha, A. O., Nazlia, O., Salwani, A., Sanep, A., Hailani, M. T., Zaher, M. Z., Azizah, J., Nor Faezah, M. Y., Choo, W. O., Abdullah, C., Sopian, B.: Holistic Islamic Virtual Banking System: SOA parameter- based semantic approach (HiCORE). Paper presented at the Information Technology (2008)
4. Raid, A., Hassan, A., Hassan, Q.: Leveraging SOA in Banking Systems Integration. Journal of Applied Economic Sciences 3(2), 4 (2008)
5. Matjaz, J., Ramesh, L., Poornachandra, S., Frank, J.: SOA Approach to Integration. BIRMINGHAM - MUMBAI: packt (2007)
6. Raid, A., Hassan, A., Hassan, Q.: Design of SOA-based Grid Computing with Enterprise Service Bus. International Journal on Advances in Information Sciences and Service Sciences 2(1), 11 (2010)
7. Rosenberg, D.: Modeling Service-Oriented Architectures (2010)
8. Abdaladhem, A.S., Partrik, F., Jacque, P.: Web Services Orchestration and Composition (2009), http://diuf.unifr.ch/drupal/softeng/sites/diuf.unifr.ch.drupal.softeng/files/file/publications/albreshne/download/WP09-03.pdf(retrieved August 12, 2012)
9. Karande, A.M., Chunekar, V.N., Meshram, B.B.: Working of web services using BPEL workflow in SOA. In: Unnikrishnan, S., Surve, S., Bhoir, D. (eds.) ICAC3 2011. CCIS, vol. 125, pp. 143–149. Springer, Heidelberg (2011)
10. Li, Z., Yan, W., Fujiang, L.: The Banking System Reference Implementation Based on SOA. Paper presented at the Fourth International Joint Conference on Computational Sciences and Optimization (2011)
11. Jenny, Chen, A., Hsu: Implementing a Banking Front End Processor in Taiwan Using SOA. Paper presented at the IEEE International Conference on e-Business Engineering, ICEBE 200 (2006)
12. Li, Q.: Bank Application System Integration based on SOA and EAI. Paper presented at the International Conference on Industrial and Information Systems (2009)

High-Performance and High-Assurance

William R. Simpson

Institute for Defense Analyses, 4850 Mark Center Drive, Alexandria, Virginia 22311 USA
rsimpson@ida.org

Abstract. Many Organizations are moving to web-based approaches to computing. As the threat evolves to higher levels of sophistication, many governmental and commercial organizations are also moving toward high-assurance. This service-based approach offers many of the advantages of the cloud-based approaches. There is a natural partitioning of entity groups similar to the Google Megastore [1] concept, but the rigidity associated with forced portioning may not be attainable. This high-assurance requirement presents many challenges to normal computing and some rather precise requirements that have developed from assurance issues for web service applications, many of which are considerably simplified by high-performance computing paradigms. The most difficult part of scaling up to higher user levels is the maintenance of the security paradigms that provide mitigation of these generic and specific threats. Not to worry, high performance computing is here. Multiple cores and server clusters provide scaling at the thread level! But, the news is not all positive.

Keywords: Security, High-Performance Computing, Scaling, Cryptography, Cloud, Virtualization.

1 Introduction

In certain enterprises, the network is continually under attack. Examples might be; banking industry enterprise such as a clearing house for electronic transactions, defense industry applications, credit card consolidation processes that handle sensitive data (both fiscal and personal), medical with concerns for privacy and statutory requirements, and content distributors worried about rights in data, or theft of content. The attacks have been pervasive and continue to the point that nefarious code may be present, even when regular monitoring and system sweeps clean up readily apparent malware. This omnipresent threat leads to a healthy paranoia of resistance to observation, intercept, and masquerading. Despite this attack environment, the web interface is the best way to provide access to many of the enterprise users. One way to continue operating in this environment is to know and vet your users, your software and devices.

2 High-Assurance Architecture Elements

Despite the obvious advantages of cloud computing, the large amount of virtualization and redirection poses a number of problems for high-assurance. In order to understand this, let's examine a security flows in a high-assurance system. The basic elements

D. Nagamalai et al. (Eds.): *Adv. in Comput. Sci., Eng. & Inf. Technol.*, AISC 225, pp. 141–147.
DOI: 10.1007/978-3-319-00951-3_14

include a user, who initially authenticates to his/her domain using a hardware token and establishes a Virtual Private Network (VPN) session; a Security Token Server (STS); and attribute stores for generating the Security Assertion Markup Language (SAML) token. The application system consists of a web application, one or more aggregation services that invoke one or more exposure services and combine their information for return to the web application and the user. The exposure services retrieve information from Authoritative Data Sources (ADSs). Each communication link in Fig. 1 will be authenticated end- to-end with the use of public keys in the X.509 certificates provided for each of the active entities. The requester initially authenticates to the service provider. Once the authentication is completed, a TLS connection is established between the requester and the service provider, within which a WS-Security package will be sent to the service.

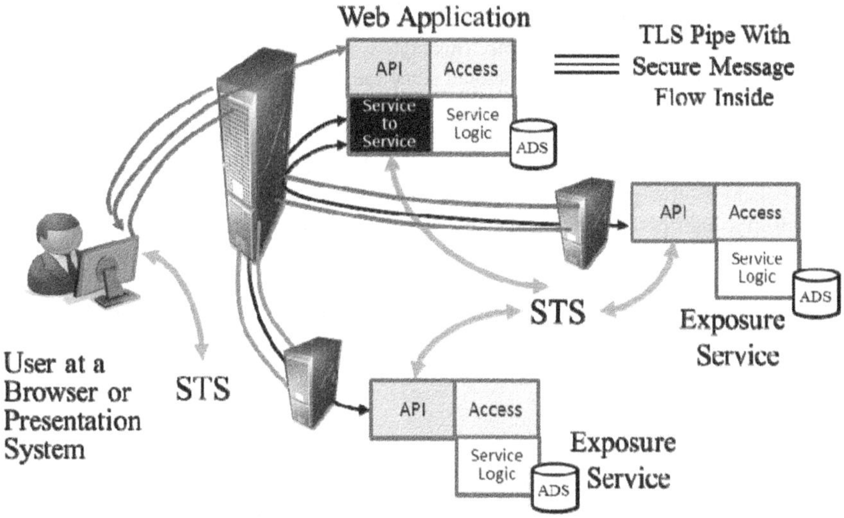

Fig. 1. High-Assurance Security Flows

The WS-Security package contains a SAML token generated by the STS in the requestor domain. The primary method of authentication will be through the use of public keys in the X.509 certificate, which can then be used to set up encrypted communications (either by X.509 keys or a generated session key). Session keys and certificate keys need to be robust and sufficiently protected to prevent exploitation. The encryption key used is the public key of the target (or a mutually derived session key), ensuring only the target can interpret the communication. The problem of scale-up and performance is the issue that makes cloud environments so attractive. The cloud will bring on assets as needed and retire them as needed. The trick is to maintain the security paradigm as we scale up. A load balancer monitors activity and posts a connection to an available instance. In this case all works out since the new instance has a unique name, end-point, and credentials with which to proceed. All of this, of course needs to be logged in a standard form and parameters passed to make it easy to reconstruct for forensics.

3 High-Assurance Scaling Issues

Among the many scaling issues in a large-scale enterprise, three stand out as related to the high-assurance, high-performance issues:

- Load balancing of security components - A Security Token Server (STS) is involved at the initiation of every session (500,000+ users and multiple services). Further, the STS uses the services of an Enterprise Attribute Store.
- Load balancing of non-security components - Many Services will need load balancing to meet users' needs.
- All session are bi-laterally authenticated by PKI and encrypted (TLS).

4 Enterprise Attribute Store

The Enterprise Attribute Store (EAS) is more than the data repository of information that is used for authorization claims. It must gather data, compute claims, provide for delegation and provide claims information about entities to trusted and authorized requesters. In doing this it must meet all of the high-assurance security requirements and processes. The goal is to provide each application, service or software element with the information needed to make access control decisions from a defined and trusted authoritative source. Each independent instance of a virtual or real machine or virtual or real service must be uniquely named [2] and provided a PKI Certificate for authentication. The Certificate must be activated while the virtual machine is in being, and de-activated when it is not, preventing hijacking of the certificate by nefarious activities. Extensions of the thread mechanism by assigning resources to the operating system may preserve this functionality. The thread mechanisms are inherently parallel computation and may benefit from high performance computing methodology.

5 Normal Scaling of the STS

Scale-up to higher levels of users will require a number of different schemes. The most critical, since it involves every request, will be the STS and EAS. Example data needed for load-balancing calculations are provided below:

- Test data are still being developed; however, assume testing indicates 100 token requests can be satisfied in 1 second by the current STS using normal threading. Improved versions of the STS or processors may reduce these requirements
- Requirement for an enterprise the size of the USAF would require 1667 SAML [3] tokens per second at peak load [500,000 users in 5 minutes]
- Assume a 25% throughput loss in multiple clustering.
- Need 23 STS

The STS is a trusted component and can be load balanced in the traditional manner. The clusters share naming, PKI credentials and end-point identities.

6 High-Performance Computing Benefits

Individual requests and user sessions are inherently parallel in nature and well-suited to the high-performance environment by spreading the threads across multiple cores. The goal is for a globally based enterprise with three primary compute centers (US, East, and West). A high-performance computing environment with a tenfold improvement (16 cores) would satisfy that by increasing the thread count and maintain security without load balancing in the traditional sense. Further, it would be scalable to 6-8 million users with another tenfold improvement and with as little as 64 core systems. Care should be taken to provision adequate bandwidth/memory/storage for the scale of operations.

6.1 Scaling of Services

Multiple sessions are inherently parallel in nature and well-suited to the high-performance environment by spreading the threads across multiple cores. The scaling of services is dependent upon the usage of the service itself. In the enterprise, the services that will get the largest utilization are in the EAS. Here too, care should be taken to provision adequate bandwidth/memory/storage for the scale of operations.

6.2 Globally-Distributed Data Bases

Google has developed a scalable, multi-version, globally distributed, and synchronously-replicated database called Spanner [4]. It is the first system to distribute data at global scale and support externally-consistent distributed transactions. Spanner's main focus is managing cross-datacenter replicated data. Utilizing a process such as Spanner will allow for placement of the EAS data stores in three global centers (US, East, and West) for use by the STS.

6.3 The Matter of Encryption

For the high-assurance process all communications are encrypted using TLS 1.2 for confidentiality. Private keys are stored in Hardware Security Modules (HSMs). All Cryptography is done locally (no enterprise cryptographic services). Software Cryptographic Suites used are FIPS 140 approved.

6.3.1 The High-Performance Dilemma

The locality of cryptographic services makes scaling through high-performance multi-core thread management easier and eliminates the need for front-end load balancing. However, maintaining confidentiality with such computing capability available to an adversary is problematic. The encryption methodology must be standard and published so that all elements of the enterprise can obtain the proper software and hardware to perform the needed operations. Rogue agents (including insider threats) may be present and to the extent possible, we should be able to operate in their presence, although this does not exclude their ability to view and export some activity. Key extraction from encrypted data is inherently a parallel function. For example, all of the possible keys can be distributed to many cores and each can spend a few cycles attempting to decrypt encrypted packets. When one recognizes the output the key is discovered. This parallel decomposition is shown in figure 2.

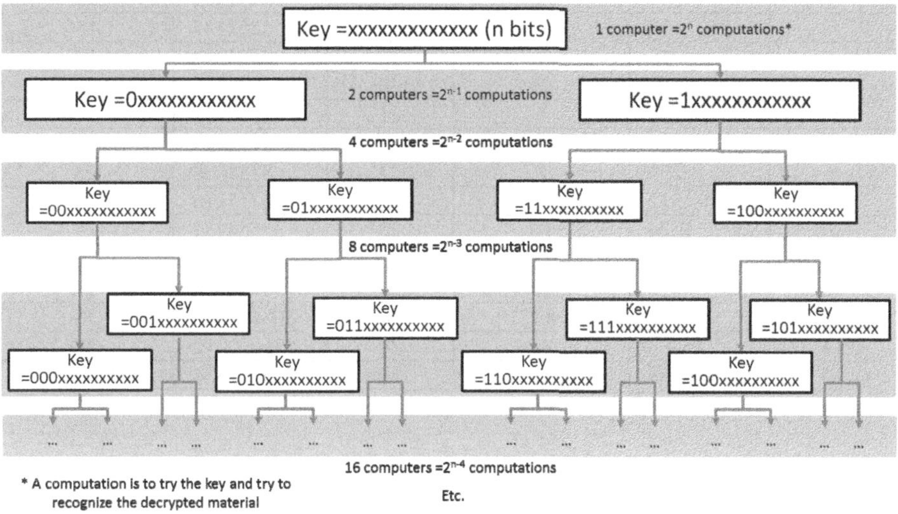

Fig. 2. Encryption key discovery is inherently parallel

6.3.2 The Mathematics of Parallel Decomposition of Key Discovery

For a given key length (m), the number of possible keys is given by 2^m and this represents the sequential computational burden for trying each possible key. Further, the first n bits may be distributed among cores (c) for processing at the scale of $n=2^c$. The remaining bits (m-n) may be sequentially processed with a total processing burden on each processor equal to $2^{(m-n)}$. This results in a confidentiality effectiveness loss of n bits. Note that this loss is independent of the original key length (m).

If the key length is only 10 bits, then the possible keys that encrypted the packet is 1024. If this is spread over 1024 cores, then a single decrypt cycle (a decrypt cycle includes applying the decryption process and applying a recognition algorithm to the answer. This could be 10 machine cycles or less) will reveal the key and provide an adversary a way to overcome confidentiality. If the key length is only 12 bits, then the possible keys that encrypted the packet is 4096. If this is spread over 1024 cores, then 4 decrypt cycles will reveal the key and provide an adversary a way to overcome confidentiality. This latter is equivalent to weakening the cryptographic key by ten bits. Table 1 provides a summary of the discovery process.

Table 1. Loss of Bit Effectiveness

distributed key bits (n)	values spread over cores (c)	confidentiality effectiveness loss (bits)	distributed key bits (n)	value spread over cores (c)	confidentiality effectiveness loss (bits)
1	2	1	16	65536	16
2	4	2	17	131072	17
3	8	3	18	262144	18
4	16	4	19	524288	19
5	32	5	20	1048576	20
6	64	6	21	2097152	21
7	128	7	22	4194304	22
8	256	8	23	8388608	23
9	512	9	24	16777216	24
10	1024	10	25	33554432	25
11	2048	11	26	67108864	26
12	4096	12	27	134217728	27
13	8192	13	28	268435456	28
14	16384	14	29	536870912	29
15	32768	15	30	1073741824	30

From table1, it can be seen that high-performance computing over 500k cores is equivalent to a loss of 19 bits in the strength of encryption. This leads to a race that technology has delivered to us, and the response is to increase the bits in the encryption process. The increased bit length adds computational burden to the normal computational process increasing the need for high-performance computing. This race can only be broken by developing a non-parallel decomposable decryption process, which is not currently available.

7 Summary

This paper has reviewed the basic approaches to scale up in high-assurance computing environment, and some of the stresses that can be created while others are relieved. Virtualization must be very carefully reviewed to ascertain if the security paradigm can be maintained. Extensions of the thread mechanism by assigning resources to the operating system may preserve this functionality. The individual mechanism for virtualization will determine whether this can be accomplished. Notably the extensive use of virtualization and redirection is severe enough that many customers who need high-assurance have moved away from the concept of cloud computing. Figure 6 provides a summary of how a user addresses an individual web application in a scaled-up system. This processes described in this paper are part of a broad-scale, high-assurance enterprise stand-up. Aspects of the enterprise processes are shown in [5-11].

References

1. Baker, J., et al.: Megastore: Providing Scalable, Highly Available Storage for Interactive Services, Google, Inc. In: 5th Biennial Conference on Innovative Data Systems Research (CIDR 2011), Alomar, California, USA, January 9-12 (2011)
2. Standard for Naming Active Entities on DoD IT Networks, Version 3.5 (September 23, 2010)
3. OASIS Identity Federation, 2011b, Profiles for the OASIS Security Assertion Markup Language (SAML) V2.0, `http://www.oasis-open.org/committees/tc_home.php?wg_abbrev=security` (accessed on February 19, 2011)
4. Corbett, J.C., et al.: Spanner: Google's Globally-Distributed Database, Google, Inc., To appear in Proceedings of OSDI (2012)
5. Chandersekaran, C., Simpson, W.R., Trice, A.: The 1st International Multi-Conf. on Eng. and Tech. Innovation. Cross-Domain Solutions in an Era of Information Sharing I, 313–318 (2008)
6. Simpson, W.R., Chandersekaran, C.: A Multi-Tiered Approach to Enterprise Support Services. In: 1st International Conference on Design, User Experience, and Usability, part of the 14th International Conference on Human-Computer Interaction (HCII 2011), Orlando, FL, p. 10 (July 2011)
7. Simpson, W.R., Chandersekaran, C.: Information Sharing and Federation. In: 2nd International Multi-Conference on Engineering and Technological Innovation, Orlando, FL, vol. I, pp. 300–305 (July 2009)
8. Chandersekaran, C., Simpson, W.R.: The Case for Bi-lateral End-to-End Strong Authentication. In: World Wide Web Consortium (W3C) Workshop on Security Models for Device APIs, London, England, p. 4 (December 2008)
9. Chandersekaran, C., Simpson, W.R.: An Agent Based Monitoring System for Web Services. In: The 16th International Command and Control Research and Technology Symposium: CCT 2011, Orlando, FL, vol. II, pp. 84–89 (April 2011)
10. Simpson, W.R., Chandersekaran, C.: A Multi-Tiered Approach to Enterprise Support Services. In: 1st International Conference on Design, User Experience, and Usability, Orlando, FL, p. 10 (July 2011)
11. Simpson, W.R., Chandersekaran, C.: Enterprise High Assurance Scale-up. In: Proceedings World Congress on Engineering and Computer Science 2012, San Francisco. Lecture Notes in Engineering and Computer Science, vol. 1, pp. 54–59 (October 2012)

Automatically Language Patterns Elicitation from Biomedical Literature

Seyed Ziaeddin Alborzi

Computer Engineering School, Nanyang Technological University, Singapore
Seyed1@e.ntu.edu.sg

Abstract. The amount of research articles being published over the years has been overwhelming and number continues to rise with each day. This rapid growth combined with the unstructured nature of text written in natural languages has created the need to develop tools and methods that aid the process of information extraction, making it more accessible and utilizable. In this work, we present an approach for language pattern acquisition from the biomedical literature. In our method, all possible patterns are generated (candidates' enumeration), and those patterns which have a match in the training corpus are selected. Equipped with genes and proteins names glossaries plus keywords database, we achieved a recall rate of 52.2% with precision of 40.9%, identifying 321 gene ontology terms.

Keywords: Bioinformatics, Natural Language processing, language pattern, Pattern acquisition.

1 Introduction

The increasing use of electronic research article has led to a massive increment of databases containing various scientific domains publications. The electronically accessible database, PubMed [1], has appeared as the most pertinent one for the biomedical society. To date, PubMed comprises more than 21 million citations for biomedical literature [2] and about 2 million full-text articles (available via PubMed Central [3]). The size and fast growth of PubMed combined with the unstructured context call for methods and tools that aid the process of extracting information, increasing the usability and accessibility.

Ontologies provide controlled terminologies which may help stated procedure if it is possible to map a text passage to an ontology concept in the biomedical domain the Gene Ontology (GO) [4] has evolved as the actual standard. It provides a structured and controlled vocabulary of terms describing features of genes and gene products.

The GO is being used for the process of gene function annotation which involves two tasks. First task is recognizing genes, gene products and gene functions in text and second task is, associating both using GO terms. This process is performed to sustain the Gene Ontology Annotation (GOA) database [5] which contains associations of

D. Nagamalai et al. (Eds.): *Adv. in Comput. Sci., Eng. & Inf. Technol.*, AISC 225, pp. 149–158.
DOI: 10.1007/978-3-319-00951-3_15

genes and gene products to GO terms that reference the PubMed article as well as endorse the annotation.

Normally, GO annotation is carried out via manual curation by an expert, the so-called curator, who must read the whole paper to extract related genes, gene products and their functional description. Then, he has to map this information to a couple of GO terms. While it has been demonstrated that the task of identifying genes and gene products can be fulfilled through the help of high-precision tools [6; 7], the latter task of identifying GO terms in free text has yet to be solved in a satisfying manner.

Automatic construction of language patterns is not a novel task, but the previous methods have been limited to just a few types of concepts. The real challenge of this research lies in the number of target concepts used. We took Gene Ontology, , as the source of concepts types for which we collected language patterns. GO has many complex concepts and they share a common vocabulary. We chose frequently used terms in the vocabulary which refer to relations (e.g., 'regulation', 'export'), and will construct language patterns for the concepts referred to by these terms. The resultant language patterns are used to identify the GO concepts in text.

Three of major ways developed to extract information from documents. Include Statistical methods, Computational linguistics methods, Frame-based methods [8].

Statistical methods are based on the frequency of occurrence of words in large corpora of text that has been previously organized in line with some form of external knowledge [8]. In [9], keywords are extracted from Medline abstracts in order to qualify the function of previously classified protein families. In [10], statistical methods assist in the annotation of experimental results obtained from DNA expression arrays. The distribution of extracted terms in order to classify them in relation to articles linked to OMIM database of human diseases is used in [11]. In 2008, a survey from Credant Technologies reported that in six months 3,000 laptops and 55,000 cellular phones were left in London taxis [1].

Computational linguistics methods use parsers and grammars to extract syntactic information and internal dependencies within individual sentences [8]. The representation of model of the patterns in the system proposed by [12] was based on the paths from predicated nodes in dependency trees. A framework for text mining is presented in [13], called DISCOTEX (DISCOvery from Text EXtraction), using a learned information extraction system to transform text into more structured data which is then mined for interesting relationships. In [14], a robust method based on Bayesian networks was introduced to extract pattern acquisition of protein-protein interaction. Moreover, an algorithm for extracting PPIs from literature, which consisted of two phases, proposed in [15]. One of the most efficient methods introduced in [16]. It is the combination of diverse lexical, syntactic and semantic information in feature-based protein-protein interaction extraction using SVM.

A third type of approach combines features of the two previous methods with a set of previously defined templates for possible textual relationships, called frames-based method [8]. The use of frame-based methods to extract large set of biological information from scientific literature based on automatic detection of protein-protein interaction extracted from scientific abstracts is shown in [17]. In addition, a method for extraction of information from biomedical literature is proposed in [18] that presented a methodology for extracting information on PPI from the scientific literature. Another approach for relation identification of proteins from biomedical literature is based on using a dynamic programming algorithm to compute distinguishing patterns by aligning relevant sentences and key words that described protein interactions [19]. [20] co-trained a decision list learner whose feature space covered the set of all syntactico-semantic patterns with an Expectation Maximization clustering algorithm that uses the text words as attributes.

2 Materials and Methodology

Related Resources: GRO is intended to represent gene regulation in a formal way rather than extremely fine-grained classes as found in ontologies such as the Gene Ontology (GO) (created for database annotation purposes) and various relevant databases. The main purpose of the ontology is to support NLP applications. [22]. Table 1 illustrates 63 GRO concepts of our system.

Table 1. 63 GRO concepts used by the system

GeneticImprinting	**GeneExpression**	**DNARegion**
TranscriptionInitiation	Methylation	Spliceosome
ModificationOfMolecularEntity	G1_SPhase	RibosomalDNA
SelenocysteineIncorporation	CarbonCatabolite	CellCycle
TranslationalAttenuation	TranslationInitiation	DevelopmentalProcess
RNAInitiatedSilencingComplex	OxidativeStress	GeneSilencing
PreMicroRNA	MicroRNA	Destabilization
RNAPolymerase_III_Promoter	RegulatoryProcess	Stabilization
TranscriptionFactor	SmallInterferingRNA	OrganismalProcess
RNAPolymerase_I_Promoter	ChromatinSilencing	TranslationTermination
TranscriptionFactorActivity	DosageCompensation	CellularProcess
RegulationTranslationalFidelity	GeneRegion	Transcription
RNAPolymerase_II_Promoter	MessengerRNA	RNAInterference
RNAElongation	Maintenance	Translation
Localization	Chromatin	Process
Decrease	Virus	Stress
Producing	Epigenetic	DNA
RNA	Binding	Protein
Gene	G1Phase	Splicing
SPhase	Nucleus	Mitosis

The Gene Ontology project standardizes the representation of gene and gene product attributes across species and databases [4; 23].

The Gene Ontology Annotation (GOA) database provides high-quality electronic and manual annotations to the UniProt Knowledgebase (Swiss-Prot, TrEMBL and PIR-PSD) using the standardized vocabulary of the Gene Ontology (GO). [24].

Finally we use ABNER, a software tool for molecular biology text analysis. [25; 26; 27].

Corpus Construction: For each GRO concept, the system provides a scientific abstract repository as the primary source in extracting language patterns. Based on the association between Gene Regulation Ontology concepts, Gene Ontology terms and Gene Ontology Annotations, system is able to pair GRO concepts with several abstracts. Figure 1 shows the association between GRO, GO, PMID and abstracts.

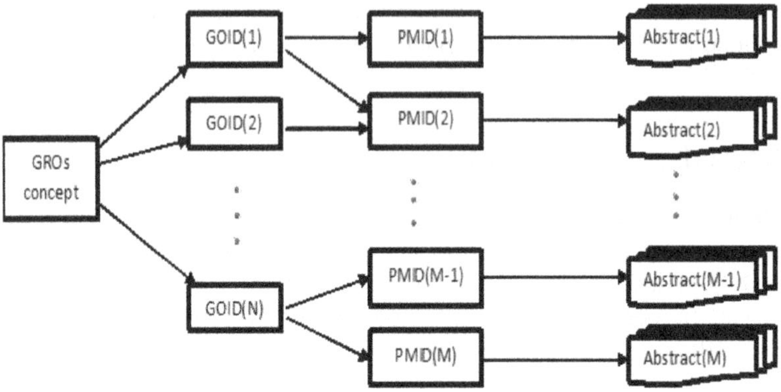

Fig. 1. Association between GRO, GO, PMID and abstracts

Pattern Acquisition: Our final goal is seeking patterns for GRO concepts and testing their quality by applying and finding the GO terms in text. But some scientists don't use the ontology vocabularies; hence our system needs to find as many keywords as possible for each GRO concept. .This expands pattern coverage by establishing a dictionary for each concept. With finding more keywords for GRO concepts and adding them to dictionaries, the system is able to unearth language patterns from abstracts which they don't comprise the main word of the specified concept. For instance, many scientists are exploiting "Generate" instead of "Produce" in their publication, or some scientists prefer to use "Alteration" instead of "Modification".

In order to find more candidate words the system excerpts the result intersection of 3 ideas. First of all, synonyms and antonyms of the concepts are extracted from several resources. In the second idea is to increase the number of possible keywords, troponym and hyponym of each concept are used [28]. In the third plan, the system discovers keywords for each specific concept based on the frequency of words inside the related abstracts to the concept and the frequency of words inside the unrelated abstracts. Table 2 is expressed this process.

Table 2. Finding keywords based on their frequencies

$$FinalScore = \frac{Count1}{Log(Count2 + 1) + 1}$$

Count 1: Frequency of X in the related repository
Count 2: Frequency of X in the unrelated repository
* If FinalScore of each word is greater than system threshold that word can be a potential keyword in the text for the concept.

Now System has words from 3 different dictionaries formed by above 3 ideas. These potential keywords are combined to produce the ultimate keywords for the system, performed by union of first and second ideas followed by intersection with the result of third idea. Table 3 shows us discovered keywords of "RegulatoryProcess" concept based on illustrated ideas.

Table 3. Finding keywords based on their frequencies

RegulatoryProcess keywords:			
Regulation	Deregulation	Modulate	Adapt
Form	Activate	Control	Develop
Interact	Bind	Cleavage	Mutate

In candidate's enumeration method, system exploits pre-generated patterns. These patterns are fed in the system as inputs to find appropriate patterns. In the first step, all possible patterns that appears in the text are generated.. These patterns are all possible combination of used POS (part-of-speech) tags. For example, a system with limitation of patterns' length by two tagtypes (Noun, Verb), creates following patterns in table 4.

Table 4. Instances of pre generated patterns

Noun Noun	Verb Verb
Noun Verb	Verb Noun

Our system assumes no pattern can be longer than six. It means patterns with more than six residues are ignored in this approach. In the next step System matches the generated patterns against sentences of abstracts. When extracting patterns from text for each GRO concept, patterns without its keywords will be ignored and others will be stored in its patterns repository.

Two advantages of this approach are its speed and the ability of extracting patterns with only one time occurrence from abstracts. Table 5 illustrates candidates' enumeration algorithm.

Filtering the unwanted result is one of the most crucial parts of the system in order to have efficacious patterns since useless patterns increase the false positive amount of the system. Filtering of the outcomes is based on the following rules:

1. System removes all language patterns which start or end with "preposition", "coordinating conjunction" and "determiner" from patterns repositories.
2. For each GRO concept, system sustains frequent patterns in relevant abstracts of that GRO concept.

Table 5. Candidates' enumeration algorithm

1- Generate all possible patterns
2- Apply patterns to abstracts of each GRO concept
3- Ignore patters without concept's keywords
4- Store other patterns

3 Results and Discussion

Evaluation Method: System needs an evaluation method that validates the quality of the extracted patterns; also this method evaluates how well the language patterns can identify Gene Ontology terms in texts. In the first step, system needs to create 2 sets a GO term IDs and PubMedIDs pair, each for system and truth. Truth set comprise all pairs of GOID-PubMedIDs from GOA and system set includes all possible pairs of GOIDs-PubMedIDs from GO term repository and PubMedIDs repository. In the next step, system gets each pair of GOID-PubMedID and retrieves GRO concepts of the GO term. Then it tries to match patterns of each concept on the corresponding abstract of the PubMedID according to following rules:

1. Finding at least one pattern for each GRO concept.
2. To eliminate the ambiguity, matched patterns of GRO concepts cannot be the same.
3. If system finds any protein, gene, RNA or DNA in the patterns pool, it can replace the element with any entity-name from gene-protein glossary generated by ABNER.
4. Matching method consider the adjacency between matched patterns that is nearer patterns express the term in the text with higher possibility. In order to consider the proximity between patterns, system computes distances between matched patterns. In calculating distance, every line gap increases the distance score by 1. At the end, total occurrence of patterns in the training corpus divided by the distance score is equal to adjacency score. If the adjacency score is higher than the threshold, system recognize the GO term in the text. An example of distance calculating is expressed in figure 2.

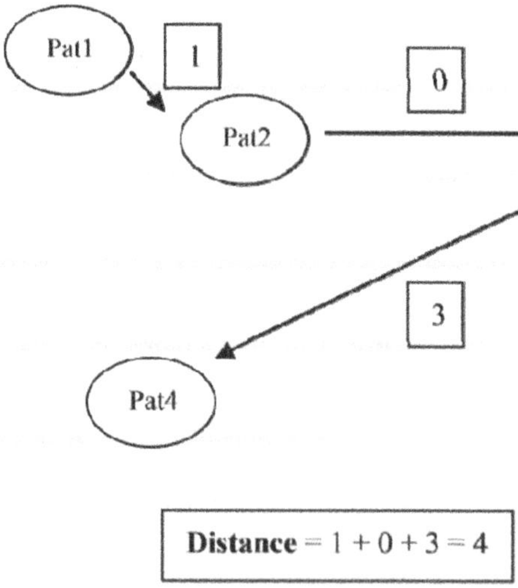

Fig. 2. An example of calculating distance score

We extracted language patterns for 63 GRO concepts, determining 321 GO ontology terms in the text. Moreover, our system was trained and tested on 3931 abstracts. It achieved recall and precision rates of 52.2% and 40.9 respectively. Guadan and his colleagues [21] in their research achieved recall rate of 34% and precision rate of 34%

for identifying GO terms texts with their approach. Evidently, our system accomplished better performance in recognizing gene ontology terms.

Figure 3 illustrates the performance of the system by changing the pattern pruning threshold. To score patterns, the system uses following equation:

$$Score = \frac{\sqrt{Freq(A)}}{Freq(A')} \tag{1}$$

Freq(A) is the frequency of the pattern in related abstracts and Freq(A') is the frequency of the pattern in unrelated abstracts. The optimum score for the system is 2 and patterns with equal or greater scores will be stored in the patterns repository. Moreover, if system adds a pattern with zero occurrence in the A', the performance of the system increases by approximately 2 percent.

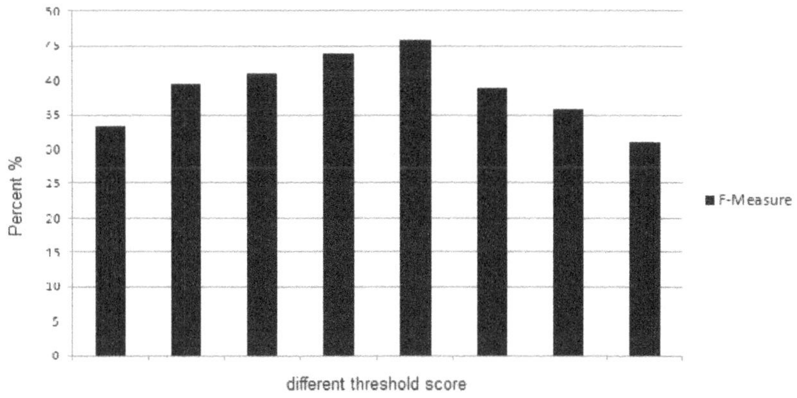

Fig. 3. F-measure changes by alteration of the threshold

Result of applied patterns on test data for two GRO concepts are illustrated in table 6.

Table 6. Example result of two GRO concepts

Splicing:		
splicing control of cell death	splicing complex	mRNA splicing
cTNT splicing	pre-mRNA splicing	IR splicing
splicing regulator	splicing of pre-mRNA	splicing of cTNT
splicing regulation	splicing pattern	splice-site pairing
5' splice site	3' site of splicing	spliced exon
3' splice site	splicing regulation	FGFR2 splicing
splicing of cardiac troponin T	splicing intermediate	spliced mRNA
splicing of CD44	Epithelial Splicing	RNA splicing
MessengerRNA:		
circularization of the mRNA	mRNA stability	growth factor mRNA
stability of TH mRNA	pre-mRNA substrate	virus mRNA export
degradation of TH mRNA	E1A pre-mRNA	mRNA silencing
spliced beta-thalassemia mRNA	pre-mRNA-splicing	binding mRNA
SRPK1 inhibits splicing	mRNA-destabilizing	mRNA binding
binds to casp-2 pre-mRNA	stabilize the pre-mRNA	cTNT splicing
splicing of Fas pre-mRNA	mRNA stabilization	CELF activity

4 Conclusion and Future Work

In this research paper a robust method for automatically extracting patterns for 63 gene regulation ontology concepts from biomedical texts is proposed and implemented. Our method extracts information from biomedical texts and identifies 321 GO terms by discovering language patterns which represent ontology concepts. The method has the capabilities of determining keywords for each GO concept and discovering novel and useful patterns in biomedical literature with high coverage. After calculating the f-measure of the system, 45.8%, and comparing it with similar methods, our method showed a better performance in construction of language patterns.

Future Work: The results of this research point to several interesting directions for future work:

1. We applied our approach to identify 321 gene ontology terms in the text; system can be applied on other logical definition like Cross-products of gene ontology.
2. We used syntactic patterns in our system; by the application of parser and grammars and then well organizing the result, system can achieve better outcome and performance.
3. Betterment in the keywords and gene-protein glossary can improve the final result of the system.

References

1. Putnam, N.C.: Searching MEDLINE free on the Internet using the National Library of Medicine's PubMed. Clin Excell Nurse Pract. 2(5), 314–316 (1998)
2. NLM Systems: Data, News and Update Information. PubMed Update. Internet (April 18, 2011), `http://www.nlm.nih.gov/bsd/revup/revup_pub.html#med_update`
3. Vastag, B.: NIH launches PubMed Central. J. Natl. Cancer Inst. 92(5), 374 (2000)
4. The Gene Ontology Consortium: The Gene Ontology (GO) database and informatics resource. Nucleic Acids Res. 32(Database issue), D258–D261 (2004)
5. Camon, E., Magrane, M., Barrell, D., Lee, V., Dimmer, E., Maslen, J., Binns, D., Harte, N., Lopez, R., Apweiler, R.: The Gene Ontology Annotation (GOA) Database: sharing knowledge in Uniprot with Gene Ontology. Nucleic Acids Res. 32(Database issue), 262–266 (2004)
6. Hirschman, L., Colosimo, M., Morgan, A., Yeh, A.: Overview of BioCreAtIvE task 1B: normalized gene lists. BMC Bioinformatics 6(suppl. 1), S11 (2005)
7. Morgan, A., et al.: Overview of BioCreative II Gene Normalization. Genome Biology 9(suppl. 2), S3 (2008)
8. Blaschke, C., Hoffmann, R., Oliveros, J.C., Valencia, A.: Extracting information automatically from biological literature. Comparative and Functional Genomics 2(5), 310–313 (2001)
9. Andrade, M.A., Valencia, A.: Automatic extraction of keywords from scientific text: Application to the knowledge domain of protein families. Bioinformatics 14, 600–607 (1998)

10. Blaschke, C., Oliveros, J.C., Valencia, A.: Mining functional information associated to expression arrays. Functional and Integrative Genomics (2000) (in press)
11. Andrade, M.A., Brok, P.: Automatic extraction of information in molocular biology. FEBS Lett. 476, 12–17 (1998)
12. Sudo, K., Sekine, S., Grishman, R.: Automatic pattern acquisition for Japanese information extraction. In: HLT 2001 Proceedings of the First International Conference on Human Language Technology Research (2001)
13. Mooney, R., Nahm, Y.: Text Mining with Information Extraction. In: Proceedings of the 4th International MIDP Colloquium, Multilingualism and Electronic Language Management, pp. 141–160 (September 2003)
14. Chowdhary, R., Zhang, J., Liu, J.S.: Bayesian inference of protein–protein interactions from biological literature. Bioinformatics 25(12), 1536–1542 (2009)
15. Bui, Q., Katrenko, S., Sloot, P.: A hybrid approach to extract protein–protein interactions. Bioinformatics 27(2), 259–265 (2011)
16. Liu, B., Qian, L., Zhou, G., Zhu, Q.: Exploiting dependency information for feature-based protein-protein interaction extraction. In: Jiang, L. (ed.) ICCE 2011. AISC, vol. 111, pp. 267–272. Springer, Heidelberg (2011)
17. Blaschke, C., Andrade, M.A., Ouzounis, C., Valencia, A.: Automatic extraction of biological information from scientific text: Protein-protein interactions, pp. 60–66. AAAI Press (1999)
18. Ono, T., Hishigaki, H., Tanigami, A., Takagi, T.: Automated extraction of information on-protein-protein interactions from the biological literature. Bioinformatics 17(2), 155–161 (2001)
19. Huang, M., Zhu, X., Hao, Y., Payan, D.G., Qu, K., Li, M.: Discovering patterns to extract protein-protein interactions from full texts. Bioinformatics 20(18), 3604–3612 (2004)
20. Surdeanu, M., Turmo, J., Ageno, A.: A hybrid approach for the acquisition of information extraction patterns. In: Proceedings of the EACL 2006 Workshop on Adaptive Text Extraction and Mining (ATEM 2006). ACL (2006)
21. Gaudan, S., Jimeno Yepes, A., Lee, V., Rebholz-Schuhmann, D.: Combining Evidence, Specificity, and Proximity towards the Normalization of Gene Ontology Terms in Text. EURASIP Journal on Bioinformatics and Systems Biology 2008, Article ID 342746
22. Beisswanger, E., Lee, V., Kim, J., Rebholz-Schuhmann, D., Splendiani, A., Dameron, O., Schulz, S., Hahn, U.: Gene Regulation Ontology (GRO): Design principles and use cases. Studies in Health Technology and Informatics. Studies in Health Technology and Informatics 136, 9–14 (2008)
23. Ashburner, M., Ball, C., Blake, J.A., Botstein, D., Butler, H., Cherry, M., Davis, A.P., Dolinski, K., Dwight, S., Eppig, J.T., Harris, M.A., Hill, D.P., Issel-Tarver, L., Kasarskis, A., Lewis, S., Matese, J.C., Richardson, J.E., Ringwald, M., Rubin, G.M., Sherlock, G.: Gene Ontology: tool for the unification of biology, The Gene Ontology Consortium. Nature Genet. 25, 25–29 (2000)
24. Camon, E., Magrane, M., Barrell, D., Lee, V., Dimmer, E., Maslen, J., Binns, D., Harte, N., Lopez, R., Apweiler, R.: The Gene Ontology Annotation (GOA) Database: sharing knowledge in Uniprot with Gene Ontology. Nucleic Acids Res. 32(Database issue), D262–D266 (2004)
25. Settles, B.: ABNER: an open source tool for automatically tagging genes, proteins, and other entity names in text. Bioinformatics 21(14), 3191–3192 (2005)

26. Settles, B.: Biomedical Named Entity Recognition Using Conditional Random Fields and Rich Feature Sets. In: Proceedings of the International Joint Workshop on Natural Language Processing in Biomedicine and its Applications (NLPBA), Geneva, Switzerland, pp. 104–107 (2004)
27. Lafferty, J., McCallum, A., Pereira, F.: Conditional Random Fields: Probabilistic Models for Segmenting and Labeling Sequence Data. In: Proceedings of the International Conference on Machine Learning (ICML), Williamstown, MA, USA, pp. 282–289 (2001)
28. Miller, G.: WordNet: A Lexical Database for English. Communications of the ACM 38(11) (November 1995)

A Routing Algorithm to Guarantee End-to-End Delay for Sensor Networks

Dong Li[1,2], Yang Liu[1], Peng Zeng[1], and Haibin Yu[1]

[1] Department of Industrial Control Network and System
Shenyang Institute of Automation Chinese Academy of Sciences
Shenyang, Liaoning, 110016, China
[2] University of Chinese Academy of Sciences
Beijing, 100049, China
lidong@sia.cn

Abstract. In this paper, we state the phenomenon of Pay Bursts Only Once in communication networks and propose a routing algorithm to guarantee end-to-end delay. Our idea is based on consider an end-to-end route as one concatenation system rather than the aggregation of several separate systems. We provide an algorithm to compute the delay bound under Pay Bursts Only Once phenomenon and compare it with other algorithms. The evaluation result show the algorithm is efficient in this case.

1 Introduction

Sensor Networks contain many distributed sensors to monitor physical conditions, such as temperature, pressure, etc. A Sensor Network typically consists of several sensor nodes and one gateway. Each sensor node has embedded processors, limited memory, low-power radio and one or more sensors. Due to the low-power and limited energy, the sensor nodes transmit messages to the gateway in a multi-hop network. The routing problem, which is the process of selecting paths to optimize the network, is a basic and important issue in sensor networks.

However, a phenomenon called the Pay Bursts Only Once (PBOO) is known that the end-to-end delay is always smaller than the sum of delay of every hop of the route [1]. It is to say that traditional delay guarantee routing algorithms which have the aim to minimum the sum of delay of every hop cannot generate the minimum end-to-end delay. Network Calculus [2][3] is a set of mathematical results which give insights into man-made systems such as concurrent programs, digital circuits and communication networks. Network calculus gives a theoretical framework for analyzing performance guarantees in networks and it is suitable in sensor networks. Network calculus can formulate the Pay Busts Only Once phenomenon as a mathematical problem and explain the phenomenon well, so we propose a routing algorithm which introduces some of the results of Network Calculus to get more accurate end-to-end delay in sensor networks.

D. Nagamalai et al. (Eds.): *Adv. in Comput. Sci., Eng. & Inf. Technol.*, AISC 225, pp. 159–166.
DOI: 10.1007/978-3-319-00951-3_16

The paper is organized as follows. In Section 2 we summarize routing algorithms of minimize delay of previous works. Section 3 gives a brief introduction of Network Calculus, which is a mathematical tool we use in this paper. We state the problem we solved and some reasonable assumptions in Section 4, and provide the algorithm to guarantee end-to-end delay. The evaluation is presented in Section 5. We conclude and propose future work in Section 6.

2 Related Works

Many studies on routing algorithms of providing end-to-end delay guarantee have been proposed. Early studies consider the delay of each hop as routing metrics and apply traditional route generating methods. [4] measures the end-to-end delay of paths and chooses the smallest one. Some other studies focus on Maximum Tolerable Delay Jitter. The definition of Delay Jitter is the difference between the upper bound and the absolute minimum of end-to-end delay [5]. The former incorporates the queuing delay at each node and the latter is determined by the propagation delay and the transmission time of a packet [6]. The transmission time between two nodes is simply the packet size in bits/the channel capacity. This metric can also be expressed as delay variance [7].

AAQR [7] provides soft guarantees of bounded delay and jitter, with the assurance of throughput.[8] focuses on providing delay-constrained routes for data sessions. SPEED [9] is another QoS routing protocol for sensor networks that pro-vides soft real-time end-to-end guarantees. The protocol requires each node to maintain information about its neighbors and uses geographic forwarding to find the paths. [10] takes MAC delay into account. The MAC delay is defined as the time used to transmit a packet from one node to another, including the buffer time and the time to acknowledge the packet. This provides a good indication of the amount of traffic at the relevant nodes.

The referenced studies focus on routing algorithms to minimize the end-to-end delay, some of them take other routing metrics into account, such as throughput, packet loss rate, etc. However, these studies ignore the phenomenon of Pay Bursts Only Once, and the primary cause is maximum delay or average delay used in these studies cannot sufficiently indicate the influence of the network over time.

3 Backgrounds on Network Calculus

As mentioned before, maximum delay or average delay cannot lead to a precise end-to-end delay of a route, so we must find another tool to represent the network. Network Calculus is the tool to analyze flow control problem in networks with particular focus on determination of bounds on worst case performance. It has been successfully applied to calculate network performance, such as delay and throughput. Network calculus is also considered as a system theory for deterministic queuing system. Network calculus focuses on quantitated worst case while traditional queuing theory on average case or equilibrium behavior.

Some basic theorems are provided, whose details can be found in [3].

Theorem 1. Delay Bound. Assume a flow $R(t)$, constrained by arrival curve α, traverses a system S that offers a service curve β. At any time t, the delay $d(t)$ satisfies:

$$d(t) \leq \sup_{s\geq 0} \{\inf_{\tau\geq 0}\{\alpha(s) \leq \beta(s+\tau)\}\} = h(\alpha,\beta)$$

$h(\alpha,\beta)$ is also called horizontal deviation between α and β.

Theorem 2. Concatenation of Nodes. Assume a flow $R(t)$, constrained by arrival curve α, traverses systems S_1 and S_2 in sequence where S_1 offers a service curve β_1 and S_2 offers a service curve β_2.Then the system S combined by S_1 and S_2 in concatenation offers the following service curve:

$$\beta = \beta_1 \otimes \beta_2$$

Theorem 3. If β_1 and β_2 are convex then $\beta_1 \otimes \beta_2$ is convex. In particular, If β_1 and β_2 are convex and piecewise linear, $\beta_1 \otimes \beta_2$ is obtained by putting end-to-end the different linear pieces of β_1 and β_2, sorted by increasing slopes, as shown in Figure 1

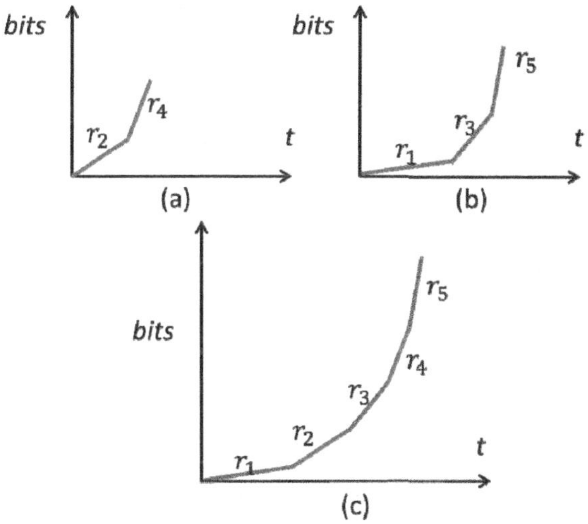

Fig. 1. Concatenation of convex and piecewise linear functions

In Figure 1, (a) and (b) are two convex and piecewise linear functions, with slope r_2, r_4 in (a) and r_1, r_3, r_5 in (b). Suppose that $r_1 < r_2 < r_3 < r_4 < r_5$. The concatenation of the two functions is shown in (c), which is obtained by putting end-to-end pieces of segments from r_1 to r_5.

4 A Routing Algorithm under PBOO Phenomenon

As mentioned before, a phenomenon called the Pay Bursts Only Once is known that the end-to-end delay is always smaller than the sum of delay of every hop of the route. This phenomenon can be well explained by network calculus with the help of Theorems in Network Calculus. Consider the concatenation of two nodes offering each a rate-latency service curve $\beta_{R_i,T_i}(i = 1,2)$), as is commonly assumed with networks. Assume the input is constrained by $\gamma_{r,b}$. Assume that $r < R_1$ and $r < R_2$. We now calculus the delay bound in two ways: 1. Calculate the maximum delay of two nodes respectively as D_1+D_2; 2. Calculate the concatenation of two nodes using Theorem 2 as D_0.

It is easy to see that $D_0<D_1+D_2$. In other words, the bounds obtained by considering the global service curve are better than the bounds obtained by considering every node in isolation. In this paper, we want to generate a routing algorithm under Pay Bursts Only Once phenomenon. The problem can be described as: Given a directed graph $G=(V,E)$ with a service curve β_e for each edge $e \in E$ and a flow constrained by arrival curve α with a source and a destination. Compute a path $e_1, e_2, \ldots, e_p$ from the source S to the destination D such that the maximum delay is minimal, where $\beta = \beta_{e_1} \otimes \beta_{e_2} \otimes \ldots \otimes \beta_{e_p}$.

We assume the arrive curve is concave and the service curves are convex, which is an usual assumption in real networks. Even if the arrive curve is not fulfilled, the arrival curve can always be replaced by its sub-additive closure [3]. For the sake of the convenient calculation, we assume the arrival curve and the service curves are piecewise linear, like β in Figure 2.

Now, let us take a deep look into PBOO phenomenon. The delay through one node has a part of the delay due to the burst of the input flow. We see that D_1+D_2 contain twice the burst delay, whereas D_0 contains it only once. We sometimes say that "we pay bursts only once". We see that this increase of burstiness does not result into an increase of the overall delay. From the analyzing of PBOO phenomenon; we can see the reason why traditional routing algorithms based on delay of every hop cannot work well under PBOO phenomenon. Next, we go through some mathematical derivation, from which we can generate a new routing algorithm.

First, we re-define the mathematic term tangency of a line to a piecewise linear curve. The tangent line is a geometric line that touches the piecewise linear curve at one point or a piece of segment but does not intersect it. As showed in Figure 2, two lines with slope k and k' are both the tangent line of piecewise linear curve.

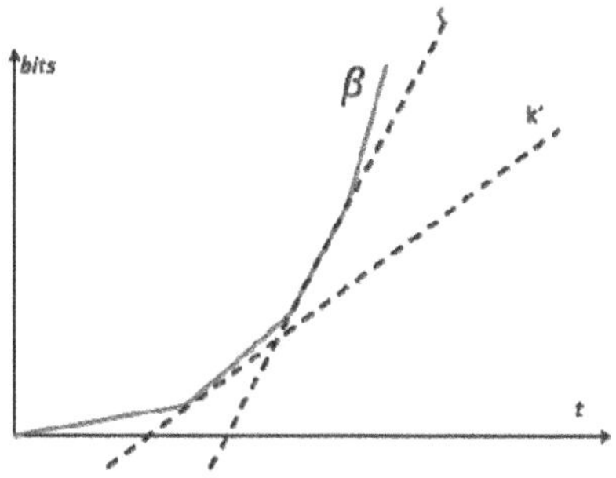

Fig. 2. Tangent line of a piecewise linear curve

For a given path P, we can compute the concatenation of nodes by calculate $\beta_p = \beta_{e_1} \otimes \beta_{e_2} \otimes \ldots \otimes \beta_{e_p}$, where $e_1, e_2, \ldots, e_p$ combine the given path P. According to previous discussion, the maximum delay of the path is the horizontal deviation between α and β, which can be expressed as the horizontal distance between two parallel lines who are the tangent line of α and β, respectively. For a given slope k, we use $H_k(\alpha,\beta)$ represents the horizontal distance between two parallel lines with slope k who are tangency to α and β. It can be proofed that for all parallel lines tangent to α and β, the minimal distance is $h(\alpha,\beta)$.Thus, we have $h(\alpha,\beta) = \min\{H_k(\alpha,\beta)\}$, as showed in Figure 3.

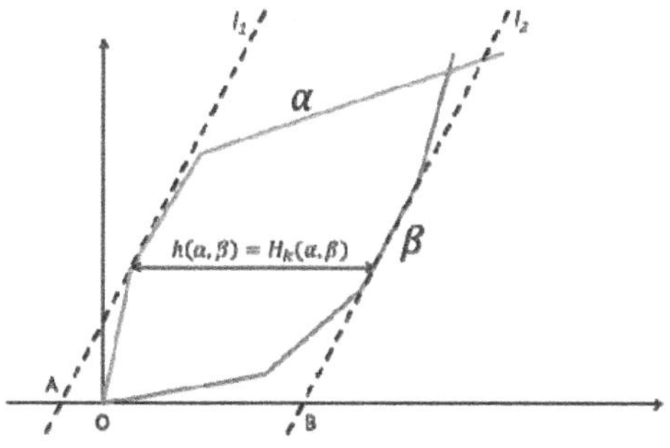

Fig. 3. Maximum delay and horizontal distance between parallel lines

Since we have the assumption of piecewise linear curves, k must be one of the slopes of piecewise segments of α and β, which we call $K_{\alpha,\beta}$ for convenience. As discussed before, the goal of minimizing the bound of delay is

$$Delay = \min_p D(\alpha, \beta_p) = \min_p \{ \min_{k \in K_{\alpha,\beta}} H_k(\alpha, \beta_p) \}$$

The two minima yields can be switched

$$Delay = \min_{k \in K_{\alpha,\beta}} \{\min_{p} H_k(\alpha, \beta_p)\}$$

Now, we focus on $\min_{p} H_k(\alpha, \beta_p)$ under a specific k. Let take a look at Figure 3 again. Suppose that the line l_1 is the tangent to α, crossing x-axis at A, and the line l_2 is the tangent to β, crossing x-axis at B. It is clear that A is on the left side of origin O and B is on the left side. We can get $H_k(\alpha, \beta_p) = |AO| + |OB|$, where |AO| can be compute directly depending on slope k and arrival curve α. From Theorem 3, we know that $|OB| = |OB_{e1}| + |OB_{e2}| + \ldots + |OB_{ep}|$, where B_i is the intersection of line l_i that is tangent to β_i and has the slope of k. We can get |OB| by employing classic Dijkstra algorithm with a graph G' with the same topology but the edge of which is $|OB_i|$. Thus, we can get a path with minimal delay bound under slope k. Going through that minimal delay bound for every k in $K_{\alpha,\beta}$, the minimal delay can be found.

5 Evaluations

The evaluation is implemented under the toolbox for network calculus. The toolbox offers a class containing some function which are often used in network calculus, and closed operations of piecewise affine functions. The can be finitely described which enables people to propose some algorithms for each of the Network Calculus.

The implementations are tested on a multi-hop hub-based sensor network with 6 nodes. The assumption hub-based means the network has a gateway as manager (such as WirelessHART, WIA-PA[11], etc.), and implies a centralized routing algorithm can be employed such as Dijkstra algorithm. The topology is shown in Figure 4, which is an acyclic graph. The routing flow has a source node 1 and a destination node 6. The edges indicate the transmitting capacity, which are described as piece-wise linear functions.

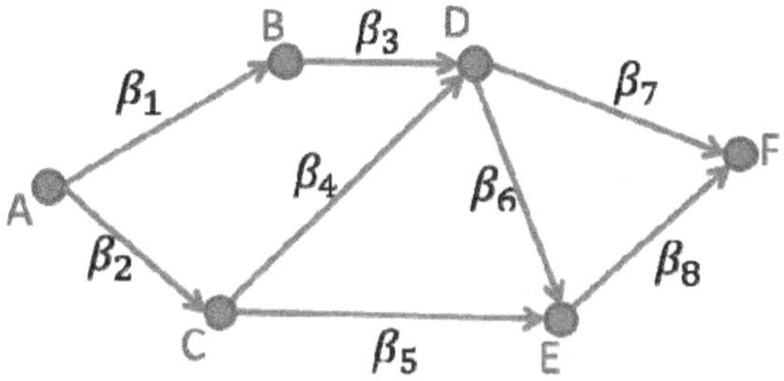

Fig. 4. The topology of implementations

The input flow has the arrival curve of $\alpha=(0.5t+2)\delta_0(t)$, while the definitions of each edge are: $\beta_1=\max(0,0.1t-5)$, $\beta_2=\max(0,5t-10)$ $\beta_3=\max(0,0.2t-4)$, $\beta_4=\max(0,2t-10)$, $\beta_5=\max(0,t/3,2t-20)$, $\beta_6=\max(0,t/10,t/5-20)$, $\beta_7=\beta_8=\max(0,t/3-2,2t-20)$, the unit of time is timeslot and the unit of flow is packet (one timeslot of a typical sensor network such as Zigbee or WIA-PA is 15.625ms and the maximum size of one packet is 127 bits).

We implement the algorithm introduced in section 4 and a minimal delay algorithm as comparison. The minimal delay algorithm computes the maximum delay of each edge and generates a route with minimal end-to-end delay using Dijkstra algorithm.

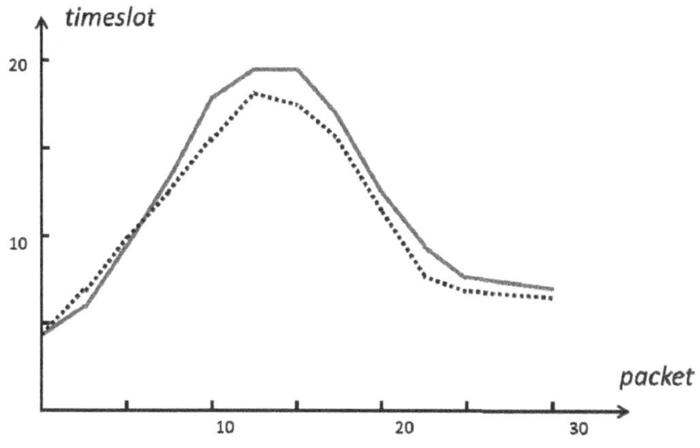

Fig. 5. The delays of two algorithms

The result is showed in Figure 5, where the red solid line implies the delay of first 30 packets in minimal delay algorithm and the black dotted line reveals the situation in our algorithm. The route of our algorithm is Route 1: A->C->E->F, with delay bound 18 timeslots (dotted line). The comparison algorithm generates Route 2: A->C->D->F, with end-to-end delay bound 19 timeslots (solid line). Considering we have $\beta_7=\beta_8$, so the difference between Route 1 and Route 2 is the choice of β_4 or β_5. If we judge that in an isolate way, β_4 is better because it has the delay bound of 6 timeslot while β_5 introduce 8 timeslots. However, when we consider the whole network, β_5 is better while concatenating with β_7. This is a numerical example of Pay Bursts Only Once phenomenon. From discussion before, we can tell our algorithm can deal well with PBOO phenomenon, while traditional algorithms cannot.

6 Conclusions

Network Calculus is a powerful tool to analyze the performances in communication networks. In Previous work, the phenomenon of Pay Bursts Only Once has been introduced and can be explained well with Network Calculus. However, routing algorithms do not consider the influence of PBOO phenomenon and cannot get the minimal end-to-end delay. We provide an algorithm which can generate the minimal delay route under PBOO phenomenon, by considering the route as a concatenation system rather than separate systems. The disadvantage of our algorithm is the time complexity is higher than traditional algorithms. Our future work will focus on reducing the time complexity and extend the algorithm to multi-path for increasing route reliability.

Acknowledgments. The authors acknowledge the financial support of the Strategic Priority Research Program of the Chinese Academy of Sciences under Grant No.XDA06020500, the Important National Science and Technology Specific Project under Contact No.2010ZX03006-005-01, the National High Technology Research and Development Program of China under 863 Program No.2011AA040103, and the Natural Science Foundation of China under Contact No.61233007.

References

1. Fidler, M.: Extending the network calculus pay bursts only once principle to aggregate scheduling. Quality of Service in Multiservice IP Networks, 19–34 (2003)
2. Chang, C.S.: Performance Guarantees in Communication Networks. Springer, New York (2000)
3. Le Boudec, J.Y., Thiran, P.: Network Calculus: A Theory of Deterministic Queuing Systems for theInternet. Springer, New York (2001)
4. Chen, S., Nahrstedt, K.: Distributed Quality-of-ServiceRouting in Ad Hoc Networks. IEEE JSA 17, 1488–1505 (1999)
5. Bashandy, A.R., Chong, E.K.P., Ghafoor, A.: GeneralizedQuality-of-Service Routing with Resource Allocation. IEEE JSAC 23, 450–463 (2005)
6. Hanzo, L., Tafazolli, R.: A survey of QoS routing solutions for mobile ad hoc networks. IEEE Communications Surveys & Tutorials 9, 50–70 (2007)
7. Wang, M., Kuo, G.-S.: An Application-Aware QoS RoutingScheme with Improved Stability for Multimedia Applications inMobile Ad Hoc Networks. In: Proc. IEEE Vehic. Tech. Conf., pp. 1901–05 (2005)
8. Zhang, B., Mouftah, H.T.: QoS Routing for Wireless AdHoc Networks: Problems, Algorithms, and Protocols. IEEE Commun. Mag 43, 110–117 (2005)
9. He, T., Stankovic, J.A., Lu, C., et al.: SPEED: A stateless protocol for real-time communication in sensor networks. In: Proceedings of International Conference on Distributed Computing Systems, pp. 46–55 (2003)
10. Fan, Z.: QoS Routing Using Lower Layer Information in AdHoc Networks. In: Proc. Pers.Indoor and Mobile Radio Commun.Conf., pp. 135–139 (2004)
11. Liang, W., Zhang, X., Xiao, Y., et al.: Survey and experiments of WIA-PA specification of industrial wireless network. Wireless Communications and Mobile Computing 11, 1197–1212 (2010)

Optimized Multiple Description Coding for Temporal Video Scalability

Roya Choupani[1,2], Stephan Wong[2], and Mehmet Tolun[3]

[1] Computer Engineering Department, Cankaya University, Ankara, Turkey
roya@cankaya.edu.tr
[2] Computer Engineering Department, Delft University of Technology, Delft, The Netherlands
{rchoupani,j.s.s.m.wong}@tudelft.nl
[3] Computer Engineering Department, TED University, Ankara, Turkey
mehmet.tolun@tedu.edu.tr

Abstract. The vast application of video streaming over the Internet requires video adaptation to the fluctuations of the available bandwidth, and the rendering capabilities of the receiver device. On the other hand, the available video coding standards are designed for optimum bit rate which makes them susceptible to packet losses. A combination of video adaptation methods and error resilient methods can make the video stream more robust against networking problems. In this paper, an optimization for combining scalable video coding with multiple description coding schemes have been proposed. Our proposed method is capable of creating balanced descriptions with optimum coding efficiency.

Keywords: Scalable Video Coding, Multiple Description Coding, Video Coding.

1 Introduction

Human's perception of his surrounding world is essentially dependent on visual information. This dependency has reached a higher level with the progress in advanced technologies, particularly in communications networks which makes it possible for video to be widely utilized in our daily life. Video as a sequence of frames, however, involves a huge amount of data. Hence, the storage and communicating video requires very large capacities which make video compression a necessity. However, the variations in the physical characteristics of the communication networks and the rendering capabilities of the receiver display device require adaptations to be made to the compressed video. Meanwhile, these adaptations should be fast to be applicable in real-time video streaming while preserving the quality of the video as much as possible. Adaptability of video to the transmission bandwidth or displaying capabilities of the recipient device is the objective of scalable video coding (SVC) methods. This adaptability however, does not require long processing and is performed by utilizing only some parts of the video data and simply ignoring the remaining parts in a flexible way. Meanwhile, this adaptability can not only handle the bandwidth fluctuation of the communication channels but also enables video rendering on older devices

D. Nagamalai et al. (Eds.): *Adv. in Comput. Sci., Eng. & Inf. Technol.*, AISC 225, pp. 167–176.
DOI: 10.1007/978-3-319-00951-3_17

by allowing them to utilize the bit-stream partially to display the video in lower quality. The flexibility however, comes with the cost of sacrificing coding efficiency to some extent. SVC however is not capable of handling the packet loss problems because almost all video coding standards are based on eliminating temporal redundancy by encoding the differences between consecutive frames instead of the frame itself. This scheme can reduce the coded video size in a very high rate however, it creates a dependency chain between the frames. A frame cannot be decoded if its previous (reference) frames is not available. This characteristic requires utilization of error concealment methods such as multiple description coding. In this paper we introduce an optimized method for decomposition of a video into multiple descriptions. Our proposed method has scalability and error resilience properties. In the following sections we introduce the basic concepts of scalable coding of video and multiple description coding. Then we describe our proposed method details followed by experimental evaluation results.

2 Scalable Video Coding

In SVC methods, the video stream is represented by a main bit-stream which consists of several sub-streams. Each sub-stream represents video in a lower spatial resolution, lower temporal resolution, or lower bit-per-pixel quality [2]. The reconstructed video by using all sub-streams is in its highest quality. In order to reconstruct the video in lower spatial, resolution, or bit-per-pixel quality, some sub-streams from the main bit-stream are left out. To adapt the data size to the changes in the bit rate of the communication channel, a unit in a video stream such as a frame or a macro-block, is divided into a set of smaller parts. A measure of the number of items comprising a unit is called its granularity [25]. The first item of this set contains the basic and coarsest part of the data and the remaining items contain refinements to the basic item [24],[23]. The scheme of gradual refining of a unit or increasing the granularity of a unit is called Fine Granularity Scalability (FGS) [23], [22]. It is clear from the definitions that a gradual increase in the frame size, bit rate or frame rate is achieved through adapting the granularity of a stream to the bit rate capability of the communication channel. The FGS scheme defines the video content in a multi-layered format [21], [20]. A higher quality for a video is achieved through increasing the number of layers decoded at the receiver side. This scheme leads to placing the layers comprising a video in an ordered sequence where the base layer is always at the first position. The base layer contains the minimum data required while remaining layers include refinements to the data carried by the base layer. This makes scalability possible, as a receiver can receive some of these layers and ignore the rest depending on its current bit rate capacity. Scalability in video is achievable through signal-to-noise ratio (SNR), spatial, and temporal changes. Bit-per-pixel or signal-to-noise ratio scalability is a technique to decompose a video sequence into two layers at the same frame rate and the same spatial resolution, but different quantization accuracy. The decomposition can be performed in pixel domain by putting more significant bits in the base layer and less significant bits in the enhancement layers. The decomposition can also be performed in the DCT domain. In this case the low frequency coefficients of the DCT are put in the base layer and the high frequency coefficients are put in the enhancement layer(s) [19], [18]. Spatial scalability is a technique to code a video sequence into multi-layers at the same frame rate, but different spatial resolutions [1]. The first layer (the base layer) is coded at the lowest spatial resolution. The base layer is created by down-sampling the frames. The difference between up-sampled base layer and the original frame is coded as the enhancement layer. In case that the video is coded in more layers, this procedure is repeated on the base layer yielding a new base layer in lower resolution, and an enhancement layer [17], [16]. This strategy of creating multi-layer spatial SVC, makes layer k dependent on all layer from 1 up to k-1. An important consideration for coding efficiency is motion compensation in each layer. Two strategies are followed for motion compensation. Temporal scalability is a technique to code a video sequence into two layers at the same spatial resolution, but different frame rates [15], [14]. The base layer is coded at a lower frame rate. The enhancement layer provides the missing frames to form a video with a higher frame rate. Coding efficiency of temporal scalable coding is high and very close to non-scalable coding [14]. Figure 1 depicts the structure of temporal scalability with two layers.

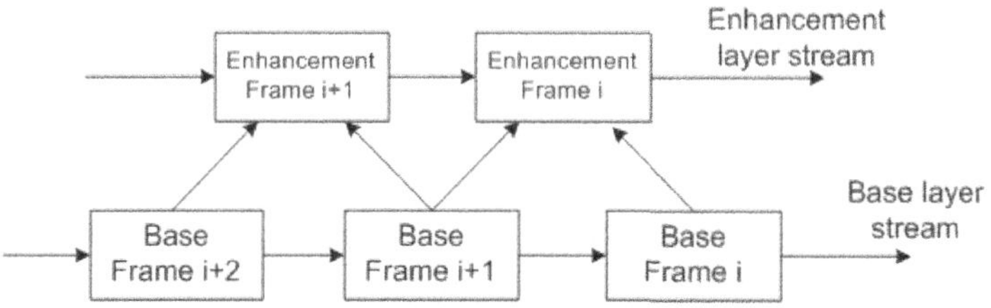

Fig. 1. Typical Structure of a Temporal Scalability Decoder

Considering the two layer structure depicted in Figure 1, the enhancement frame i is the successor of the base layer frame i in the original sequence. Enhancement frame i-1, base frames i or i+1 can be used as a reference frame for enhancement layer frame i. Therefore, complying with the restrictions in video coding standards before H.264, only P-type predicted frames are used in the base layer [3]. The enhancement layer predicted frames can be either P-type, or B-type referencing a P-type frame from the base layer or the enhancement layer. Motion compensation in the based layer utilizes only the base layer information so no drift error is expected here [4]. However, with moving some of the frames to enhancement layer(s), the distance between consecutive frames in the base layer is increased. This increase can cause a slight decrease in the coding efficiency.

3 Multiple Description Coding

A multiple description coder (MDC) for video coding divides the video data into some bit-streams called descriptions which are then transmitted separately over different network channels [13]. Generally, descriptions have the same importance and data rates, even though this is not a necessary requirement. Each description can be decoded independently from other descriptions. This means that the loss of some of these descriptions does not affect the decoding of the rest [12]. The accuracy of the decoded video depends on the number of received descriptions [9]. Figure 2 depicts the basic framework for a multiple description encoder/decoder with two descriptions.

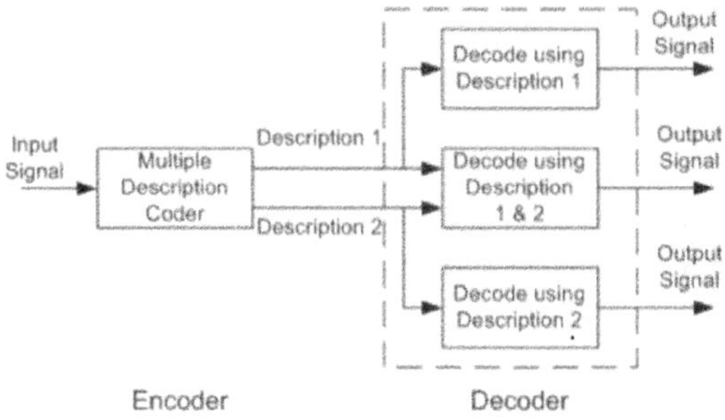

Fig. 2. Multiple descriptions coding block-diagram

In case of a failure in one of the channels, the output signal is reconstructed from the other description. In contrast to descriptions, in a multi-layer coded video, layer i cannot be decoded if layer i-1 is not present [7]. This means that in order to decode a multi layer video using m layers out of a total of n layers, the available layers should be the lowest layers. However, the descriptions utilized for decoding a video are not necessarily from any order since the main goal of MDC is delivering video (although in a lower quality) when parts of video data are lost [8]. In order to reconstruct video in presence of data loss or corruption, redundancy should be added to the bit-stream. This redundancy is in the form of repeated bits (duplicated blocks), or inefficiency in the encoder when the bit-stream is encoded by a rate below channel capacity. When a frame or a block of a frame is missing, the decoder estimates it by utilizing its adjacent data that was received correctly. The adjacency can be in the spatial or temporal domain. Recovering data completely or partially when some parts of data is lost and masking the data loss effect is called error concealment. MDC schemes are among the techniques that are commonly utilized for error concealment. Even if the descriptions are designed as non-overlapping sets, or partitions, it does not necessarily mean that there is no redundancy in the data. Given that each partition is encoded independently from other partitions, the spatial or temporal correlation between the data in different partitions is not utilized and hence the redundancy is not eliminated [6]. On the other hand, the preserved spatial or temporal correlation can be used for estimating the lost data for error concealment [5]. This helps to create a scalable video resilient to packet losses [10][11].

4 Optimizing Temporal MDC

Decomposing a video into several descriptions by putting the frames in different descriptions is the main idea utilized in temporal scalability with multiple descriptions. For instance, using two descriptions, the odd numbered frames are put in the first description while the even numbered frames are assigned to the second description. The main drawback of this scheme is that in order to make descriptions independent from each other, a frame and its reference frame should be in the same description. Hence, the reference frame may not be necessarily the most similar frame to the current frame, and the most similar frame may have been assigned to the other description. This drawback reduces the coding efficiency because the temporal redundancy is not completely eliminated. Meanwhile, the decomposition of a video sequence into multiple descriptions should create balanced descriptions. This requirement is based on the assumption that the transmission networks used for delivering descriptions can be subject to bandwidth fluctuations and data loss. In presence of data loss, the video is reconstructed using the delivered descriptions. Hence, to minimize the video quality degradation in all cases, the required condition is the dependency of the reconstructed video quality on the number of delivered descriptions regardless of which description is lost. In this section we present an optimization to solve this problem.

The proposed method presented here assumes only two descriptions (D1 and D2) however, it is readily extendable to more number of descriptions. Meanwhile, in our proposed method, we have assumed that each frame can have only one reference frame. Assuming that a GOP includes n frames, the proposed method starts by encoding frames F1 and F2 by using intra-frame coding. These two frames are the first frames of each of the descriptions in our proposed method. This assumption can be relaxed by a few minor changes in our proposed method. The method starts by considering frames F3 and F4. Since these frames are encoded by using inter-frames coding, their differences with their reference frames are computed. The proposed method considers reference frames from both descriptions and finds the differences for F3 and F4. The differences are summed up for each frame as given in Equation 2.

$$Diff_{total} = \sum_{i\in\{blocks\}} MAD_{v,w}(B_i\,, B_{Ri}) \tag{1}$$

$$MAD_{s,t}(F_1, F_2) = \frac{1}{mn}\sum_{i=1}^{m}\sum_{j=1}^{n}|F_1(i,j) - F_2(i+s, j+t)| \tag{2}$$

where MAD is mean absolute difference Equation 1, Bi and B_{Ri} are a block from the current frame and its most similar area from the reference frame respectively, and v,w indicates the amount of displacement by the current block to reach to its most similar area in the reference frame (motion vector). The proposed method assigns each frame to the descriptions with smaller total difference as computed in Equation 1. Figure 3 depicts the decomposition of a GOP with 16 frames into two description with optimized assignment of the frames to descriptions as proposed in our method.

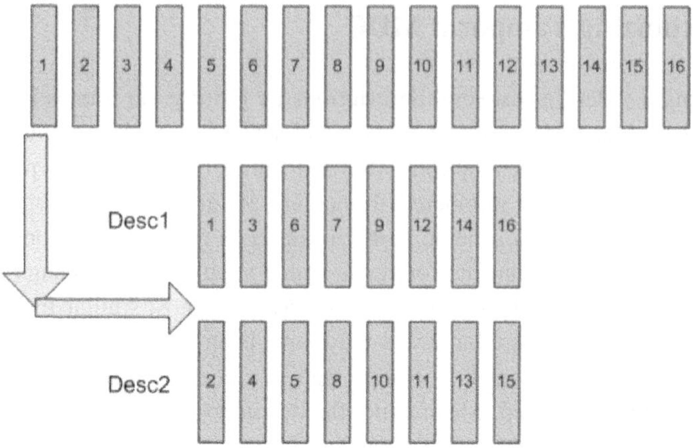

Fig. 3. A sample decomposition of a GOP into two descriptions in the proposed method

As depicted in Figure 3 the worse case of having frames with consecutive sequence numbers puts only two adjacent frames in a description. Algorithm 1 defines how the assignment of the frames to the descriptions is carried out.

Algorithm 1. Assigning frames to descriptions in the proposed method

1. Encode the frames 1 and 2 of a GOP using intra-frame coding and assign them to description 1 and 2 respectively
2. While NOT end of the GOP DO

   a. Get next two frames (F_i and F_j)
   b. Find the motion compensated difference of each block of F_i and F_j in description 1 and 2 as Diff(i,1), Diff(i,2), Diff(j,1), Diff(j,2)
   c. IF Diff(i,1) + Diff(j,2) < Diff(i,2) + Diff(j,1)THEN

      Assign frame F_i to description 1 and F_j to description 2

   d. ELSE

      Assign frame F_i to description 2 and F_j to description 1

As shown in Algorithm 1, the proposed method preserves balance in the creation of the descriptions by grouping the frames of a GOP in pairs and assigning each frame of a pair to one of the descriptions. In case of extending the method to multiple descriptions, the grouping will be in n where n is the number of descriptions. In case that the size of GOP is not divisible by n, some descriptions will contain one frame less than the others (worst case).

5 Experimental Results

To evaluate the proposed method experimentally we have utilized the sequences 'Foreman', 'Stefan', and 'City'. The specifications of the sequences are given in Table 1.

Table 1. The Video Sequences Utilized in Experimental Evaluations

Sequence Name	Resolution	Frame Rate	Length (frames)
Foreman	352 × 288	30	300
Stefan	768 × 576	30	300
City	704 × 576	60	600

The comparison is considered to measure the effect of the proposed optimization. Therefore, as a benchmark, the frames of a GOP of 32 frames are decomposed as odd and even number frames into two descriptions. Subsequently, the same frames are decomposed into descriptions with the optimization proposed in our method. The coding efficiency in terms of bit-per-pixel is computed for each sequence for with optimization and without optimization cases as provided in Table 2.

Table 2. Performance Comparison of the Proposed Method with Odd-Even Decomposition of Video into Descriptions

Sequence Name	Optimized using the Proposed Method (bpp)	No Optimization (bpp)
Foreman	1.181	1.2212
Stefan	1.237	1.291
City	1.062	1.076

As it is shown in Table 2, in all three sequences the bit-per-pixel values have been improved, although the improvements does not result in a major reduction in bit-per-pixel rates. Besides to the increased coding efficiency, the proposed method preserves the balance in the descriptions in terms of the number of frames, by decomposing the GOPs evenly, and avoids creating large timing gaps between adjacent frames in a description.

Our second experiment compares the coding efficiency of the odd-even decomposition of the frames with the coding efficiency the proposed method at different bit rates. Figure 4 depicts the results of our comparison using Foreman video sequence. The comparison indicates that the impact of the proposed method is higher in high bit rates. The closeness of the results in low bit rates is the result of the fact that in low bit rates much of the differences between the frames which correspond to high frequencies are eliminated. It is also important to note that the main goal of combining MDC with SVC is avoiding jitter in video transmission when a sudden bandwidth fluctuation occurs at high bit rates. Hence, the proposed method suitability for these types of applications are verified once more.

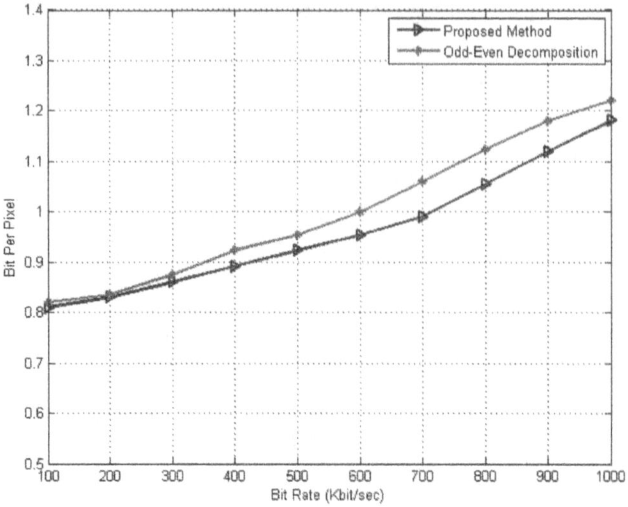

Fig. 4. Odd-Even Temporal Decomposition vis-à-vis Proposed Method Decomposition

6 Conclusion

A new method for handling the data loss during the transmission of video streams has been proposed. Our proposed methods are based on combining SVC with MDC where the video is decomposed into temporal sub-streams. In our proposed method, the error resilience of the video is increased. The proposed method has the capability of being used as scalable coding methods in which any data loss or corruption is reflected as reduction in the quality of the video. However, except for the case when all descriptions are lost, the video streams do not experience jitter at play back. In our method, an improvement is proposed for the well-known method of temporal decomposition of video into multiple descriptions by putting odd and even numbered frames in different descriptions. This method improves the coding efficiency of the odd-even temporal decomposition method by grouping the frames in pairs and assigned to the descriptions so that the differences with the reference frames are minimized. The proposed method preserves the balance in the descriptions and avoids creating large timing gaps between adjacent frames in a description.

References

1. Segall, A., Sullivan, G.J.: Spatial scalability. IEEE Transaction on Circuits Systems for Video Technology 17(9), 1121–1135 (2007)
2. Schwarz, H., Marpe, D., Wiegand, T.: Overview of the scalable video coding extension of the h.264/avc standard. IEEE Transaction on Circuits and Systems for Video 17(9), 1103–1120 (2007)

3. Hewage, C., Karim, H., Worrall, S., Dogan, S., Kondoz, A.: Comparison of stereo video coding support in mpeg-4 mac, h.264/avc and h.264/svc. In: Proceeding of the 4th Visual Information Engineering Conference, pp. 25–27 (2007)
4. Choupani, R., Wong, S., Tolun, M.: A drift-reduced hierarchical wavelet coding scheme for scalable video transmissions. In: First International Conference on Advances in Multimedia (MMEDIA), pp. 68–73 (2009)
5. Choupani, R., Wong, S., Tolun, M.: Multiple description scalable coding for video transmission over unreliable networks. In: 9th International Workshop on Embedded Computer Systems: Architectures, Modeling, and Simulation, Samos-Greece, pp. 58–67 (2009)
6. Franchi, N., Fumagalli, M., Lancini, R., Tubaro, S.: A space domain approach for multiple description video coding. In: ICIP 2003, vol. 2, pp. 253–256 (2003)
7. Tillo, T., Olmo, G.: A low complexity pre-post processing multiple description coding for video streaming. In: IEEE International Conference on Information and Communication Technologies, ICTTA 2004 (2004)
8. Akyol, E., Tekalp, A.M., Civanlar, M.R.: A flexible multiple description coding framework for adaptive peer-to-peer video streaming. IEEE Journal of Selected Topics in Signal Processing 1, 231–245 (2007)
9. Goyal, V.K.: Multiple description coding: Compression meets the network. IEEE Signal Processing Magazine 18, 74–94 (2001)
10. Choupani, R., Wong, S., Tolun, M.: Multiple description coding for SNR scalable video transmission over unreliable networks. Multimedia Tools and Applications (June 2012), doi: 10.1007/s11042-012-1150-9
11. Choupani, R., Wong, S., Tolun, M.: Unbalanced multiple description wavelet coding for scalable video transmission. Journal of Electronic Imaging 21(4), 43006 (2012), doi:10.1117/1.JEI.21.4.043006
12. Venkataramani, R., Kramer, G., Goyal, V.K.: Multiple description coding with many channels. IEEE Transaction on Information Theory 49, 2106–2114 (2003)
13. Wang, Y., Reibman, A.R., Shunan, L.: Multiple description coding for video delivery. Proceedings of IEEE 93, 57–70 (2005)
14. Katata, H., Ito, N., Kusao, H.: Temporal-scalable coding based on image content. IEEE Transaction on Circuits and Systems for Video Technology 7, 52–59 (1997)
15. Domanski, M., Luczak, A., Mackowiak, S.: Spatial-temporal scalability for mpeg video coding. IEEE Transaction on Circuits and Systems for Video Technology 10, 1088–1093 (2000)
16. Tan, W., Zakhor, A.: Real-time internet video using error resilient scalable compression and tcp-friendly transport protocol. IEEE Transaction on Multimedia 1(2), 172–186 (1999)
17. Delp, E.J., Salama, P., Asbun, E., Saenz, M., Shen, K.: Rate scalable image and video compression techniques. In: Proceedings of the 42nd Midwest Symposium on Circuits and Systems, pp. 635–638 (1999)
18. Mathew, R., Arnold, J.F.: Layered coding using bit-stream decomposition with drift correction'. IEEE Transaction on Circuits and Systems for Video Technology 7, 882–891 (1997)
19. Wilson, D., Ghanbari, M.: Exploiting interlayer correlation of SNR scalable video. IEEE Transaction on Circuits and Systems for Video Technology 9, 783–797 (1999)
20. Jiang, H.: Experiment on post-clip FGS enhancement. ISO/IEC JTC1/SC29/WG11, MPEG00/M5826 (2000)
21. Li, W., Chen, Y.: Experiment result on fine granularity scalability. ISO/IEC JTC1/SC29/WG11, MPEG99/M4473 (1999)

22. Gharavi, H., Partovi, M.H.: Multilevel video coding and distribution architectures for emerging broadband digital networks. IEEE Transaction on Circuits and Systems for Video Technology 6, 459–469 (1996)
23. Ghanbari, M.: Two-layer coding of video signals for VBR networks. IEEE Journal of Selected Areas Communications 7, 771–781 (1989)
24. Puri, A., Chen, T.: Multimedia Systems, Standards, and Networks. Marcel Dekker, New York (2000)
25. Weiping, L.: Overview of fine granularity scalability in mpeg-4 video standard. IEEE Transaction on Circuits and Systems for Video Technology 11(3), 301–317 (2001)

An Estimate Frequency Assignment Approach for GSM New Cell Neighbours

Pınar Tanrıverdi and H. Ali Mantar

Computer Engineering Department
Gebze Institute of Technology Institution, Gebze Turkey
pinartanriverdi@turkcellteknoloji.com.tr,
hamantar@bilmuh.gyte.edu.tr

Abstract. Operators have to increase their network capacity to meet the traffic caused by increased mobile user and use of services. To increase capacity new cells are added. As a traditional method, for every added new cell, frequency is chosen manually. Several approaches have been developed by operators for frequency assignment. In most of them, along with new cells, neighbour cell frequencies need to be redefined. This causes service failures and decreases service quality.

In this paper, we propose an algorithm for automatic frequency planning for new cells. The proposed algorithm gives smart ideas for frequency assignment in an effective and efficient way.

Keywords: GSM frequency planning, neighbour cell, frequency interference.

1 Introduction

GSM world users are increasing day by day and operators are expanding their network with new cells or developing new capacities to current cells for maintaining service quality or not to fail. This expansion is bringing new frequency assignment process up. Most effective frequency assignment is important for operators to use their new investments productively.

Generally, like wireless communication, GSM technology is performed between two points called transmitter and receiver. Transmitter produces propagations called electrical carrier frequency. Receiver catches these frequencies and turns them to voice or data according to their type. If two transmitters use the same carrier frequency, this causes an interference. The Interference level changes according to many different parameters. It depends on factors like distance between transmitter and receiver, geographic position of the transmitter, signal power, signal direction and weather conditions. Accordingly, if interference level become too high, received signal can drop below of specific signal-to-noise ratio (SNR) which is unacceptable for quality [1]. For this reason, frequency assignment process is performed with an intelligent plan due to limited frequency range and hardware limitations. The complexity of this problem is known as NP – Complete. Current studies are trying to accomplish best results with accepting error margin of problem solutions.

D. Nagamalai et al. (Eds.): *Adv. in Comput. Sci., Eng. & Inf. Technol.*, AISC 225, pp. 177–186.
DOI: 10.1007/978-3-319-00951-3_18

In [2], the authors propose a model that depends on new frequency plan developing with fixed assignment approach to increase quality of critic cells by optimizing current frequency plan. It is based on planning frequency with the data collected with drive test. This approach uses fixed frequency assignment approach. The evaluation results shows that Drive Test resulted data is more powerful than progressing with assumptions on network. In [3], Integer Programming algorithm is used for solving frequency assignment problem. It is mentioned that 75% recovery at KPI data.

Another work named HiPlan is presented in [4]. In this work, real data on network and genetic algorithm are used. Within genetic algorithm, for area in which frequency planning is made partial C/I (Carrier to Interference Ratio) based cost function is given as a condition.

In [5], an idea of completely automated frequency planning method for new cell is presented. With this new method, the pre-existing system is not required to change frequencies of the other cells. Frequency planning study is based on neighbourship information listing of new cell. No handover measurement report is available because new cell is at installation process. Instead of this information, information in survey report of new cell is used to list neighbourship information.Traditional approaches describes neighbourship only depending on distance information. However, in this study, neighbourship information is excluded; instead distance between new cell and its neighbours with antenna angles are taken into account to determine potential neighbours.

In literature frequency reuse model often proposed for frequency planning topic. This causes blind spots. Alternate Row Antenna Rotation technique is proposed for recovering the failure of 4x12 frequency plan [6]. In other words, the paper includes optimization works performed after frequency planning.

Lately automatic frequency planning applications are prepared for GSM network planning based on Interference Matrix. Most of the algorithms used for interference matrix are using MMR (Mobile Measurement Report) data. MMR report is prepared according to signal level measurement of the cell, which is providing service to mobile user, and the six neighbour cells. This data is not enough for a clear AFP [7]. Focused on interference matrix preparation in which CIR simulation and traffic grid mapping data are used. According to this approach, cell name providing service, latitude and longitude information related to CIR and traffic information will be used in interference matrix development process. CIR information that creates a difference would come without a current value. It is aimed to increase sensitivity with forecasting CIR and traffic values. It is claimed that AFP, in which interference matrix developed by new method used, is giving more genuine frequency plan [7].

GSM system needs frequency planning for two type carrier frequency channel. These are Broadcast Control Channel (BCCH) and Traffic channel (TCH) which is only traffic carrier [8]. The differences between two strategies are investigated in the study. In the first strategy; frequency hopping system is only used for TCH or only for BCCH channels in order to have no interference between TCH and BCCH. In the second strategy; frequency hopping system is analyzed including frequency groups assignment for TCH and BCCH. As a result, no matter how the network traffic's condition is, both strategies' control strategies do not affected. [8] We can say that, this paper includes optimization works after frequency planning.

In our study, we use the neighbour categorization method presented in [5]. We add some upgrades and included categorization method of new cell potential neighbours and relationship with this neighbour. Then, we use interference ratio data as parameter in planning. This data is taken Turkcell operator. After categorization of new cell neighbours and interference possibility ratios, similar to cost calculation function mentioned in [9], we used constants instead of ICDM data to develop cost calculation method.

The rest of the paper is organized as follow. Section 2 presents the method suggested for listing potential neighbours of new cell and categorization of relationship between them. How interference ratio values are reached for neighbourship information list is explained in Section 3. Section 4 details cost function for frequency assignment design.

2 Neighbourship Classification

We use a three step approach to categorize neighbour cells and relationship between them. The new cell's angle and coordination information are the main relationship factors.

2.1 Neighbourship Information

We draw a circle while taking new cell as a central point. While deciding circle's radius size, we choose a cell which can be reference to new cell and distance between this reference cell and its farthest neighbour gives us our radius size. This reference choosing process is risky. There is the possibility of obtaining an irrelevant radius by choosing only one reference in case of new cell has city centre on one side and forest lands on the other side. To prevent this type of situations, more than one reference cell can be chosen and average of these references can be taken.

While choosing reference cell it is important to choose the closest and parallel area to new cell. We will call this distance information as "r". To give an example; blue coloured cell is our new cell; we reference the closest and parallel green cell. For the situations, more than one reference is necessary. We will take red cell as reference.

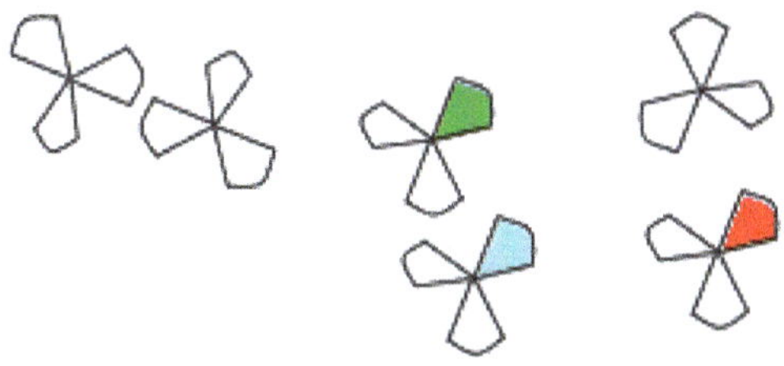

Fig. 1. Sample Reference cell choose

By accepting new cell as centre, we draw three circles with radius of "r/3", "2r/3", "r" and categorize neighbours according to which circle they are in [5].

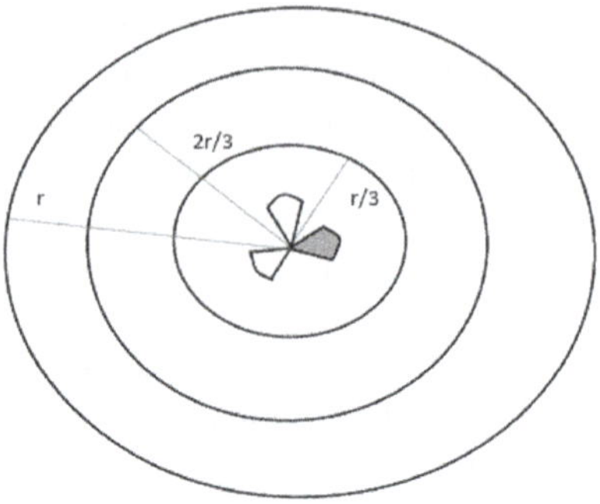

Fig. 2. Layering process

2.2 Categorization Process of Other Cells within Big Circle

If we take blue cell as our new cell, green and yellow cells will be "inward" because the angle between them is less than 90° and red cell will be "outward" because the angle is more than 90° [5]. Instead of 90° value, we can use different values based on operators' statistical data. We preferred 90° in our study as mentioned in [5].

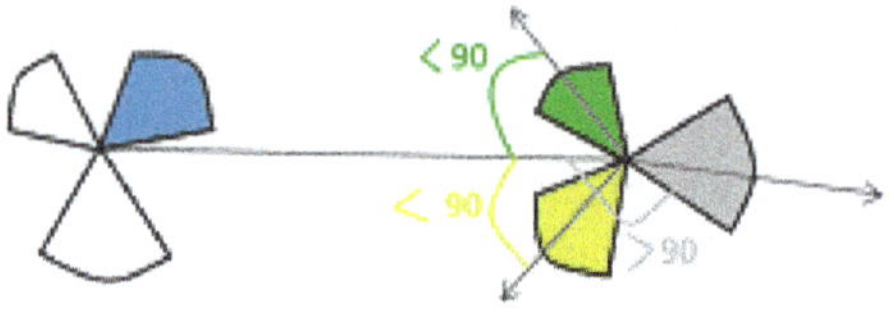

Fig. 3. Inward – outward sample

2.2.1 Frontward – Backward – Sideward

We split circle into four equal pieces according to cell's direction and categorized other cells according to in which piece they are in [5]. We accept blue cell as our new cell. According to this approach, green cell will be frontward, yellow cell will be sideward and purple cell will be backward.

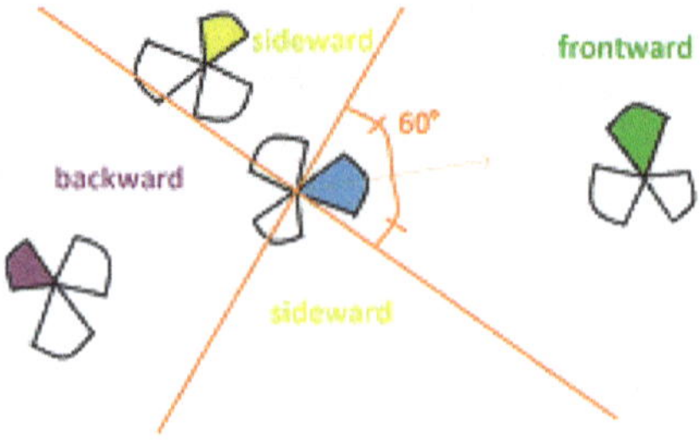

Fig. 4. Frontward – sideward –backward sample

2.2.3 Opposite Neighbour Cell

It is the process of categorization of neighbour cells to new cell of which intersection is estimated highly. If we imagine a line between new cell and neighbour cell and calculate their angles, we call opposite cell if the total is smaller than 45° [5]. 45° value can be change according to operators' statistical data. We use 45° as mentioned in [5]. In order to include a cell to opposite cell classification, it also has to be in first layer. This restriction is added differently from [5].

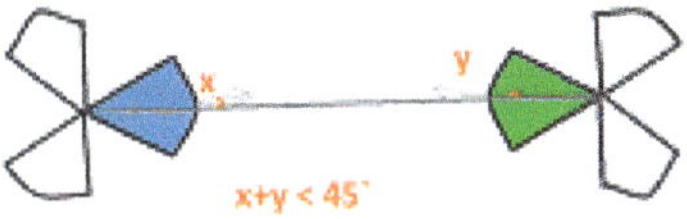

Fig. 5. Opposite cell sample

As seen in Figure 5, blue and green cells are opposite neighbour cell to each other.

2.3 Neighbourship Classification Results

After all these categorization processes, the category in which the potential neighbour cells in will be equal one of the classification types given below. The accuracy of neighbour list and classification can be different according to geographic conditions, environmental conditions like city centre or rural area in which new cell will be developed.

Table 1. Classification Types of Cell Neighbourship Relationship

Classification Type	Layer	inward / outward	Frontward/ backward /sideward
1	Layer 1	inward	frontward
2			backward
3			sideward
4		outward	frontward
5			backward
6			sideward
7	Layer 2	inward	frontward
8			backward
9			sideward
10		outward	frontward
11			Backward
12			Sideward
13	Layer 3	inward	Frontward
14			Backward
15			Sideward
16		outward	Frontward
17			Backward
18			Sideward
19	OPPOSITE CELL		

3 Estimated Interference Value Determination of Neighbourship Relationship Classifications

Data support for 1000 cells are obtained from Turkcell GSM Operator in Turkey. While choosing for sample cell, macro cell property and more than 35 neighbours are taken into account. In order to accomplish data set of chosen 1000 cell, the model mentioned in Section 2 is used. Estimate neighbour class for chosen 1000 cell and get real attempt ratio between them. Then obtained constant value for each class type will be use frequency estimation. This table 3.1 should be update by periodically for increase of sensitivity actual network data.

Table 2. Estimated Interference Values of Neighbourship Relationship Classification

Classification Type	Class Count	Attempt Ratio by class type	Constant for Class Type
1	4482	0,022103971	2210
2	5571	0,006471011	647
3	5696	0,012793188	1279
4	5592	0,012616237	1262
5	5459	0,000890273	89
6	5724	0,003151642	315
7	11479	0,003584807	358
8	12160	0,000710526	71
9	12482	0,001407627	141
10	11806	0,001565306	157
11	12127	0,000101427	10
12	12393	0,00025821	26
13	16097	0,00056905	57
14	18614	0,000109595	11
15	15949	0,000210045	21
16	16992	0,000224223	22
17	18515	0,0000156629759	2
18	16371	0,0000476452263	5
19	846	0,036631206	3663

4 Cost Function Design

The cost function design is based on calculating cost in the assignment of Bcch and Tch frequencies to new cell. While calculating this, using constants from Table 4.1 and Constant for Class Type in Table 3.1 from third section is thought.

Table 3. Constants used in cost calculation

Constant	Assumed Value	Definition
Cosite Co- Bcch	10000000	Used to prevent same Bcch frequencies in the same site.
Cosite Adj-Bcch	1000000	Sequenced Bcch frequencies of cells in the same site will cause big quality problems, so this is used to prevent it.
Cosite Co- Tch	10000000	Used to prevent same Tch frequencies in the same site.
Cosite Adj-Tch	1000000	Sequenced Tch frequencies of cells in the same site will cause big quality problems, so this is used to prevent it.
Neighbour Co- Bcch	40	Used to reduce the possibility of same Bcch frequencies at neighbour cells.
Neighbour Adj-Bcch	10	Used to reduce the possibility of sequenced neighbour cell and Bcch frequency.
Neighbour Co- Tch	4	Used to reduce the possibility of same neighbour cell and Bcch frequency.
Neighbour Adj-Tch	1	Used to reduce the possibility of sequenced neighbour cell and Tch frequency.

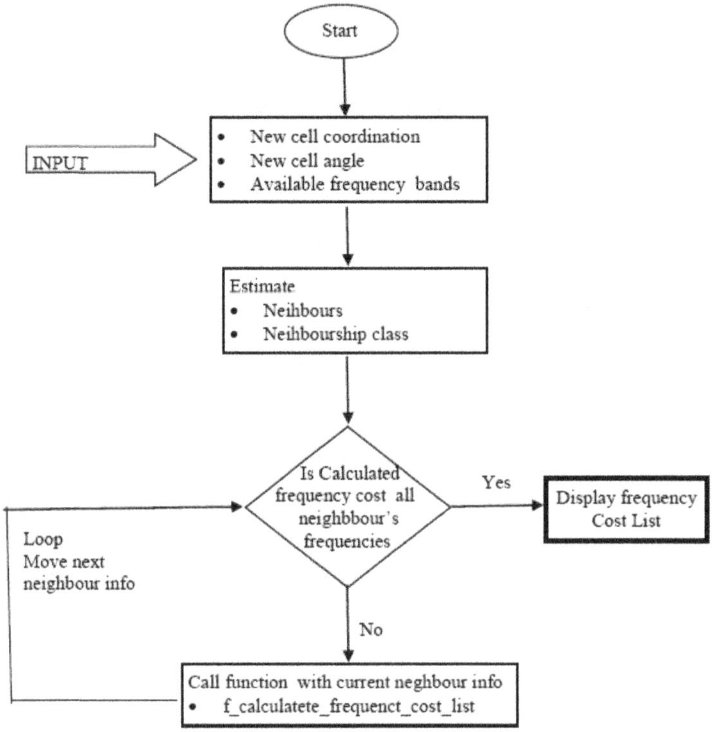

Fig. 6. Flow diagram of frequency assignment to new cell approach

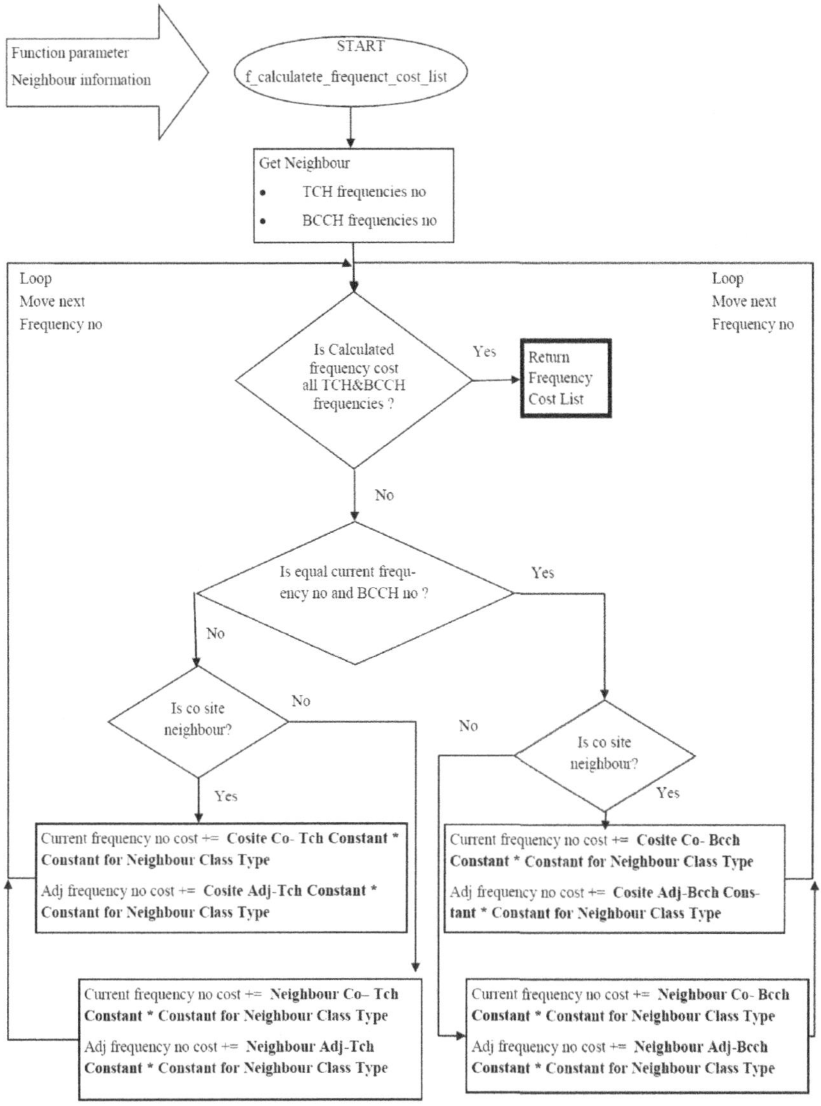

Fig. 7. Detailed flow diagram of cost calculation function

5 Conclision

We propose a new algorithm for assigning frequency to new cell. Generally, previous studies provided different genetic, integer programming, algorithms etc. and implementation for frequency planning for all cells within a specific area chosen in GSM network using data sets developed with traffic, drive test or propagation prediction models. In some studies new approaches to determining data sets, in other words interference matrix, is presented. These approaches are used in planning of current cells' frequency rather than new cell planning.

No traffic data presented for new cell is forming disadvantage. We tried to eliminate this disadvantage by excluding values and potential interference constants obtained with neighbourship descriptions and classification approaches. By taking real interference values within Turkcell network as interference value references, cost constants based on neighbourship classification is tried to be established. Thus, more sensitive frequency planning for new cells can be obtained with traditional fixed frequency assignment approach. In the next study, we aim to make implementation of this topic and share the results.

References

1. Catharinus, A.M., Koster, A.: Frequency Assignment: Models and Algorithms. Arie Koster (1999)
2. De Pasquale, A., Magnani, N.P., Zanini, P.: Optimizing Frequency Planning In the GSM Systems. CSELT, Italy (1998)
3. Chang, Y., Subramaniau, S.: Advanced frequency planning techniques for TDMA and GSM networks. In: IEEE Global Telecommunications Conference 1998, Sydney, pp. 934–938 (1998)
4. Atamanesh, M., Farzaneh, F.: Frequency Planning of GSM cellular Communication Network in Urban Areas Including Traffic Distribution, A Practical Implementation. In: Electromagnetic Compatibility and 19th International Zurich Symposium on Electromagnetic Compatibility, APEMC 2008, May 19-23, pp. 891–894. IEEE CNF (2008)
5. Su, C., Lan, L., Yu, C., Gou, X., Zhang, X.: A new method of frequency planning for new cells in GSM. In: 2nd Conference on Environmental Science and Information Application Technology (2010)
6. Zhang, M., MacDonald, V.H.: Generalized cell planning technique applied for 4x12 frequency-reuse. In: Proc. PIMRC 2003, vol. 2, pp. 1884–1888 (September 2003)
7. Gao, Y., Song, J., Zhang, Q., Ma, Y., Wang, J., Feng, Z.: A novel interference matrix generation algorithm in GSM networks. In: International Conference on Communications and Networking in China 2009, Xian, pp. 1–5 (2009)
8. Kronestedt, F., Frodigh, M.: Frequency planning strategies for frequency hopping GSM. In: Proc. VTC 1997, Phoenix, Arizone, USA, May 4-7, vol. III, pp. 1862–1866 (1997)
9. Kayacan, S., Toker, L.: Implementation of Mobile Measurement-based Frequency Planning in GSM. Ege Universty Master Dissertation (2007)

Design of Optimal Digital Fir Filter Using Particle Swarm Optimization Algorithm

Pavani Uday Kumar, G.R.C. Kaladhara Sarma, S. Mohan Das, and M.A.V. Kamalnath

Department of ECE,
AVR & SVR College of Engineering & Technology, Nandyal, A.P, India
udayaccess@yahoo.com, {grcksrgm,mohantech418,kamalmav}@gmail.com

Abstract. Designing a good performance FIR filter is a core problem in signal processing field over the years in determining the sample value in the transition zone. Obviously, Genetic Algorithm method cannot guarantee that interpolator is the optimal sampling point. This method is complex in structure which takes longer time in operation & suffers from local optimal solutions. In this paper PSO method is used to determine the frequency response of Digital FIR low pass filter, consequently the optimal filter coefficients are obtained with fast convergence speed and also error function is minimized, when compared with the errors obtained from windowing techniques. PSO algorithm is implemented in FIR filter in an efficient way to improve the stop band attenuation such that the samples are interpolated near the discontinuity and reduce errors. The performance of this PSO is compared with the conventional window techniques have been verified via computer simulations using Matlab.

Keywords: PSO, LMS Error, Filter Ripples, Windowing Methods, Swarm.

1 Introduction

Particle Swarm Optimization is a population based stochastic optimization technique developed by Dr.Eberhart and Dr. Kennedy in 1995, inspired by social behavior of bird flocking or fish schooling[1]. In PSO the potential solutions, called particles, fly through the problem space by following the current optimum particles [3]. Each particle keeps track of its coordinates in the problem space which are associated with best solutions (fitness) it has achieved so far. This value is called pbest. When a particle takes all the population as its topological neighbors. The best value is called global best (gbest). The particle swarm optimization concepts consist of, at each time step, changing the velocity of each particle towards its pbest location. PSO has no potential evolution operators such as crossover and mutation[7].

2 Quantitative Analysis of PSO and Fir Low Pass Filter

The frequency response of a linear-phase FIR filter is given by

$$H(\omega) = \sum_{i=1}^{k} h(n) e^{-j\omega n} \tag{1}$$

D. Nagamalai et al. (Eds.): *Adv. in Comput. Sci., Eng. & Inf. Technol.*, AISC 225, pp. 187–196.
DOI: 10.1007/978-3-319-00951-3_19

Where h(n) is the real valued impulse response of filter, N+1 is the length of filter and ω is frequency according to the length being even and odd and the symmetry being an even and odd four types of FIR filters described.

The linear phase is possible if the impulse response h(n) is either symmetric(i.e. h(n) =h(N-n) or is anti symmetry h(n)=-h(N-m) for 0<=n<=N.

In general the frequency response [5] for type1 FIR filter can be expressed in the form

$$H(e^{j\omega}) = e^{-jn\omega/2}\tilde{H} \tag{2}$$

Where amplitude response $\tilde{H}(\omega)$, also called the zero response, is given by

$$\tilde{H}(\omega) = h(N/2) + \sum_{n=1}^{N/2} h(N/2-n)\cos(\omega n) \tag{3}$$

The amplitude response for the type 1 linear phase FIR filter [5] (using the notation N=2M) is expressed as

$$\tilde{H} = \sum_{k=0}^{M} a(k)\cos(\omega k) \text{ Where a(0)=h(M) and a(k) = 2h(M-k), 1<=k<M}$$

The amplitude response for the type 2linear phase FIR filter is given as

$$\begin{aligned} A(\omega) &= \sum_{i=1}^{k} \{W(\omega_i)[\sum_{k=0}^{M} a(k)\cos(\omega_i k) - D(\omega_i)]\}\wedge 2 \\ \partial \in / \partial a(k) &= 0 \end{aligned} \tag{4}$$

$$A(\omega) = \sum_{k=1}^{2M+1/2} b(k)\cos[\omega(k-1/2)$$

Where b(k)=2h[2m+1/2-K], 1<K<2M+1/2

The amplitude response for the case of type 3 linear phase FIR filter is given as

$$A(\omega) = \sum_{k=1}^{M} c(k)\sin(\omega k) \text{ Where c (k) = 2h (M-k) 1<=k<=M}$$

The amplitude response for the case of type 4 linear phase FIR filter is given as

$$\tilde{H}(\omega) = \sum_{k=1}^{2M+1/2} d(k)\sin[\omega(k-1/2)$$

Where d(k)=2h[2M+1/2-k] 1<=k<=2M+1/2

The design of a linear phase FIR filter with least mean square error criterion, we find the filter coefficients a(k) such that error is minimized. Corresponding to the coefficients the filter coefficients are obtained as shown by the equations.

The Least mean square design function for this design is given as

$$\epsilon = \sum_{i=1}^{k} W(\omega_i)[A(\omega_i - D(\omega_i)] \tag{5}$$

For type 1 FIR filter the amplitude response A(w) is a function of a(k) to arrive at the minimum value of ε, we set

$$\partial \in / \partial a(k) = 0 \qquad 0<=k<=M$$

Which results in a set of (M+1) linear equations that can be solved for a(k)

For the type 1 the expression for mean square error [5] is ε expressed as

$$\sum_{i=1}^{k} \{W(\omega_i)[\sum_{k=0}^{M} a(k)\cos(\omega_i k) - D(\omega_i)]\}\wedge 2 \tag{6}$$

A similar formation can be derived for the other three types of linear phase FIR filters. This Design approach can be used to design a linear phase FIR filter with arbitrarily shaped desired response. Where D(w)is the frequency response.

3 PSO Algorithm

Particle Swarm Optimization (PSO) algorithm is a population based optimization algorithm[6]. Its population is called a swarm and each individual is called a particle. Each particle flies through the solution space to search for global optimization solution. The implementation of PSO algorithm[2][8] for optimizing the filter coefficients is given as follows.

Step 1:

Error function is to be minimized is expressed in equation

Step 2:

Initial population (swarm) is generated where each particle in the swarm is a solution vector containing M=5 elements, then initial population can be expressed as follows

$$U_i^0 = [U_{i1}^0, U_{i2}^0, U_{i3}^0, U_{i4}^0, U_{i5}^0]$$

Particle U_{ij}^0 of particle U_j^0 is generated from uniform distribution [0 , $U_{ij,\max}^0$].

Step 3:

Initial velocities of each particle are written as follows

$$V_i^0 = [V_{i1}^0, V_{i2}^0, V_{i3}^0, V_{i4}^0, V_{i5}^0] \quad i=1, 2 ,.....5$$

Each elements V_{ij}^0 of V_j^0 is selected a random digits for example, between

[0, 0.1 $x_{i,\max}$]

Step 4:

Set iteration [4] count K=1.

Step 5:

Calculate Error value by using equation (5).

$$\in = \min[\in_1, \in_2, \in_3, \in_4, \in_5]$$

and corresponding particle is the gbest is the particle which leads to Є.

Step 6:

Update velocities of each particle using following relation where $Pbest_i^k$ is the best previous position of the ith particles, $gbest^{k-1}$ is particle which leads to ε

$$V_i^k = W * V_i^{k-1} + c1 * r1(pbest_i^{k-1} - U_i^{k-1}) + c2 * r2(gbest_i^{k-1} - U_i^{k-1})$$

the position among all particles. r1 and r2 are random digits between [0,1]. c1 and c2are acceleration constants and W an inertia weight typically selected in the range 0.1 to 2.W can be calculated as follows.

Where W_{max} and W_{min} are maximum and minimum values of inertia weight, respectively. K_{max} is the total number of iterations and K is the current iteration.

Step 7:

Update position of each individual particle as

$$U_i^k = U_i^{k-1} + V_i^k$$

Where i=1, 2....M.

Step 8:

Update $Pbest_i^k$ and $Gbest_k$

$$Pbest_i^k = U_i^k \quad \text{if} \quad C(U_i^k) < C(Pbest_i^k)$$

$$= Pbest_i^k \quad \text{if}\ C(U_i^k) >= C(Pbest_i^k)$$

Out of all particles which give min error gives the Gbest.

Step 9:

Repeat 5 to 9 for maximum no of times to get the least possible error.

4 Results of PSO Algorithm for 8 Particles with 8 Elements

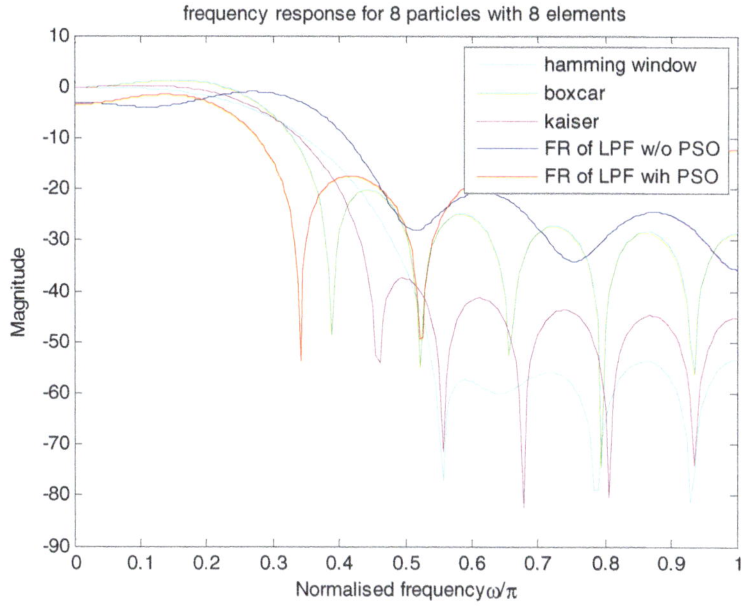

Fig. 1. Frequency response of first Order low pass FIR digital filter

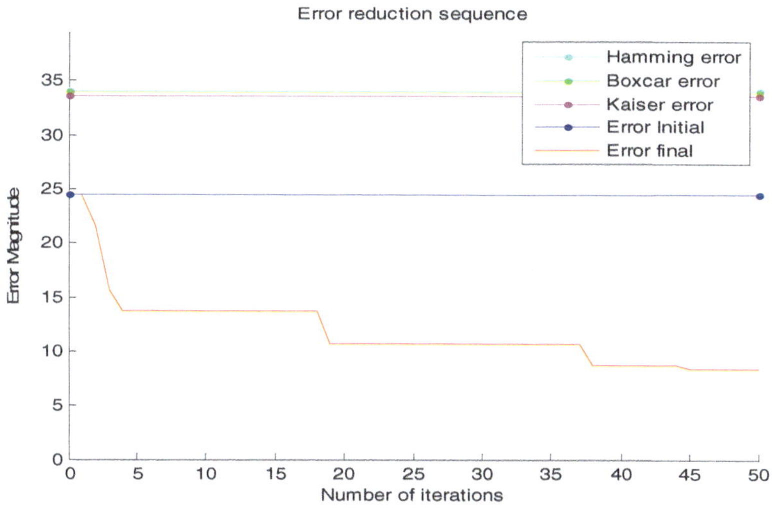

Fig. 2. Number of Iterations (N=50) VS MSE

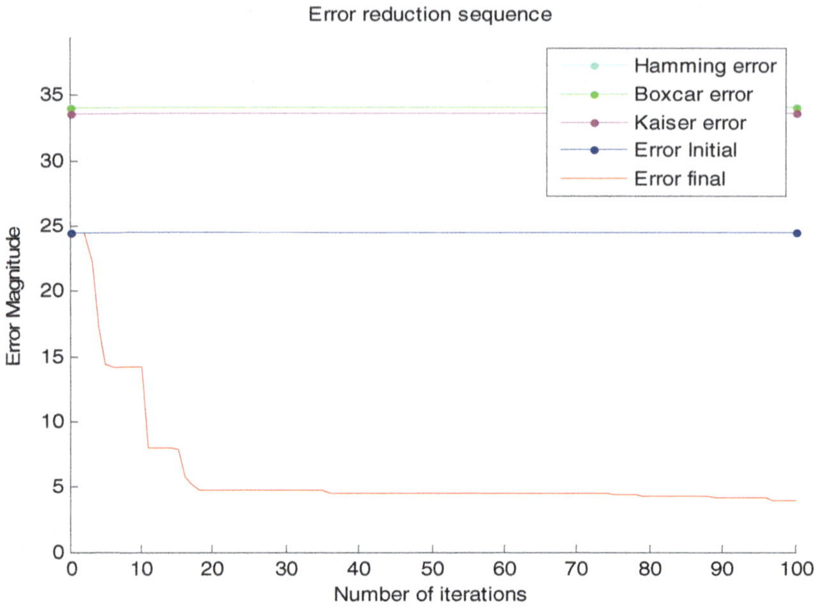

Fig. 3. Number of Iterations (N=100) VS MSE

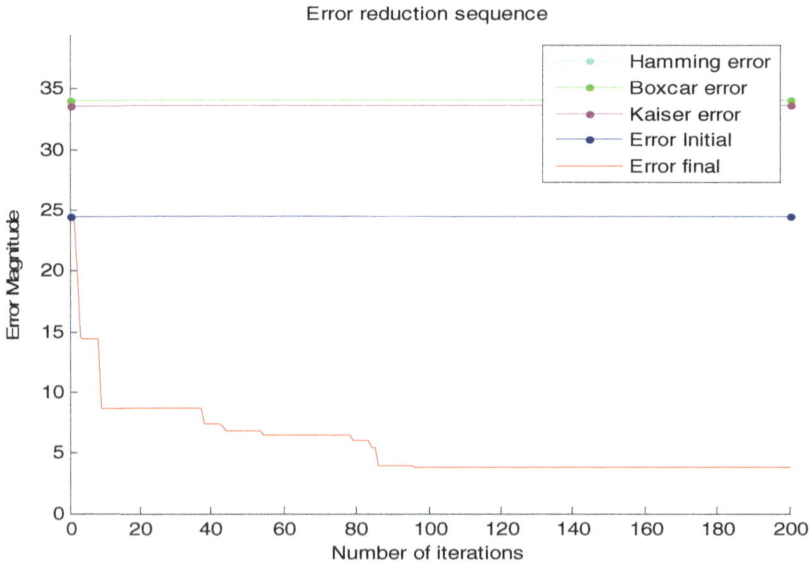

Fig. 4. Number of Iterations (N=200) VS MSE

5 Comparisons of Window Based Fir Low Pass Filter for Different Specifications

The following table shows the different performance parameters like pass band gain, stop band gain, cutoff frequency & error in the pass band & stop band for the conventional window based FIR filters.

Table 1. Filter specifications for different window methods

Specifications	Filter Length	Pass band gain (H1)dB	Stop band gain (H2)dB	Cutoff frequency (ωc)	Error in pass band & stop band
Hamming window	7	-0dB	-70dB	0.3π	16.958
	11	-0dB	-80dB	0.3π	17.698
	15	-0dB	-80dB	0.3π	24.497
Rectangular window	7	-0dB	-88dB	0.3π	27.486
	11	-0dB	-60dB	0.3π	32.588
	15	-0dB	-76dB	0.3π	34.051
Kaiser window	7	-0dB	-80dB	0.3π	26.750
	11	-0dB	-88dB	0.3π	31.957
	15	-0dB	-82dB	0.3π	33.624

In the above Table 1 different filter specification are calculated for conventional window based methods and they are tabulated. For all the window based methods a common cut off frequency is proposed in designing the digital FIR low pass filter.

In this the pass band gain , stop band gain for different windows are tabulated in Table 1 and ripples in pass band & ripples in stop band are calculated and compared among the Hamming window, Rectangular window & Kaiser window as shown in the Table 1. These ripples are again compared with the ripples which is calculated using PSO algorithm method as shown in Table 2 and these ripples are reduced when compared to the window based techniques in designing digital FIR low pass filter .

Table 2. Comparison of main lobe & side lobe of window based methods[9] and PSO Algorithm

Techniques	Main Lobe	Side Lobe	Pass band Gain	Stop band Gain	Error in pass band & stop band
PSO	-2dB	-22 dB	-3 dB	-58 dB	**3.7616**
Hamming	0 dB	-60 dB	0 dB	-80 dB	24.497
Rectangular	0 dB	-20 dB	0 dB	-76 dB	34.051
Kaiser	0 dB	-36dB	0 dB	-82 dB	33.624

In this PSO algorithm is compared with the window method. The following table Shows Accuracy parameters like main lobe, side lobe, pass band gain, stop band gain & execution time & its Comparison with windowing techniques. This comparison is shown in table 2.

Table 3. Comparison of number of iterations and Error reduction using PSO

Number of iterations	Initial error without PSO(dB)	Final Error reduced using PSO (dB)	Program execution time
50	**24.4659**	**5.3642**	**3.012755S**
100	**24.4659**	**3.9453**	**6.582142S**
200	**24.4659**	**3.7616**	**5.573579S**

In this analysis as the number of iterations are increased then the ripples in pass band and stop band is reduced using PSO algorithm which is shown in the Table 3. PSO algorithm not only used in optimizing the filter coefficients in digital FIR filter but also in various areas like Training of neural networks, Identification of Parkinson's disease, Extraction of rules from fuzzy networks, Image recognition Optimization of electric power distribution networks, Structural optimization, Optimal shape and sizing design, Topology optimization, Process biochemistry, System identification in biomechanics.

6 Conclusions

The information mechanism in PSO is significantly different, because the whole population moves like a one group towards an optimal area. In PSO, only gbest gives out the information to others. PSO converges to the best solution quickly and gives minimum error which is compared with the conventional technique whose filter specifications are tabulated in the table 1 & the error reduction for different particles are shown in the figures(2,3,4) & its quantitative information is tabulated in table 3. The frequency response of PSO is better than the conventional methods at the same time as the number of iterations increases error is reduced & the information is furnished in the table 3.

As the Number of coefficients are increased then probability of reduction in error is getting reduced which can be shown in figure 2 and observed in table 3 .In this table 3 the Initial error is shown in 2nd column & Final error is shown in 3th column. As the number of iterations increases then reduction of error is consequently reduced. and The frequency response of multiple particles with multiple elements are compared with the conventional window based methods in which pass band ripples and stop band ripples are reduced particularly in PSO algorithm.

The main lobe and side lobe parameters of the conventional methods & PSO algorithm are tabulated in table 2 in which the transition width of the PSO algorithm is getting narrow reaching to the ideal condition as shown in figure1. As PSO is a purely random algorithm in which time taken to compute the algorithm is large compared to the conventional methods. However PSO does not have any genetic operators like crossover and mutation. So, PSO is better than genetic algorithm

References

1. Zhu, Y., Huang, C.: Hybrid Opti- mization design of FIR filters. In: IEEE the Conference Neural Networks & Signal Proessing, Zhenjing, June 8-10 (2008)
2. Li, H., Zhang, A., Zhao, M., et al.: Particle Swarm Optimization Algorithm for FIR Digital Filters Design. Acta Electronica Sinica 33(7), 1338–1341 (2005)
3. Li, J., Xiao, X.: Multi-swarm and Multi - Best Particle swarm Optimization Algorithm, June 25-27. School of sciences, wuhan university (2008)
4. Li, B., Xiao, R.Y.: The particle swarm Optimization algorithm: How to select the Number of iteration. School of mathematical Sciences, south china university

5. Mitra, S.K.: Digital signal processing: A Computer based approach, 3rd edn. McGraw-Hill, New York (2006)
6. Kennedy, J., Ebhert, R.C.: Particle swarm optimization. In: Proceedings of 1995 IEEE International Conference on Neural Networks, vol. 4 (1995)
7. Angeline, P.: Evolutionary Optimization versus Particle Swarm Optimization: Philosophy and Performance Differences. In: Porto, V.W., Waagen, D. (eds.) EP 1998. LNCS, vol. 1447, pp. 601–610. Springer, Heidelberg (1998)
8. Arya, R., Choube, S.C., Arya, L.D.: System Reliability Enhancement using Particle Swarm Optimization (PSO). 89 (2008) IE(I) Jrnl–EL
9. Babu, P.R.: Digital Signal Processing, 4th edn.

Cooperative Spectrum Sensing for Cognitive Radio Networks Application: Performance Analysis for Realistic Channel Conditions

Waleed Ejaz[1], Najam ul Hasan[1], Muhammad Awais Azam[2], and Hyung Seok Kim[1,*]

[1] Dept. of Information and Communication Engineering, Sejong University Seoul, Republic of Korea

[2] University of Engineering and Technology, Taxila, Pakistan

{waleedejaz,hasan}@sju.ac.kr, awais.azam@uettaxila.edu.pk, hyungkim@sejong.edu

Abstract. Cognitive radio is a key technology to overcome spectrum scarcity by using spectrum opportunistically. It can be applied to maritime wireless networks to provide more bandwidth and reduce communication cost. Spectrum sensing is a primary issue to develop cognitive radio networks. There are few challenges for spectrum sensing in maritime networks which are different from terrestrial networks, for example, sea's surface channel properties. High probability of detection is required to achieve better network performance. In this paper, centralized cooperative spectrum sensing schemes are compared for maritime cognitive radio networks. The simulation results show that exiting schemes are well suitable for lower sea states but fail for higher sea states and we need to devise some advanced algorithms for spectrum sensing in the maritime wireless networks.

1 Introduction

Satellite communication has been used for long range ship-to-ship and ship-to-shore communication in recent years [1]. However, narrowband ultra high frequency (UHF) and very high frequency (VHF) based communication systems are used in near water ports ship-to-shore communication. The use of satellite links for internet access along with voice calls to and from ships is very expensive in comparison with land communication. There is huge scope for researchers to improve existing techniques and explore new research directions to reduce communication costs while satisfying the data rate requirements at sea.

It is hard to find a dedicated spectrum for maritime wireless networks due to congested bandwidths [2]. The maritime wireless network devices on shore may coexist with other radio devices installed for terrestrial networks. In addition, it is also necessary to synchronize frequency bands around the world as ships travel between countries and from continent to continent. The importance of traffic can change both the operational requirements for each ship and its network-wide goals. Therefore, network management in maritime networks should be efficient and well distributed.

* Corresponding author.

D. Nagamalai et al. (Eds.): *Adv. in Comput. Sci., Eng. & Inf. Technol.*, AISC 225, pp. 197–206.
DOI: 10.1007/978-3-319-00951-3_20

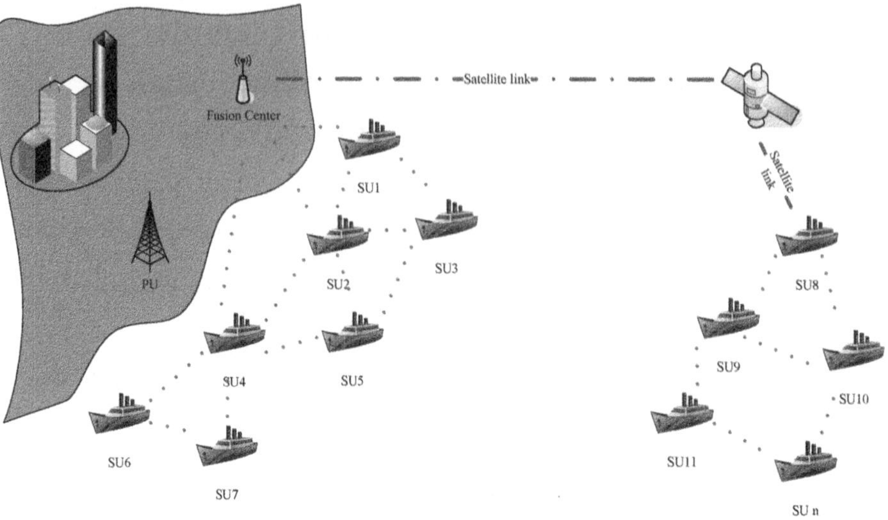

Fig. 1. Architecture of maritime cognitive radio network

Some new communication systems for maritime networks have been proposed in the last few years. A communication system for maritime networks was developed in Norway, which was the first digital VHF network of data rate 21 and 133 kbps with a coverage range of 130 km [3]. This system operates in the licensed VHF channel, which results in narrow bandwidth and slow communication speed. An IEEE 802.16e-based communication system named WISE-PORT (Wireless-broadband-access for SeaPORT) has been proposed in [4]. It can access up to 5Mbps, with a coverage distance of 15 km, but it still requires enhancement. In order to provide high speed and low communications costs, the mesh/ad hoc network based on the IEEE 802.16d mesh technology was proposed in a project called TRITON [5]. The authors developed a prototype that operates at 2.3 and 5.8 GHz.

In [2], the authors concluded that there was plenty of spectrum that was underutilized at sea. This inefficient utilization of spectrum can be mitigated with the help of Cognitive Radio (CR), using licensed spectrum bands opportunistically. The advantages of CR networks are an alleviation of spectrum scarcity, long range communication using TV bands, reduced cost for communication, and large bandwidth. One of the most essential tasks of CR networks is spectrum sensing, in which CR needs to detect the licensed/primary user (PU), which is achieved by sensing the radio environment. If PU is absent, its spectrum is available for cognitive radio/secondary users (SUs) and is called spectrum hole/white space.

Currently, the main focus of research in spectrum sensing for cognitive radio is divided into two main streams: improving local sensing and enhancing the cooperative spectrum sensing for better data fusion results. Major local sensing techniques considered for cognitive radios are energy detection, matched filter detection and cyclostationary detection. Energy detection is the simplest technique that has a short sensing time, but its performance is comparatively poor under low signal to noise ratio (SNR) conditions. Matched filter detection is another simple technique but it requires prior knowledge about the waveform of PU. On the contrary, cyclostationary detection provides reliable spectrum sensing at the cost of its computationally complex and long sensing time [6]. In cooperative spectrum sensing, all the local sensing observations made by SUs are reported to a fusion centre and a final decision about the presence or absence of PU is conducted at the fusion centre [7]. Based on the final decision received by the fusion centre, each SU reconfigures its operating parameters.

In this paper, the performance of existing cooperative spectrum sensing schemes on land and maritime CR networks is examined. Spectrum sensing in maritime networks have to face some unique challenges in the sea environment because of the following reasons: 1) radio wave propagation over water, 2) surface reflection and 3) wave occlusions. TV, cellular and maritime spectrum is available for maritime CR networks and it should be intelligent enough to switch its operating parameters to suit sea state, geographic location/region and communication range in order to achieve better throughput in addition to quality of service (QoS). UHF is used in USA for maritime navigation services and therefore it is important to protect the primary band. To achieve this, high probability of detection and low probability of false alarm are required to detect PU precisely and for efficient utilization of bandwidth, respectively. To the best of author's knowledge, no one has yet considered spectrum sensing in maritime CR networks. Given the challenging environment of the sea, it is identified that more sophisticated spectrum sensing algorithms and fusion rules are required for maritime CR networks.

The rest of the paper is organized as follows. In Section 2, brief overview of maritime cognitive radio network and channel modeling is presented. Section 3 presents spectrum sensing in maritime cognitive radio networks. Section 4 demonstrates simulation results and their detailed analysis. Lastly conclusions are drawn in Section 5.

2 Maritime Cognitive Radio Network and Channel Modeling

Fig. 1 shows that maritime CR networks can be divided into two types 1) the ship-to-ship/ship-to-shore network (close to shore) which can have a central entity on the shore and 2) the ship-to-ship communication deep in the sea with the support of satellite communication link. Each ship is equipped with the capabilities to perform the functionality of cognitive radio, i.e., the cognitive capability and reconfigurability. The radio environment is being sensed regularly by each ship to use the spectrum opportunistically. A satellite link can be used to access the central entity in deep sea where it becomes impossible to access the central entity by using available wireless networks. In addition to terrestrial CR network's spectrum usage, a ship can switch its operating parameters according to the sea state, geographic location and density of nodes.

Recently, researchers investigated the spectrum measurements which have been carried out for CR networks on the land. Most common of those are on the television (TV) and cellular bands. On the contrary, investigations are required for spectrum allocated for maritime communications because of the environmental differences.

Environmental differences include the obvious maritime radio atmosphere, i.e., sea motion and antenna model. There is almost no obstacle in the sea and the sea surface is also flat, which causes huge path loss due to negative interference between the line of sight (LoS) path and the reflected path [1].

2.1 Sea Motion

The generation of random sea surface is the fundamental of propagation analysis for maritime CR networks. Sea movement is described by the sea state divided into 10 levels characterized by the Pierson-Moskowitz [8]. Sea states 0-3 are generally considered as calm sea conditions. Moderate sea conditions are followed in sea states 4-5. The sea state 6 and above are the worst sea conditions having high waves which cause severe degradation in the communication by effecting the moment of antenna.

2.2 Channel Model

Maritime wireless networks have to face unusual challenges because of the variable channel statistics. The radio waves are reflected by the sea surface due to which signal degrades completely along the path. In terrestrial environment, there are obstacles of different sizes which result in reflection, refraction and scattering of signal in the communication channel. The path loss in terrestrial environment is higher than in free space and defined as [9]:

$$L_T(d) = L_s(d_o) + 10 * \alpha * \log\left(d/d_o \right) + X_f \tag{1}$$

where d_o is the distance of a reference location with measured path loss L_s, d is the physical distance between transmitter and receiver, α is the path loss exponent for the radio environment and the Gaussian random contributor X_f with zero mean and standard deviation σ, which represents fast fading effects. The accurate estimation of path loss exponent is the major characterization of communication channel. Usually values of path loss exponent ranges from 1 to 4 depending on the physical terrain features.

The typical path loss exponent in terrestrial environment is comparable with the sea state 4. Path loss increases rapidly for the sea state 5 and above. Path loss in the maritime communication channel during shadowing is proposed in [10].

$$PL(h, f) = PL(d_o) + 10 * [(0.498 \log_{10}(f) + 0.793)* h + 2]* \log_{10}\left(d/d_o \right) + X_f \tag{2}$$

where f is the frequency in GHz, the observable sea height is h in meters, d is the physical distance between transmitter and receiver, do is the distance of a reference location with measured path loss PL(d_o) and the random variable X_f with zero mean and standard deviation σ which is also represented as a function of wave height:

$$\sigma_f = [0.157 f + 0.405] * h. \tag{3}$$

3 Spectrum Sensing in Maritime Cognitive Radio Networks

3.1 System Model

In this paper, the UHF band as PU's band is assumed. It is broad and can offer bandwidth of more than 100MHz opportunistically in maritime networks [2]. The communication range for these frequencies is up to 10km. Ships satisfy all the requirements for acting as SU. The base station at the shore acts as the fusion center. The system model for the performance analysis of cooperative spectrum sensing schemes is shown in Fig. 1 in which the maritime wireless network with N SUs is considered.

The ultimate goal of spectrum sensing is to determine the presence of a PU using a binary hypothesis model, i.e., the basic model for spectrum sensing by the SU, which is defined as

$$r(t) = \begin{cases} n(t) & \text{in the case of } \mathrm{H}_0, \\ hs(t) + n(t) & \text{in the case of } \mathrm{H}_1 \end{cases} \tag{4}$$

where $r(t)$ is the signal received by SU, $s(t)$ is the transmitted signal of the PU, $n(t)$ is the additive white Gaussian noise (AWGN), and h is the amplitude gain of the channel. H_0 indicates only noise, whereas the presence of a PU is H_1.

3.2 Local Spectrum Sensing

Energy detection is considered in the implementation of our local spectrum sensing technique. The output of the energy detector has a distribution that is defined as follows [11]:

$$Y = \begin{cases} \chi^2_{2M} & \text{in the case of } \mathrm{H}_0, \\ \chi^2_{2M}(2\gamma) & \text{in the case of } \mathrm{H}_1 \end{cases} \tag{5}$$

where χ^2_{2M} and $\chi^2_{2M}(2\gamma)$ represent a central chi-square distribution and a non-central chi-square distribution with $2M$ degrees of freedom and non-centrality parameter 2γ, respectively.

3.3 Centralized Cooperative Spectrum Sensing

A centralized cooperative spectrum sensing scheme with cooperative N SUs is considered. The entire N SUs forward their local decision to centralized fusion center as shown in Fig. 2. In order to get cooperative decision, linear fusion rules are applied and most commonly used fusion rules are AND, OR and majority rules. The fusion center forwards the cooperative decision to all individual SUs after applying fusion rules.

The probability of detection and probability of false alarm for k out of the N rules are given by [7].

$$Q_d = \sum_{l=k}^{N} \binom{N}{l} P_d^l (1 - P_d)^{N-l} \tag{6}$$

$$Q_f = \sum_{l=k}^{N} \binom{N}{l} P_f^l (1 - P_f)^{N-l} \tag{7}$$

where Q_d and Q_f are the probability of detection and probability of false alarm at the fusion center. P_d and P_f are the probability of detection and probability of false alarm determined by each SU. Total number of SUs participating in cooperation is n and k determines one among AND, OR and majority rules. In (6) and (7), if the value of k is taken as n or 1, k out of N rule becomes AND or OR rules, respectively. When k is larger than $n/2$, k out of N rule works as the majority rule.

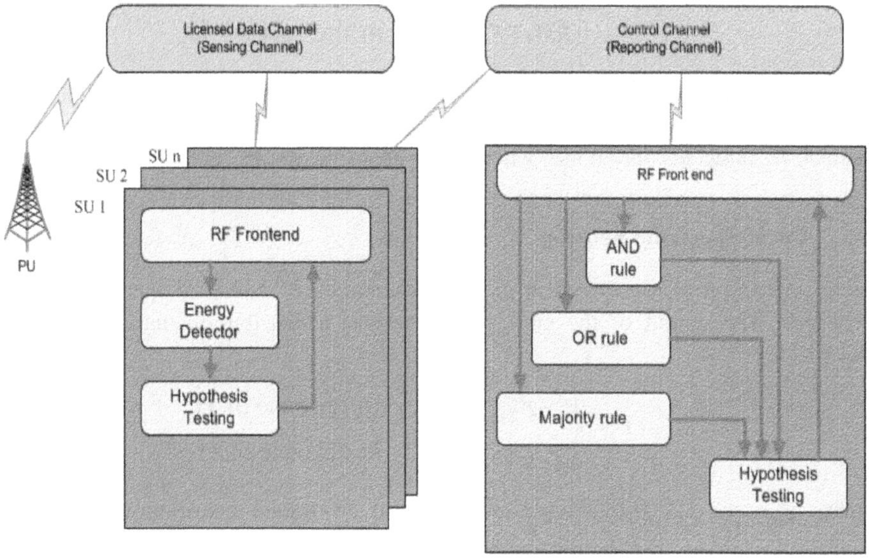

Fig. 2. Framework of centralized cooperative spectrum sensing

4 Simulation Results

In the simulations, it is assumed that there are fifteen SUs and all of them are experiencing additive white Gaussian noise (AWGN) with the same variance whereas the path loss depends on the radio environment during communication. Each SU uses energy detection for its local observations having an average SNR $\bar{\gamma}$ and the time-bandwidth product, TW. Let SUs be unaware of any relevant PU information such as the position, moving direction and velocity.

Fig. 3 shows the path loss for 2.492GHz for different radio environments including land and sea environments. Sea states considered to measure path loss are 4, 6 and 7 having wave heights 2, 4 and 11m, respectively. The reference distance is 1m and the physical distance between the transmitter and the receiver varies from 0 to 20 km. Results show that path loss in the maritime environment is comparable to one in the land environment up to the sea state 4. Severe path loss is observed in comparison with land environment at the sea states higher than 4. For the sea state 7, path loss is almost 5 times when compared with either sea state 4 or land environment. It becomes worse at higher sea states.

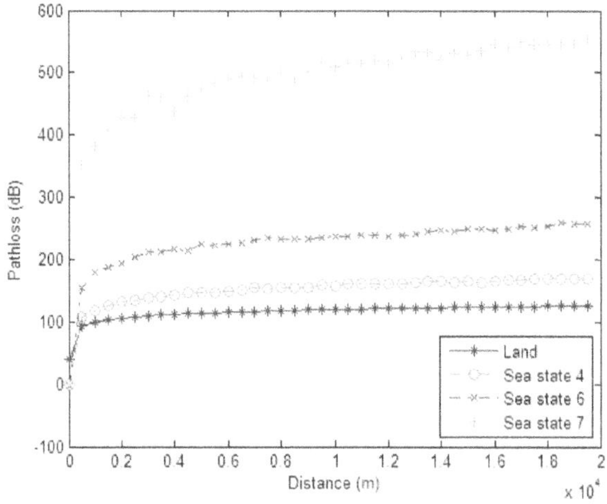

Fig. 3. Comparison of path loss model in land and different sea states

The probability of detection for the reference SU was investigated to determine its sensitivity for detecting PU's presence. SNR varies from -20 dB to 20 dB for the reference SU. According to the draft IEEE 802.22 standard [12], the probability of false alarm should be less than or equal to 0.1. Therefore, the decision threshold λ was set to maintain $P_f = 10^{-1}$. The time-bandwidth product, *TW*, was set at 5. Energy detection is used for local detection at SU. Fig. 4 shows that the detection probability at the land and sea state 4 is the same with each other because both of them experiencing almost the same path loss. In the land environment and the sea state 4, the reference SU detects the PU with 100% certainty even under a very low SNR value of -20 dB. When the sea state is 6, the probability of detection is almost zero except for very low SNR values ranging from -20dB to -15dB. This is due to inherent limitation of the energy detector because it is unable to discriminate between signal and noise energy. For the sea state 7, the probability of detection is zero over the entire range of SNR because of severe path loss.

The complementary receiver operating characteristic (ROC) curves at the land, sea state 4, sea state 6 and sea state 7 using the reference SU, AND rule, OR rule and majority rule are shown in Fig. 5. There are total 15 SUs participating in cooperation

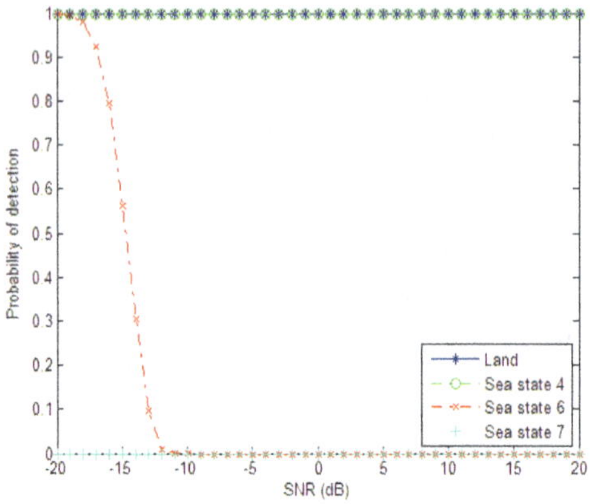

Fig. 4. The impact of SNR on the probability of detection (P_f=10^{-1}, TW=5)

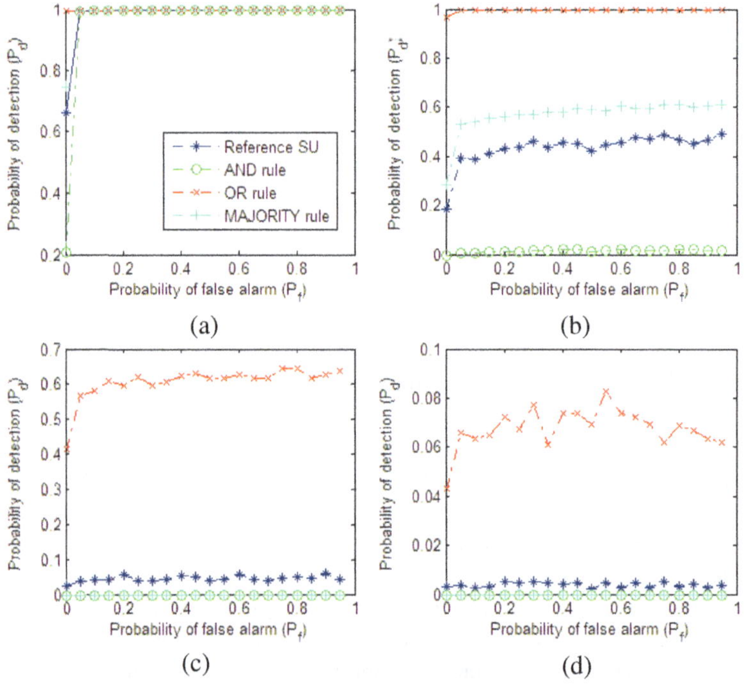

Fig. 5. Probability of detection vs. the probability of a false alarm in a 15-node network at (a) Land Network, (b) Sea state 4, (c) Sea state 6 and (d) Sea state 7

including a reference user. SNR for the reference user is considered 0 dB and for the rest of the users SNR varies from -20dB to 10dB. Distance from the PU varies from 0 to 2 km for each SU randomly. Fig. 5 (a) shows the complementary ROC for the land environment. Only in the OR rule, the probability of detection is 1 over the entire range of the probability of false alarm. The probability of detection is also high for the AND rule, reference SU and majority rule. Thus, we can say that simple cooperative fusion rules work reasonably well for the land environment. Fig. 5 (b) shows the complementary ROC for the sea environment having the sea state 4. The probability of detection for OR rule is still high because of its inherent property of fusion. For the AND rule, the probability of detection is almost zero over entire range of the probability of false alarm. The probability of detection is approximately in the range of 0.4 and 0.57 for the reference user and majority rule, respectively. The majority rule improves the detection performance in comparison with the reference SU. The complementary ROCs for the sea states 6 and 7 are shown in Fig. 5 (c) and (d), respectively. The probability of detection for the OR rule is approximately 0.6 in the sea state 6 and 0.06~0.08 in the sea state 7. Including AND and majority rules for sea states 6 and 7 leaves no improvement in the probability of detection of the reference SU.

5 Conclusion

This paper presents the performance analysis of centralized cooperative spectrum sensing schemes for maritime cognitive radio networks. Simulation results show that the relatively calm sea, i.e., sea state 4, has better detection probability with existing fusion rules, i.e., OR and majority than higher sea states does. For sea state 7 and higher, all the existing fusion rules fail and therefore advanced sensing algorithms and fusion rules are required for better performance.

Some advanced signal processing algorithms are required to improve the performance of spectrum sensing. The complex algorithm like cyclostationary feature detection needs relatively long time for sensing in comparison with energy detector. However, the performance of energy detector is poor under low SNR conditions and higher sea states. Therefore, for the future work, hybrid schemes can be considered for the spectrum sensing in maritime cognitive radio networks.

Acknowledgments. This study was supported by the CITRC (Convergence Information Technology Research Center) support program (NIPA-2013-H0401-13-1003) supervised by the NIPA (National IT Industry Promotion Agency) of the MKE (Ministry of Knowledge Economy). It was also supported by Special Disaster Emergency R&D Program from National Emergency Management Agency through Kyungil University (2012-NEMA10-002-01010001-2012) and Basic Science Research Program through the NRF funded by the MEST (2010-0025316).

References

1. Pathmasuntharam, J.S., Jurianto, J., Kong, P.Y., Ge, Y., Zhou, M., Miura, R.: High Speed Maritime ship-to-ship/ shore Mesh Networks. In: Proc. IEEE Int. Conf. on ITS Telecommunications, Sophia Antipolis, France, pp. 1–6 (2007)

2. Zhou, M.-T., Harada, H.: Cognitive maritime wireless mesh/ ad hoc networks. J. of Netw. and Comput. Appl. 35(2), 518–526 (2012)
3. Bekkadal, F.: Emerging maritime communications technologies. In: Proc. IEEE Int. conf. on ITS Telecommunications, Lille, France, pp. 358–363 (2009)
4. First in the World: Wireless Mobile Wimax Access In Singapore Seaport Now a Reality (March 6, 2008), `http://www.mpa.gov.sg/sites/global_navigation/news_center/mpa_news/mpa_news_detail.page?filename=nr080306.xml`
5. Pathmasuntharam, J.S., Kong, P.-Y., Zhou, M.-T., Ge, Y., Wang, H., Ang, C.-W., Su, W., Harada, H.: TRITON: high speed maritime mesh networks. In: Proc. IEEE Int. Symp. Pers., Indoor and Mobile Radio Commun., Cannes, France, pp. 1–5 (2008)
6. Yucek, T., Arslan, H.: A survey of spectrum sensing algorithms for cognitive radio applications. Commun. Surveys Tuts. 11(1), 116–130 (2009)
7. Akildiz, I.F., Lo, B.L., Balakrishan, R.: Cooperative spectrum sensing in cognitive radio networks: a survey. Physical Commun. 4(1), 40–62 (2011)
8. Pierson Jr., W.J., Moskowitz, L.: A proposed spectral form for fully developed seas based on the similarity theory of S. A. Kitaigorodskii. J. Geosci. Res. 69(24), 5181–5190 (1964)
9. Elliott, W.: Results of a VHF propagation study. IEEE Trans. Antennas Propag. 29(5), 808–811 (1981)
10. Timmins, I.J., O'Young, S.: Maritime Communication Channel Modeling using the Finite-Difference time domain method. IEEE Trans. Veh. Technol. 58(6), 2626–2637 (2009)
11. Urkowitz, H.: Energy detection of unknown deterministic signals. Proc. IEEE 55(4), 523–531 (1967)
12. IEEE Computer Society, IEEE Std 802.22^{TM}-2011 Part 22: Cognitive Wireless RAN Medium Access Control (MAC) and Physical Layer (PHY) Specifications: Policies and Procedures for Operation in the TV Bands, IEEE Standard for Information technology, 1-672 (2011)

HaarWavelet Based Distributed Predictive Target Tracking Algorithm for Wireless Sensor Networks

Mahsa Ghasembaglou and Abolfazl Toroghihaghighat

[1] Electrical, Computer, and IT Department Qazvin Islamic Azad University, Qazvin, Iran
mg.soft87@gmail.com, at_haghighat@yahoo.com

Abstract. Target tracking is a typical and important application of wireless sensor networks (WSNs). In the consideration of scalability and energy efficiency for target tracking in large scale WSNs, it has been employed as an effective solution by organizing the WSNs into clusters. However, tracking a moving target in cluster-based WSNs suffers many problems e.g. energy consumption and need to use complex predictive tracking. In this paper, we propose a novel scheme, called Haar wavelet-based target tracking (HWDPT). We proposed an efficient prediction algorithm for mobile targets tracking based on Haar Wavelet transform. Also, we integrated on-demand dynamic clustering into a cluster-based WSN for effective target tracking. The simulation results validate that the proposed HWDPT protocol performs better in terms of track failed ratio, energy waste ratio and algorithm complexity, respectively.

Keywords: wireless sensor network, target tracking, Haar wavelet transform, prediction algorithm.

1 Introduction

In wireless sensor networks, many inexpensive and small sensor-rich devices are deployed to monitor and control our environment [1, 2]. Each device, called a sensor node, is capable of sensing, computation, and communication. Sensor nodes form a wireless network for communication. The limited supply of power and other constraints, such as manufacturing costs and limited package sizes, limit the capabilities of each sensor node. For example, a typical sensor node has short communication and sensing ranges, a limited amount of memory, and limited computational power. However, the abundant number of spatially spread sensors will enable us to monitor changes in our environment accurately despite the inaccuracy of each sensor node.

One of important application of wireless sensors is target tracking. According to [3-15], target tracking using wireless sensor networks was initially investigated on 2002. In such scenarios, the sensor networks may be deployed for The applications of tracking include surveillance, military (tracking enemy vehicles, detecting illegal border crossings) and civilian purposes (tracking the movement of wild animals in wildlife protection), search and rescue, disaster response system, pursuit evasion games [16], distributed control [17] and other location-based services [18]. In the

D. Nagamalai et al. (Eds.): *Adv. in Comput. Sci., Eng. & Inf. Technol.*, AISC 225, pp. 207–219.
DOI: 10.1007/978-3-319-00951-3_21

literature [19] [20], various algorithms and approaches such as tree-based tracking, cluster-based tracking; prediction-based tracking and mobicast message-based tracking for target tracking are presented.

Our proposed algorithm is based on cluster based tracking. The cluster-based methods are better than other types of methods in terms of provided scalability, usage of bandwidth and facility collaborative data processing [19][20][21]. Cluster based target tracking algorithms can be further divided into two groups: static and dynamic clustering. In static approaches, clusters are formed at the time of network deployment. The attributes of each cluster, such as the size of a cluster, the area it covers, and the members it possesses, are static. These attributes of static cluster causes a boundary problem, that nodes of different clusters can't share their information. The static clustering architecture offers the sensor networks to save the important energy of the sensors for finding cluster heads. In dynamic approaches clusters are formed as the events occur in the network area. This approach does not impose any restriction on memberships. For example, a node can be a member of different clusters at different times which make this approach more advantageous for minimization of localization errors. Although dynamic clustering offers many benefits, it consumes the sensor's essential energy for choosing the next cluster head and trying to form cluster with its neighbors until the target leaves the sensor networks [19][20][22][23–26]. In this paper we use static cluster and dynamic cluster architecture to involve advantages of both of them.

The main contributions of this paper are summarized as follows: We present a novel Haar Wavelet based prediction for target tracking scheme, which estimates accurately next position of a mobile target based on its history. The proposed scheme is fully distributed with no help of global information or prediction algorithms. We conduct simulations to show the efficiency of the proposed scheme compared with other typical target tracking solutions.

2 Related Works

A lot of protocols have been proposedfor target tracking in WSNs. The cluster-based methods provide scalability and better usage of bandwidth than other types of methods.

Prediction-based tracking methods are rely on tree-based and cluster-based tracking in addition to prediction method, which allow limited number of sensors to track the moving target [20].

2.1 Cluster Based Techniques

Clusters are formed to facilitate collaborative data processing in target tracking applications. A static cluster-based distributed predictive tracking (DPT) protocol is presented in [27]. The protocol uses a predictive mechanism to inform predicted area cluster head who activates k sensors to sense the target before the arrival of the target. DPT uses a cluster based approach for Scalability. However, this algorithm has the problem of local sensor collaboration as the target moves across or along the boundary among clusters.

An exponential distributed predictive tracking (EDPT) cluster based algorithm is proposed in [28] for target tracking, which using smoothing prediction algorithm and recovery mechanism. This algorithm is static cluster based and it has the problem of local sensor collaboration too.

In [29] ZhiboWang et al. proposed a hybrid cluster-based on target tracking (HCTT), which integrates on-demand dynamic clustering into a cluster-based WSN To overcome the boundary problem. However, in this algorithm unnecessary dynamic clustering may take place frequently that in this case HCTT will consume more energy.

2.2 Prediction-Based Techniques

Prediction-based techniquesare used to predict the upcoming location of mobile target for energy saving [20][32][33]. Base Assumptions of the prediction models are heuristics information that the moving targets will stay at the current speed and direction for the next few seconds, the target's speed and direction for upcoming few seconds can be deduced from the average of the target's movement history, and to the various stages can be assigned various weights based on the history [24].

The dual prediction [33] and prediction-based energy saving (PES) [32] focus on reduction of energy consumption in the monitoring step of mobile target tracking sensor network by keeping most of the nodes in sleeping mode until waken up by an active node. The next location of mobile target is calculated at sensor node and sink from historical data [20]. The basic idea of PES is that a sensor node should stay in sleeping mode as long as possible. Thus, based on the prediction model used, the current node will predict the possible location(s) of the moving target and determine a group of sensor node(s), to help tracking the moving target after certain period of sleeping.

Prediction-based techniques will be used as a part of our project, because they predict the future position of the mobile target, in order to conserve the battery.

3 The Proposed Algorithm

In the proposed algorithm, sensor nodes are organized into static clusters according to any suitable clustering algorithm. Cluster heads, corner nodes and boundary nodes in each cluster are also formed. When a target is in the network, a static cluster sensing the target and wakes up its cluster head and number of nodes to sense and track the target. When the target approaches the boundary, the boundary nodes can detect the target and neighbor cluster head would be announced in advance for a static to static handoff. However, an on-demand dynamic cluster will be constructed when target is lost. An on-demand dynamic cluster may be destructed after target exit from dynamic cluster and enter in static cluster.

When a target is in a cluster, its cluster head selects number of nodes to senses the target. This number is determined based on density of nodes. In this paper according to [27] we use three sensors to sense and track. Cluster head of cluster, which the target is in it called current cluster head and cluster head of next location of target called next cluster head. Current cluster head selects three sensors that are more suitable and wakes up them for tracking. Sensors sense the target and send report to current cluster head. Then this cluster head using proposed prediction algorithm estimates the next location of the target and wakes up the next cluster head before arriving the target. As the target moves, static clusters using new prediction algorithm manage the tracking task. Thus when next position of the target is in current cluster, current cluster head sends an inform message to aware next cluster head about future entrance of the target. The next cluster head awakes a few sensors to track the target in new cluster. Upon a detector receive detection signal from the target, it sends a message to current cluster head and informs the target detection event. The next cluster head will be responsible for tracking task in new cluster and send a message to the current cluster head about it. The current cluster head has to check the target status when the target has not entered to new cluster. There are two situations: the target is detectable in the previous cluster or the target has been lost. The lost target can be recovered using the new proposed recovery mechanism with minimum waste energy. Following section describes the proposed scheme for target tracking in wireless sensor networks.

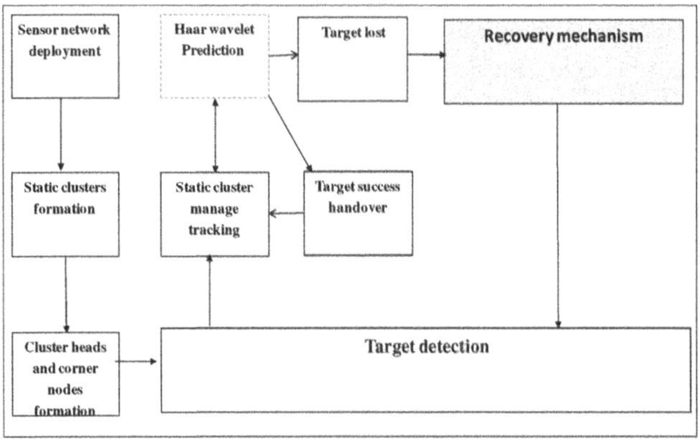

Fig. 1. Block diagram of the proposed algorithm

4 Haar Wavelet Based Prediction Algorithm

In this paper, to enhance the tracking precision and save energy, a prediction-based protocol for mobile target tracking is proposed. The proposed protocol has lower computation complexity and better performances in terms of energy consumption and tracking precision. Haar wavelet transforms are used in prediction issues in signal processing applications. In this paper, we use Haar wavelet transform to predict next position of a mobile target. Estimation of Haar wavelet module is based on eight-point vector recording previous positions of target. Suppose X[1..h] is a vector, which records track of a target. Haarwavelet equations are described in equations (1) and (2).

$$X(i) = 2^{-\frac{1}{2}}(X(2i-1) + X(2i)) \tag{1}$$

$$X(w+i) = 2^{-\frac{1}{2}}(X(2i-1) - X(2i)) \tag{2}$$

Equations (1) and (2) use only two types of very simple mathematic operations; addition and subtraction.

Our prediction algorithm runs in 4 steps and values of variables in equations (1) and (2) varying in each step as follow:

Step 1) h=8, i=1…w and w=h/2.
Step2) h=4, i=1…w and w=h/2.
Step3) h=2, i=1…w and w=h/2.

These steps are shown in "fig.2", "fig.3", and "fig.4", respectively.

Step 4) After running these three steps Haar transform is done and vector of $X[1..h]$is obtained as x31,x32,x23,x24,x15,x16,x17,x18 , which is shown in figure 4. Then we sum these values to estimate next location of the target.

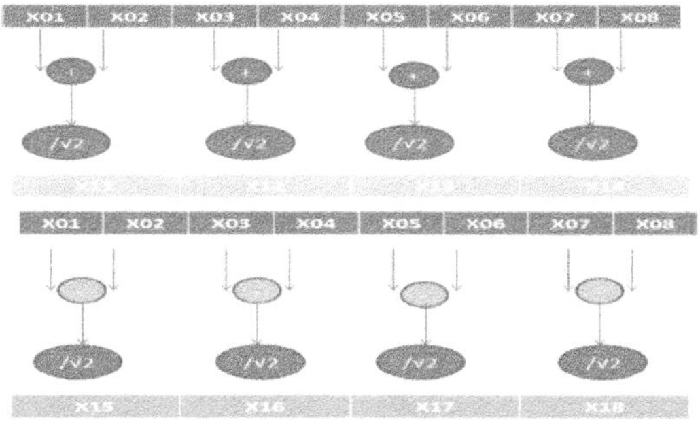

Fig. 2. First step of Haar transform based prediction

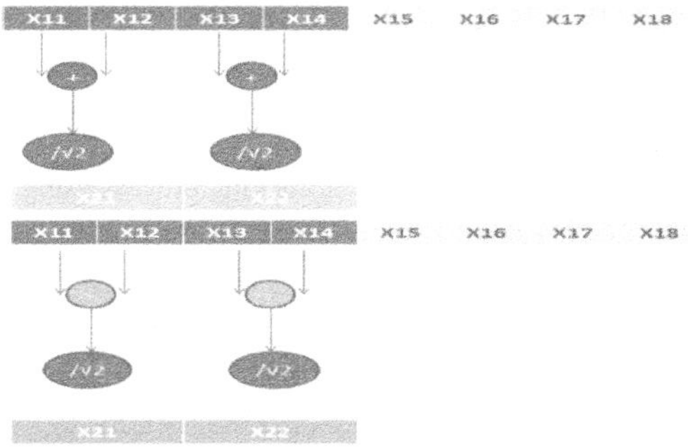

Fig. 3. Second step of Haar transform based prediction Equations

We repeat these four steps to compute x and y coordinates identically. As shown in "fig.5", suppose we have a positions vector for x coordinate of a typical target.

"Fig.6" and "fig.7" presents different steps of these transform to predict the next location of a mobile target. "fig.5" presents some errors in predicting next position of mobile target. To overcome this problem, we use a regular regression to correct prediction of Haar wavelet transform.

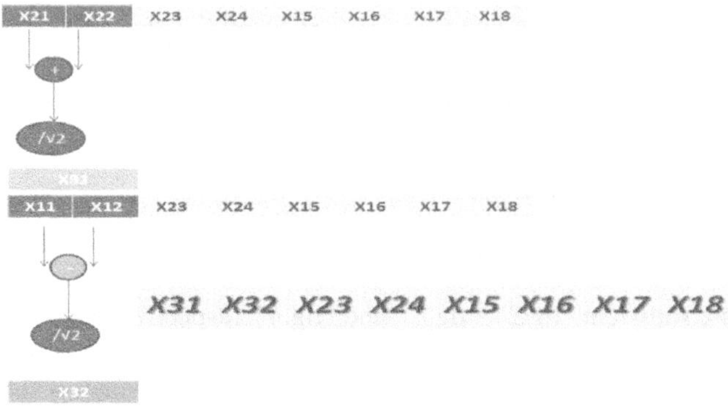

Fig. 4. Third step of Haar transform based prediction

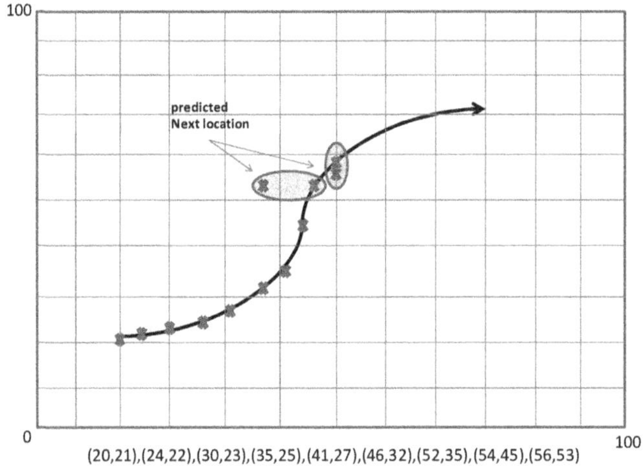

Fig. 5. Next location prediction in target tracking process using Haar wavelet

"Fig.8" presents predictor module with error corrector. Error corrector module uses history of predictor's operation to correct output of predictor module.

Suppose(X, Y) is output of error corrector module and (x, y) is outout of predictor module. We use two linear regressions to improve predictor's results. In statistics, regression analysis includes many techniques for modeling and analyzing variables to present relationship between a dependent variable and one or more independent variables. More specifically, regression analysis helps one to understand how the typical value of the dependent variable changes when any one of the independent variables is varied. The other independent variables are held fixed. The estimation target is a function of the independent variables called the regression function. In regression analysis,

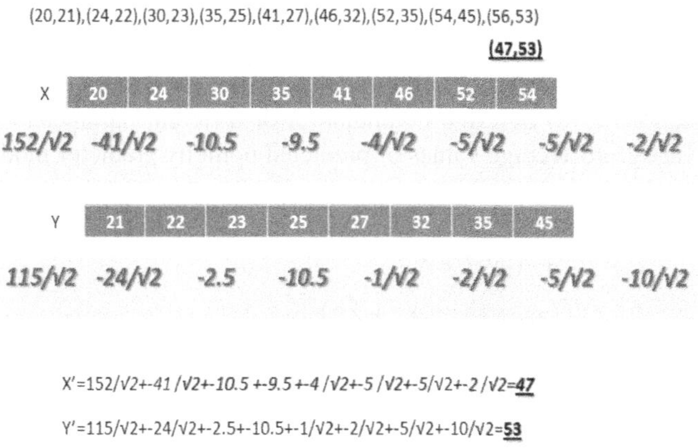

Fig. 6. Example 1 of using Haar wavelet for predicting next location

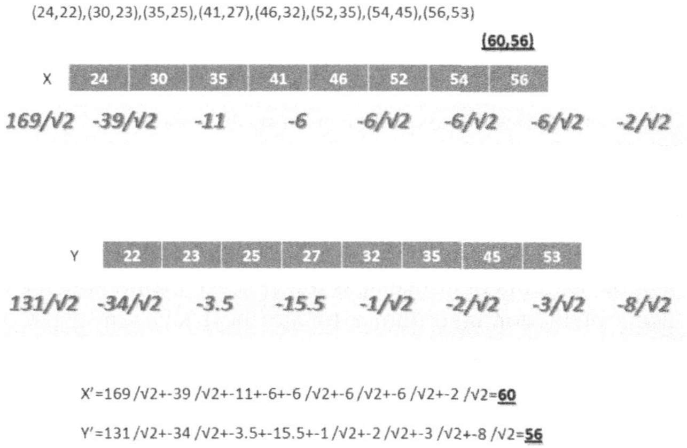

Fig. 7. Example 2 of using Haar wavelet for predicting next location

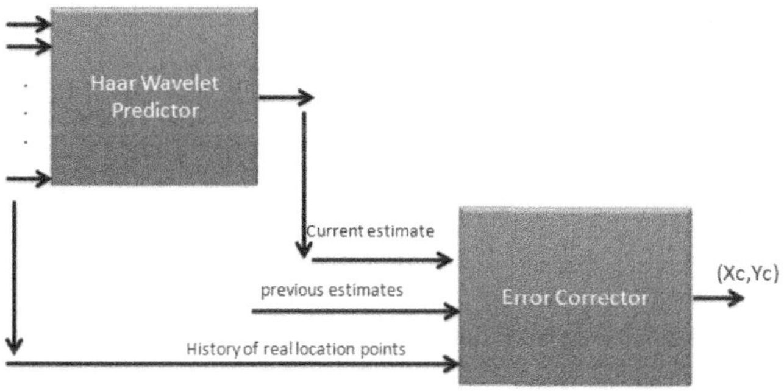

Fig. 8. Predictor and error corrector modules in tracking algorithm

it is also of interest to characterize the variation of the dependent variable around the regression function, which can be described by a probability distribution. Equations (3) and (4) describe two different regression equations for x and y point of the target location. Equations (5-8) describe computation of different parameters for equation (3) and (4). $\tilde{x}$and$\tilde{y}$ are average values of predicted point by predictor module.

$$X = \beta_0 + \beta_1\tilde{x} + ex_i \tag{3}$$

$$Y = \alpha_0 + \alpha_1\tilde{y} + ey_i \tag{4}$$

$$\beta_1 = \frac{\sum((x_i - \tilde{x}) \quad (X_i - \tilde{X}))}{\sum(x_i - \tilde{x})^2} \tag{5}$$

$$\alpha_1 = \frac{\sum((y_i - \tilde{y}) \quad (Y_i - \tilde{Y}))}{\sum(y_i - \tilde{y})^2} \tag{6}$$

$$\beta_0 = \tilde{X} - \beta_1\tilde{x} \tag{7}$$

$$\alpha_0 = \tilde{y} - \alpha_1\tilde{y}\,. \tag{8}$$

To summarize, we proposed a novel method to track mobile target in wireless sensor networks. Our HWDPT algorithm is proposed for accurate tracking by using Haar wavelet transform for predicting next location of mobile target.

5 Recovery Mechanism

Each cluster that the target is in it, manages tracking. While moving the target current cluster head using prediction algorithm estimates next location of the target and informs next cluster head. Current cluster head sets a timer to receive confirm message from next cluster head. If before expiration period of time current cluster head does not receive any message, it supposes that the target is lost. There are two cases when the target is lost: (1) the target is detectable in the previous cluster or (2) the target moves to another cluster. Our recovery scheme is based on [28] and is kept simple in order to be suitable for sensor networks.

5.1 Proposed Recovery Mechanism

1) **First level of recovery:** If current active sensors sense with low beam, they switch to high beam to sense. If sensors detect the target then they follow the tracking in normal state.

2) **Second level of recovery:** If the first level of recovery does not succeed, then Cluster head of the considered target moves with maximum speed and search all clusters that can be along with moving target thus ck sends wake up messages to all heads cj that satisfy d(cj, ck) ≤SR. Then each cj which is in this radius wakes up their member nodes to recover the lost target.

6 Experimental Results

In this section, we evaluate the performance of the proposed algorithm through simulations.

Table 1. system parameters

System parameters	Value
Network area	1000 × 1000 (m2)
n	6000
Node placement	random-waypoint (RWP)
Cluster head placement	Grid
Target moving speed/(m/s)	5 – 10 (m/s)
Time interval	5(T/s)
sensing radius of each node	100 (m)
transmission radius of each node	200 (m) to 250 (m)

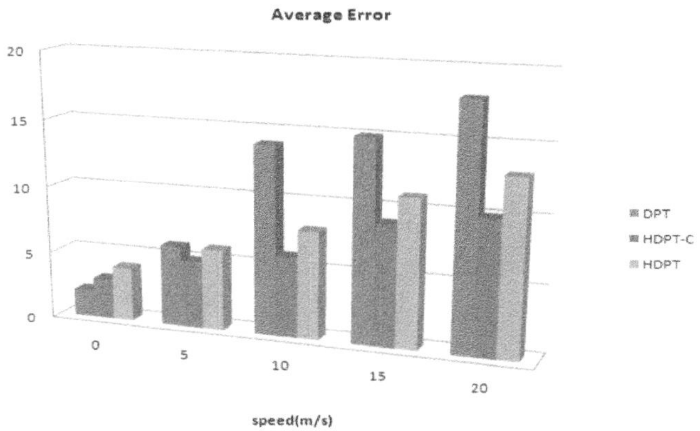

Fig. 9. Average error of prediction algorithms in different mobility

The size of a control message for exchanging the required information between cluster heads is about 20 bytes. The size of a sensing report generated by each node sensing the target is 40 bytes. Moreover, we adopt the energy consumption model presented in [34]. Our simulations consisted of three different scenarios. In the first scenario, we study error percent of the proposed prediction algorithm. The second scenario studies impact of varying of mobility and in the last scenario impact of density of nodes is probed. The simulation study mainly concentrates on two performance metrics: missing ratio, and energy consumption. The missing ratio is the probability of missing the target as the target moves in the network. The energy consumption refers to the energy consumed for transmitting control messages and sensing reports.

"Fig.9" presents error values of DPT and HWDPT protocols in different speeds, which the proposed method overcomes to DPT. As shown in "Fig.10" HWDPT protocol performs better than HCTT and EDPT protocols in high speed movement of the target due to its efficient prediction algorithm by Haar wavelet.

In the second scenario, we study performance issues (e.g. spent energy and miss rate) under different speed of mobile target. More speed leads to more error in prediction algorithms. "Fig.11" depicts the spent energy value of different algorithms. HWDPT is the proposed algorithm without recovery mechanism. The proposed HWDPT-rec has the same rate with HCTT while it spends very lower energy than HCTT.

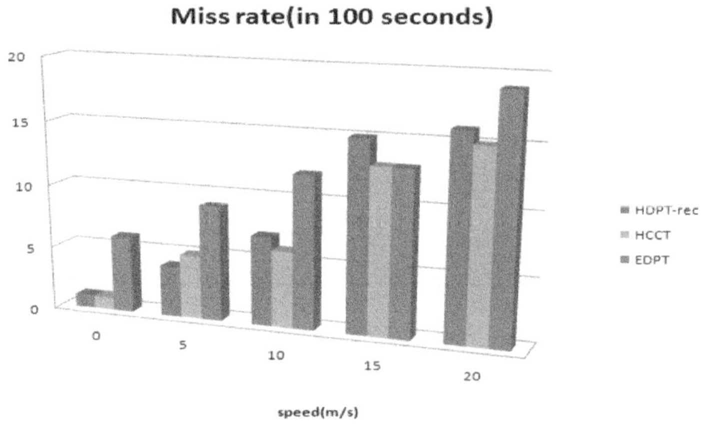

Fig. 10. Miss rate of different algorithms as function of speed

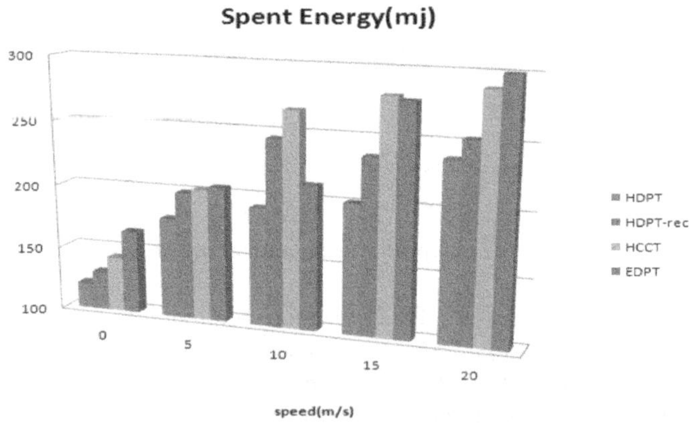

Fig. 11. Value of spent energy for different algorithms as function of speed

In the third scenario, we simulate the proposed methods under various member nodes in a wireless sensor network. Large number of sensor nodes in a typical network result in more facility to track mobile targets with lower possibility of losing targets. In other hand, large number of sensors lead to more energy consumption in different stages of tracking algorithm. "Fig.12" and "fig.13" present energy waste and miss rate for different member nodes.

Our simulation describes the proposed method has considerable improvement in spent energy in comparing to tracking algorithms while it has neglectable increase in target miss rate issue.

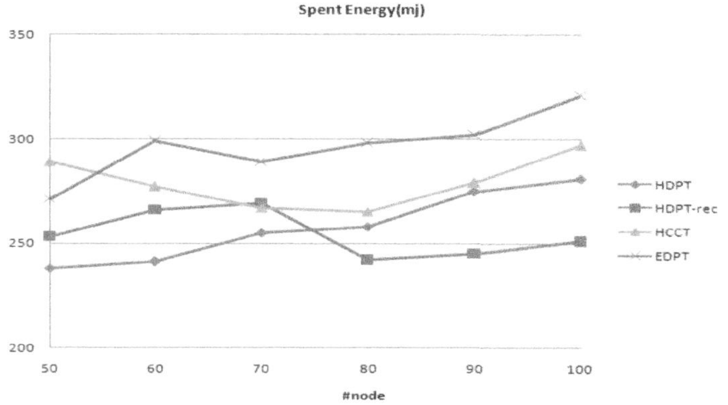

Fig. 12. Value of spent energy by tracking algorithms in different mobility

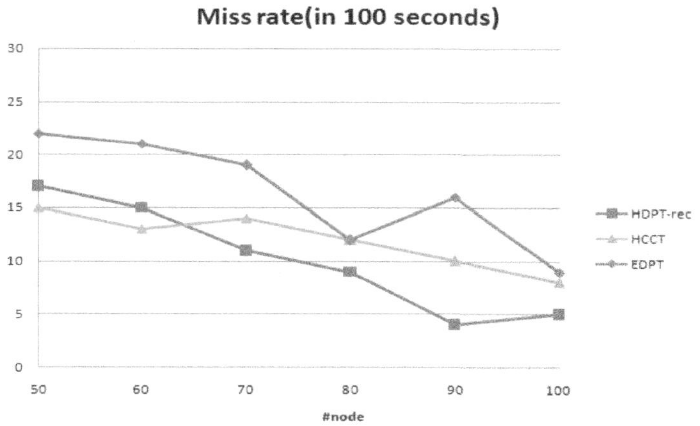

Fig. 13. Miss rate of tracking algorithms in different mobility

7 Conclusion and Future Works

In this paper, we propose a novel and fully distributed tracking scheme in cluster-based WSNs. The proposed method uses Haar wavelet transform to predict next location of a mobile target with minimum cost. Proposed algorithm increases accuracy of tracking and decreases energy consumption. Efficiency of the proposed protocol is confirmed by the simulation results.

References

1. Culler, D., Estrin, D., Srivastava, M.: Overview of sensor networks. IEEE Computer, Special Issue in Sensor Networks (August 2004)
2. Estrin, D., Culler, D., Pister, K., Sukhatme, G.: Connecting the physical world with pervasive networks. IEEE Pervasive Computing 1(1), 59–69 (2002)
3. Alhmiedat, T.A., Yang, S.-H.: A Survey: Localization and Tracking Mobile Targets through Wireless Sensors Network (2007) ISBN: 1-9025-6016-7- PGNet
4. Shorey, R., Ananda, A., Chan, M., Ooi, W.: Mobile, Wireless, and Sensor Networks. John Wiley & Sons, Canada (2006)
5. Hu, L., Evans, D.: Localization for mobile sensor networks. In: Proceedings of the Annual International Conference on Mobile Computing and Networking, MOBICOM, pp. 45–57 (2004)
6. Aslam, J., Butler, Z., Crespi, V., Cybenko, G., Rus, D.: Tracking a moving target with a binary sensornetwork. In: Proc. ACM Int. Conf. Embedded Networked Sensor Systems, SenSys (2003)
7. Brooks, R.R., Ramanathan, P., Sayeed, A.M.: Distributed target classification and tracking in sensornetwork. Proc. IEEE 91(8) (2003)
8. Chu, M., Haussecker, H., Zhao, F.: Scalable information-driven sensor querying and routing for ad hoc heterogeneous sensor networks. Int. J. High Perform. Comput. Appl. 16(3) (2002)
9. Liu, J., Chu, M., Liu, J., Reich, J., Zhao, F.: Distributed state representation for tracking problems insensor networks. In: Proc. 3rd Int. Symp. Information Processing in Sensor Networks, IPSN (2004)
10. Liu, J., Liu, J., Reich, J., Cheung, P., Zhao, F.: Distributed group management for track initiation and maintenance in target localization applications. In: Proc. Int. Workshop on Information Processing in Sensor Networks, IPSN (2003)
11. Mechitov, K., Sundresh, S., Kwon, Y., Agha, G.: Cooperative Tracing with Binary - DetectionSensor Networks. Technical report UIUCDCS-R-2003-2379, Computer Science Dept., Univ. Illinoisat Urbaba, Champaign (2003)
12. Wang, Q.X., Chen, W.P., Zheng, R., Lee, K., Sha, L.: Acoustic target tracking using tiny wireless sensor devices. In: Proc. Int.Workshop on Information Processing in Sensor Networks, IPSN (2003)
13. Zhang, W., Cao, G.: Dctc: Dynamic convoy tree-based collaboration for target tracking in sensor networks. IEEE Trans.Wireless Commun. 11(5) (2004)
14. Zhang, W., Hou, J., Sha, L.: Dynamic clustering for acoustic target tracking in wireless sensornetworks. In: Proc. 11th Int. Conf. Network Protocols (ICNP). IEEE (2003)
15. Zhao, F., Shin, J., Reich, J.: Information-driven dynamic sensor collaboration for tracking applications. IEEE Signal Proces. Mag (2002)
16. Schenato, L., Oh, S., Sastry, S.: Swarm coordination for pursuit evasion games using sensor networks. In: Proc. of the International Conference on Robotics and Automation, Barcelona, Spain (2005)
17. Sinopoli, B., Sharp, C., Schenato, L., Schaffert, S., Sastry, S.: Distributed control applications within sensor networks. Proceedings of the IEEE 91(8), 1235–1246 (2003)
18. Hightower, J., Borriello, G.: Location systems for ubiquitous computing. Computer 34(8), 57–66 (2001)
19. Li, J., Zhou, Y.: Target Tracking in Wireless Sensor Networks. Wireless Sensor Networks: Application-Centric Design (2010)

20. Bhatti, S., Xu, J.: Survey of Target Tracking Protocols using Wireless Sensor Network. In: Fifth International Conference on Wireless and Mobile Communications (2009)
21. Rapaka, A., Madria, S.: Two energy efficient algorithms for tracking targets in a sensor network. Wireless Communication Mobile Computing 12(7), 809–819 (2007)
22. Soe, K.T.: Increasing Lifetime of Target Tracking Wireless Sensor Networks. World Academy of Science, Engineering and Technology 42 (2008)
23. Lin, C.-Y., Tseng, Y.-C.: Structures for In-Network Moving Target Tracking in Wireless Sensor Networks. IEEE Broadnets (2004)
24. AL-Ghanem, W., Mahgoub, I., Ilyas, M.: Energy Efficient Cluster-Based Target Tracking Strategy. IEEE (2009) 978-1-4244-5995-7109
25. Alaybeyoglu, A., Dagdeviren, O., Erciyes, K., Kantarci, A.: Performance Evaluationof Cluster-based Target Tracking Protocols for Wireless Sensor Networks. IEEE METU Northern Cyprus Campus (2009)
26. Lee, I.-S., Fu, Z., Yang, W., Park, M.-S.: An Efficient Dynamic Clustering Algorithm for Target Tracking in Wireless Sensor Networks. Complex System and Application-modeling, Computer and Simulation, DCDIS Series B 4(S2) (2007)
27. Yang, H., Sikdar, B.: A Protocol for Tracking Mobile Targets using Sensor Networks. In: Proc.of IEEE IWSNPA, pp. 71–81 (2003)
28. Xue, L., Liu, Z., Guan, X.: Prediction-based protocol for mobile target tracking in wireless sensor networks. Journal of Systems Engineering and Electronics 22(2), 347–352 (2011)
29. Wang, Z., Lou, W., Wang, Z., Ma, J., Chen, H.: A Novel Mobility Management Scheme for Target Tracking in Cluster-Based Sensor Networks. Springer, Heidelberg (2010)
30. Li, D., Wong, K., Hu, Y.H., Sayeed, A.: Detection, classification and tracking of targets. IEEE Signal Processing Magazine 17-29 (March 2002)
31. Tsai, H.-W., Chu, C.-P., Chen, T.-S.: Mobile Target Tracking in Wireless SensorNetworks (2011)
32. Xu, Y., Winter, J., Lee, W.-C.: Prediction-based Strategies for Energy Saving in Target Tracking Sensor Networks. In: Proceedings of the IEEE International Conference on Mobile Data Management (MDM 2004), Berkeley, California, pp. 346–357 (2004)

The Design and Implementation of a DTN Naming System

Lishui Chen[1,3], Songyang Wang[2], Zhenxi Sun[2], Changjiang Yan[1], and Huijuan Rao[2]

[1] Science and Technology on Information Transmission and Dissemination in Communication Networks Laboratory, Shijiazhuang 050081, China
wangsongyang19@126.com

[2] School of Computer Science and Engineering, Beihang University Beijing 100191, China
1812458408@qq.com

[3] Department of Communication Engineering, Harbin Institute of Technology Harbin 150001, China
842337235@qq.com

Abstract. This paper presents the design and implementation of the naming mechanism (NAME), a resource discovery and service location approach for Delay/Disruption-Tolerant Network (DTN). First discuss the architecture of NAME mainly including Name Knowledge Base, Name Dissemination, Name Resolution and Name-based Routing. In the design and implementation of NAME, we introduce the simple name-specifiers to describe name, the name-tree for name storage and the efficient predicate-based routing algorithm. Future work is finally discussed for completing NAME and providing APIs for abundant applications.

Keywords: Naming Mechanism, Delay/Disruption-Tolerant Network, Name-specifiers, Predicate-based Routing.

1 Introduction

Delay/Disruption-Tolerant Network (DTN) is an architecture and a set of protocols that enable communication in environments with intermittent connectivity and long delays [1]. Bundle Protocol (BP) [2] and an architecture for DTN [3] have been previously defined. To provide its services, BP sits at the application layer of some number of constituent internets, forming a store-and-forward overlay network. Key capabilities of BP include:

- Custody-based retransmission
- Ability to cope with intermittent connectivity
- Ability to take advantage of scheduled, predicted, and opportunistic connectivity (in addition to continuous connectivity)
- Late binding of overlay network endpoint identifiers to constituent internet addresses

D. Nagamalai et al. (Eds.): *Adv. in Comput. Sci., Eng. & Inf. Technol.*, AISC 225, pp. 221–229.
DOI: 10.1007/978-3-319-00951-3_22

In a DTN, nodes and services can appear, move, and disappear dynamically. That makes nodes impossible have the whole information about the state of addresses names and routes. After a bundle is created, the destination node may have changed. In such a flexible environment, it is significant to use name attributes of nodes (such as roles, location, or sensed values) and canonical DTN endpoint identifiers to locate nodes.

Canonical EID refers to the EID of a bundle processing entity that is capable of receiving bundles addressed to that EID from other DTN nodes. Every DTN node should possess a unique EID. Naming mechanism's main target is to bind the name attributes to the canonical EID.

The main contribution of our work is the design and implementation of NAME, a naming mechanism for Delay/Disrupted-Tolerant network. The complete architecture of NAME, mainly including Name Knowledge Base, Name Dissemination, Name Resolution and Name-based Routing, is designed as a base for realizing NAME system. In the design and implementation of NAME, we introduce the simple name-specifiers to describe name, the name-tree for name storage and the efficient predicate-based routing algorithm. In future work, we will complete our DTN naming system, and design APIs for all kinds of applications such as content-based data dissemination and resource location.

The rest of this paper is organized as follows. Section 2 discusses the architecture of DTN naming system, Section 3, the design and implementation of naming mechanism NAME. Finally in Section 4, we conclude and introduce future work.

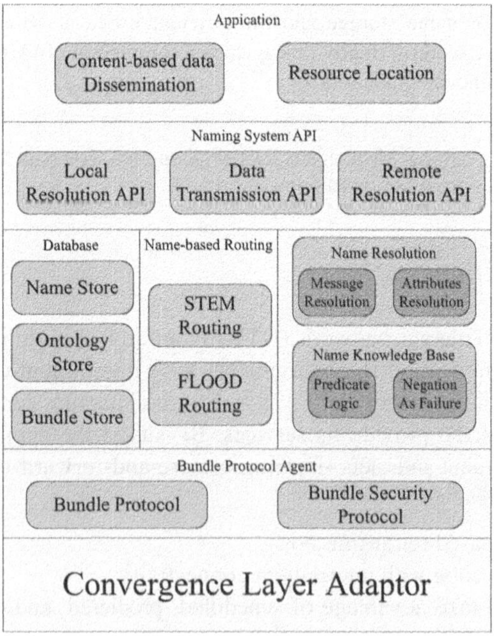

Fig. 1. The architecture of DTN naming system

2 The Architecture of NAME

In this section, we discuss the whole architecture of NAME, which mainly consists of eight modules: Applications (APP), Naming system API, Database (DB), Name-based routing, Name resolution, Name knowledge base (NKB), Bundle protocol agent (BPA) and Convergence layer adaptor (CLA). The architecture is displayed in Fig.1.

2.1 Name Knowledge Base

Each node maintains a Name Knowledge Base for storing and processing ontology as well as attributes. NKB can contain information about bindings of name attributes to locally registered application End Identifiers (EIDs), or to remote related nodes' EIDs. The Name knowledge base can range from a simple table, matching name attributes to nodes' EIDs, to a powerful deductive database engine (such as Prolog) that can use predicate logic and negation as failure to infer complex derived attributes.

The range of possible knowledge base implementation is broad and is likely to be the subject of significant research and experimentation. NKB is made of the following parts [1]:

- A simple lookup table mapping ontology names (e.g. geographic) to nodes that have advertised the corresponding ontology. When an ontology advertisement arrives it is inserted into the table.
- A lookup table matching name attributes to nodes' EIDs. The knowledge base can perform simple matches and comparisons to improve its forwarding (choice of related node) decisions. This mapping is possible if the corresponding ontology rules already reside on the current node.
- A deductive database that stores facts about name attributes as well as rules used for attributes derivation. This kind of KB can be used for table matching, as well as execute a potentially complex rule to infer the result from a given fact base.

2.2 Name Dissemination

The procedure of Name dissemination is as follows. Applications first register in the bundle protocol agent and add attribute' name and attribute's value by using name system API. The attributes' information is stored in the local name knowledge base. Based on the name information offered by applications, bundle protocol agent can infer and extent name information. The name knowledge base in every node should periodically disseminate information to other neighbor nodes.

The interaction among BPA, Resolver and Router is displayed in Fig.2.

1. An application registers in BPA
2. BPA sends name information to Resolver
3. Resolver updates the local NKB, extracts name attributes and resolves these attributes
4. Resolver sends name attributes and resolution information to Router

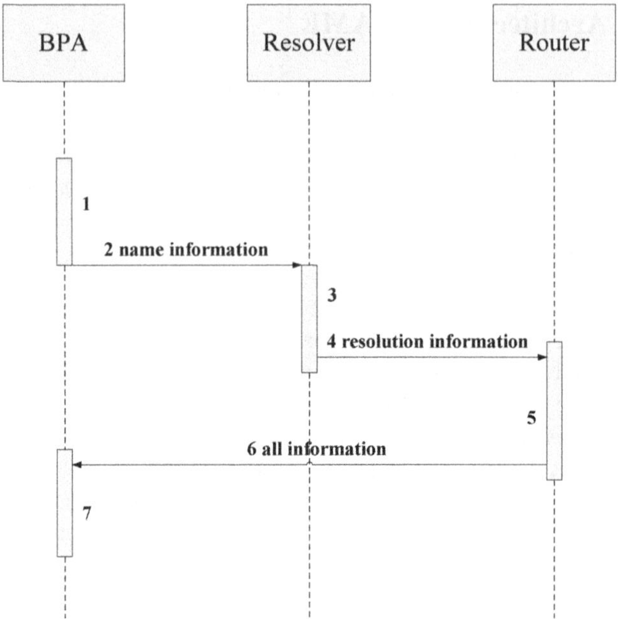

Fig. 2. The process of data dissemination

5. Router prepares to disseminate name based on the resolution information.
6. Router sends all the information to BPA
7. BPA creates a bundle and inserts it to CLA

2.3 Name Resolution

Name Resolution (also called as binding) is the process of matching a name to several destination nodes' EIDs, or to other names, or to related nodes' EIDs. In the process of name resolution, the bundle coming from local application or remote application is first stored in local DB, and then triggers the event BundleReceived. Router queries Resolver for the EIDs of the next related nodes and sends the bundle to these nodes. If some nodes cannot resolve the coming information, they should use bundle-in-bundle technique to transfer the bundle. In bundle-in-bundle technique, the nodes that are not able to resolve the current bundle could first encapsulate it in an "envelope bundle", then address this "envelope bundle" to the next resolver EID.

The interaction among BPA, Resolver and Router is displayed in Fig.3.

1. An application registers a bundle in BPA
2. Trigger the event BundleReceived
3. Resolver queries local name knowledge base and sends the information to Router
4. Router asks Resolver to resolve and infer name attributes
5. Resolver sends resolution information back to Router
6. Router sends all the information including routing approaches to BPA
7. BPA creates a bundle and inserts it to CLA

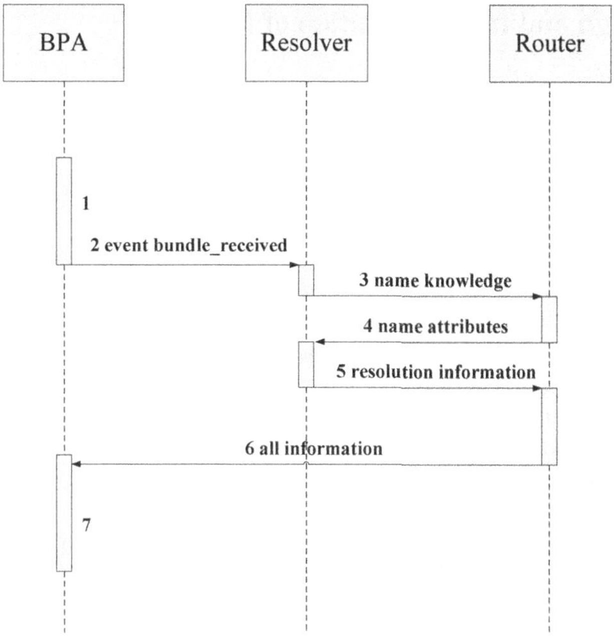

Fig. 3. The process of name resolution

2.4 Name-Based Routing

There have been numerous investigations about routing in a DTN. In a naming system, the main mission of DTN router is finding the destination nodes that match the name information in the bundle. DTN router should also make an appropriate decision to transfer a bundle based on the resolution clues. In all, the capability of DTN router directly determines the complex of a naming system.

Name is used to describe the features of destination nodes. Only by making full use of name information and other local information can router finish its function. Naming and routing are inseparable. The simplest routing is early-binding that determines the EIDs of destination nodes in the source node. The EIDs information is usually stored in the Metadata Extension Block (MEB).

2.5 Name-Based Routing

BPA of a node is the node component that offers the BP services and executes the procedures of the bundle protocol. The manner in which it does so is wholly an implementation mater. For example, BPA functionality might be coded into each node individually; it might be implemented as a shared library that is used in common by any number of bundle nodes on a single computer; it might be implemented as a daemon whose services are invoked via inter-process or network communication by any number of bundle nodes on one or more computers; it might be implemented in hardware.

3 Design and Implementation of NAME

In this section, we present the details of design and implementation of our naming system based on the above architecture of DTN naming system.

3.1 Overview

Our naming mechanism (NAME) resolves name mainly based on the geographic information. We assume that every node has gotten their geographic information (longitude and latitude) through GPS device. The name in our system is made up of role attributes, location attributes and distance attributes, for example, "generals (role) located within 2 km (distance) of 116 degrees (longitude), 112 degrees (latitude)".

In the NAME, name resolution is completed during the transmission of a bundle instead of in the source node. At the beginning, NAME determines some closer nodes by using geographic information and transfer the bundle to them. This process will not stop until the bundle reaches the target region. When the bundle is within the target region, NAME will use other attributes to find the destination nodes.

The routing state can be STEM or FLOOD [4]. When it is STEM, NAME makes use of less network resource to send the bundle to the target region. When the bundle is within the target region, the routing state will change as FLOOD to take advantage of abundant resource. FLOOD is predicate-based epidemic routing algorithm. The process of routing is displayed in Fig.4.

In the predicate-based epidemic routing algorithm, we hope the bundle could reach the target region within (longitude, latitude, distance). This limitation can be realized by deploying predicate filter. During the implementation of NAME, the information of routing state and routing control will be stored in the MEB.

3.2 Name Format

Our naming system uses name-specifiers [5] to express the meaning of a name. The source node puts name-specifies in the head of bundle to determine the destination node. Name-specifiers are designed to be simple and easy to operate. The two main parts of the name-specifier are the attribute and the value. An attribute can classify an object into a category, for example, 'color'. A value is the object's classification within that category, for example, 'red'. Attributes and values are free-form strings that are defined by applications; name-specifiers do not restrict applications to using a fixed set of attributes and values. An attribute and its associated value together form an attribute-value pair or av-pair.

A name-specifier is a tree arrangement of av-pairs such that an av-pair only depends on its parent av-pairs. For instance, in the example name-specifier shown in Fig.5, the av-pair "mission = command" depends on the "role = general"; the av-pair "longitude = 116 degrees" is dependent on the av-pair "latitude = 112 degrees", and is meaningful only when the "information" is about "location".

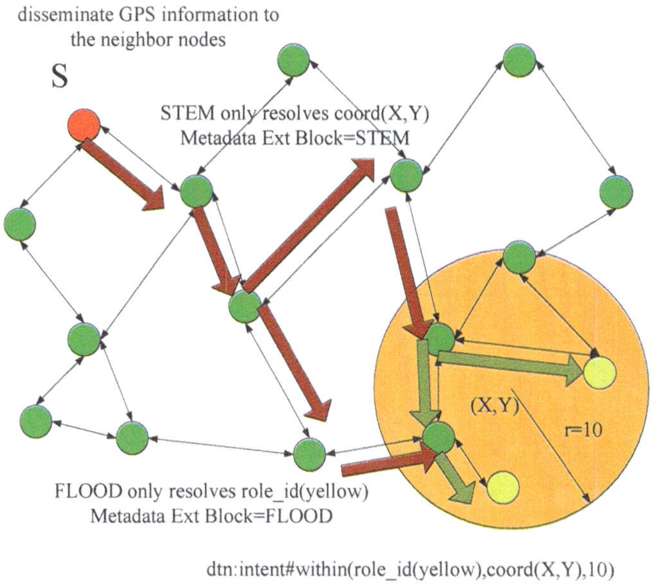

Fig. 4. The process of routing in NAME

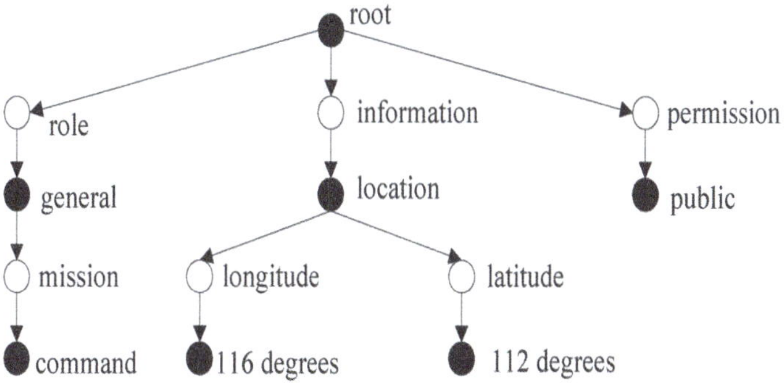

Fig. 5. Name-specifier tree

In the head of a bundle, the format of name-specifiers is shown in Fig.6.

```
[ role = general  [ mission = command ] ]
[ information = location [ longitude = 116 degrees]
                         [ latitude = 112degrees ] ]
[ permission = public ]
```

Fig. 6. The format of name-specifier in a bundle

3.3 Name Storage

We use data structure Name-trees to store the correspondence between name-specifiers and name-records. The format contains the destination node's EID, the next-hop nodes' EID and the life circle. Similar to the structure of name-specifers, Name-trees also consist of several av-pairs. The difference is that in the name-tree, an attribute can have many values. Each of these name-specifiers has a pointer from each of its leaf-values to a name-record. Fig.7 depicts an example name-tree.

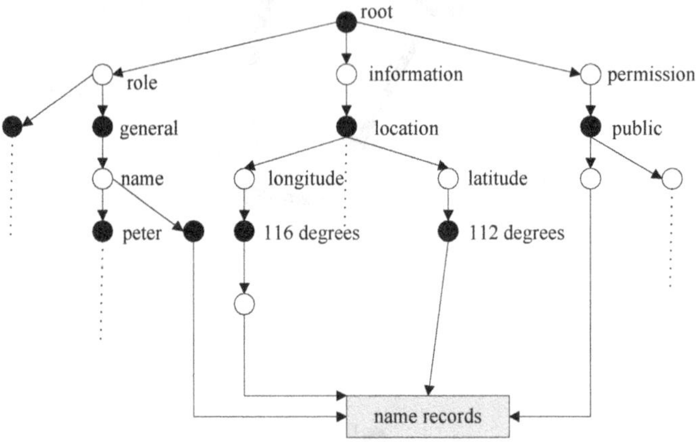

Fig. 7. Name-tree storage

3.4 NAME Routing Algorithm

The NAME Routing Algorithm is as follows:

NAME Routing Algorithm

Pick a bundle B_i in pending_bundle_table (B_i):

if dis(B_i .location, B_i .target_location) > B_i . target_radius:

B_i .route_state = STEM

for each Node N_j in node_table(N_j):

if dis(N_j .location, B_i .target_location) < min

min = dis(N_j .location, B_i .target_location)

Node_tag = min_node

forward B_i to min_node

else

change B_i .route_state from STEM to FLOOD

for each Node N_j in node_table(N_j):

if B_i have a direct link to Node N_j

forward bundle B_i to Node N_j

end

The original node first uses predicate expression "within (location, distance)" to narrow the scope of target region. In other words, if the distance between the original node and a target node is greater than target_radius which is a constant value, we will choose the routing state STEM. The bundle is to be transferred to nodes that are closer to the target node. This process will not stop until the bundle locates in the target region, then the routing state changes as FLOOD. In this routing state, the bundle will be delivered to all the nodes in the list of next hop nodes, which are within the target region.

4 Conclusion

In this paper, we first discuss the whole architecture of DTN naming mechanism. And then introduce our design and implementation of NAME. Our goals are to create an expressiveness, responsiveness, robustness and easy configuration naming mechanism. NAME uses a simple naming strategy based on attributes and values to finish locating resource and name resolution. We also design a predicate-based routing algorithm which is simple and efficient.

In future work, we will complete our DTN naming system, and design APIs for all kinds of applications such as content-based data dissemination and resource location.

Acknowledgment. This work is supported by the National Science Foundation of China under Grant No. 61073076, the Science and Technology on Information Transmission and Dissemination in Communication Networks Laboratory under Grant No.ITD-U12001, and Beihang University Innovation and Practice Fund for Graduate under Grant No. YCSJ-02-02. The authors would like to thank great support.

References

1. Basu, P., Brown, D., Polit, S., Krishnan, R.: Intentional Naming in DTN draft-pbasu-dtnrg-naming-00 (November 23, 2009)
2. Scott, K., Burleigh, S.: Bundle Protocol Specification. RFC 5050 (November 2007)
3. Cerf, V., Burleigh, S., Hooke, A., Torgerson, L., Durst, R., Scott, K., Fall, K., Weiss, H.: Delay-Tolerant Networking Architecture. RFC 4838 (April 2007)
4. Basu, P.: BBN Technologies. Intentional Naming and Deferred Binding in DTN. DTNRG meeting, IETF 71 Philadelphia (March 13, 2008)
5. Adjie-Winoto, W., Schwartz, E., Balakrishnan, H., Lilley, J.: The design and implementation of an intentional naming system. In: Proc. ACM SOSP 1999 (December 1999)

Taking Differences between Turkish and English Languages into Account in Internal Representations

Tassadit Amghar[2], Bernard Levrat[1,2], Sultan Turhan[1], and Burak Parlak[1]

[1] LERIA, Universit dAngers, Angers, France
[2] Galatasaray University, Department of Computer Engineering, Ciragan Cad. No:36 34357 Ortakoy, Istanbul, Turkey
amghar@info.univ-angers.fr

Abstract. It is generally assumed that the representation of the meaning of sentences in a knowledge representation language does not depend of the natural language in which this meaning is initially expressed. We argue here that, despite the fact that the translation of a sentence from one language to another one is always possible, this rests mainly on the fact that the two languages are natural languages. Using online translations systems (e.g. Google, Yandex translators) make it clear that structural differences between languages gives rise to more or less faithful translations depending on the proximity of the implied languages and there is no doubt that effect of the differences between languages are more crucial if one of the language is a knowledge representation language. Our purpose is illustrated through numerous examples of sentences in Turkish and their translation in English, emphasizing differences between these languages which belong to two different natural language families. As knowledge representations languages we use the first order predicate logic (FOPP) and the conceptual graph (CG) language and its associated logical semantics. We show that important Turkish constructions like gerunds, action names and differences in focus lead to representations corresponding to the reification of verbal predicates and to favor CG as semantic network representation language, whereas English seems more suited to the traditional predicates centered representation schema. We conclude that this first study give rise toideas to be considered as new inspirations in the area of knowledge representation of linguistics data and its uses in natural language translation systems.

Keywords: Knowledge representation, Effect of natural languages differences, Turkish, Reification, Conceptual graph.

1 Introduction

It is generally assumed that the representation of the meaning of sentences in a knowledge representation language does not depend of the natural language in which this meaning is initially expressed. We argue here that, despite the fact that the translation of a sentence from one language to another one is always possible, this rests mainly on the fact that the two languages are natural languages. Using translations systems

D. Nagamalai et al. (Eds.): *Adv. in Comput. Sci., Eng. & Inf. Technol.*, AISC 225, pp. 231–241.
DOI: 10.1007/978-3-319-00951-3_23

(e.g. Google, Yandex translators) makes it clear that structural differences between languages gives rise to more or less faithful translations depending on the proximity of the implied languages and there is no doubt that effect of the differences between languages are more crucial if one of the language is a knowledge representation language.

Therefore, this study aims at showing through the representations of some illustrative examples of sentences in Turkish and their corresponding translation into English, emphasizing differences between these languages, that the Turkish and English version corresponds to different structures each of them rendering the way each language has to depict the word.

As knowledge representation language we use the first order predicate logic (FOPP) and the conceptual graph language (CG) introduced by J. Sowa [REF] and its associated logical semantics. We show that important Turkish constructions like gerunds, action names and differences in focus lead to representations corresponding to the reification of verbal predicates and to favor CG as semantic network representation language. For their part, English seems more suited to the traditional representations where the logical structure of formulae renders the predicates centered representation schema of these natural languages.

The first paragraph is devoted to the representation languages and since FOPL language is commonly well known we only give a sketchy presentation of it and a more detailed one for the conceptual graphs language (J. Sowa 84).

The following paragraphs each tackles some important Turkish structures whose from which corresponding translation in English differs greatly and the consequences on the representation in each language. We will in turn analyze the representations of gerunds, and action names constructions, structures induced by the lack of auxiliary to be and to have and finally expression of moods, and other feelings.

We conclude that this first study furnishes interesting ideas to be considered as new inspirations in the area of knowledge representation of linguistics data and its uses in natural language processing. This a primary work in this direction in the semantic field but we are currently studying other aspects of Turkish language (mor phology, syntactic construction) which also deserves specific treatments.

2 Methodology

Knowledge representation languages are classically classified into two families: The logical languages on one hand and the labeled graph languages on the other hand.

- **The logic based language family:** This language family contains standard and non standard logic languages, but often, if soundness of reasoning is crucial for the application, classical logical languages are preferred due to a well grounded semantics and rather good tractability properties.
- **The graph modeling language family:** Also known as semantic nets [Findler, 1970] they are mainly inspired by psychological works [Collins&Quillian,1969]. If their expressive power is greater than the preceding representation languages, they loose in counterpart some formal properties in the definition of their semantic.

Nevertheless, to characterize this last one, a logical semantic in terms of first order predicate formulae is generally associated with them when it is possible.

2.1 First Order Logic

In this paper, we started to adapt a first order logical formalism and its reification to create a model the tuples in Turkish and English sentences. The relationship between the agent and the object was designed through the verbs. In an sentence like X is doing Y, the predicative first order expression will be written as do(X,Y) and X Y işini yapıyor as yapmak(X,Y), respectively. This relationship is generally represented as It exists R, R(x1,x2) similarly for both languages. On the other hand, the reification procedure re-expresses the same sentence in a distributive way.

2.2 Conceptual Graph Formalism

CGs are a knowledge representation language well known for its ability to cope with natural language data due to the direct mapping to language they permit [Sowa, 2000]. They currently conveniently serve as an intermediate language in applications between computer-oriented formalisms and natural language. We only focus here on the component of CGs necessary to I U@the presentation of the main ideas underlying our study

Definition 1: A conceptual graph is a finite, connected, bipartite graph with nodes of a first kind called *concepts* and nodes of the second kind called *conceptual relations*.

Definition 2: Every conceptual relation has one or more arcs, each of which must be attached to a concept. If the relation has n arcs, it is said to be n-adic and its arcs are labeled 1,2,...,n

In the graphical representation, concepts are rectangles and relations are circles

Axiome 1: There is a set T of type labels and a function *type*, which maps concepts and conceptual relations into T. T is a partially ordered by a subsumption relation $\leq$. Practically T is often taken as being a complete lattice in case of simple inheritance assumption between types. If a is a concept, *type*(a)=ta if ta denotes the lowest type a belongs to.

Axiome 2: There is a set I={i1, i2, …, in} of individual makers and a *referent* application from the concept set to $I \cup \{*\}$ where *referent*(a)=ij denotes an a particular individual ij. In this case *a* is said to be an *individual concept.* If referent(a)=* a is said to be a generic concept.

There is different notation of GCs, a linear and a graphical notation. More over there is a logical semantic associated with GCs. Each conceptual graph is associated with a formula in FOPL. Concept types correspond to unary predicates. For example, **[human** :*] *has exist (x) human(x),* as associated semantic and **[human :"Burak"**], has human(Burak) as associated formulae. (individual markers correspond to logical constants, generic marker correspond to existentially quantified variables. *n*-ary relation correspond to *n*-ary predicates for example : [**human** :*] <---(owns)---> [**car** :*] *has exist(x), exist (y) (human(x) & car(y) & owns(x,y)* as logical semantic.

GCs permits to define lambda abstraction a way to define types from existing ones. Generally new types defined by lambda abstractions use the Aristotelian way to define types. They can be used to abstract CGs to have a general view or to expand an existing CGs by replacing a defined type by its associated lambda abstraction.

Two operations expansion and contraction consist in replacing a defined type in a graph by the body of its definition in its associated lambda abstraction for expansion and replacing the body of the definition of a defined type by its name as it is defined in its lambda abstraction.

For example, if we have a type person in the type set we can define the type Turkish learner as a person which learns Turkish by the following lambda abstraction:

Define type Turkish_learner (x) is
[**human** :*]←(agt)→[**to learn** :*] ← (obj) → [language : "**Turkish**"] ;

3 Results

In this section, we try to focus on the examples to demonstrate what the theoretical methods cited on the previous sections mean in the sentence and how these representation forms convert to an analysis. We examine our methods predicate logic and conceptual graph on Turkish and English sentences.

Our supports are linguistic differences between Turkish and English languages and the hypothesis that it would be more accurate to use different methods while identifying the computer based representation form of these two languages.

In a typical Turkish sentence, the verb includes all the information about the subject, time and other qualities. For this reason, a representation model which describes the entities with process logic should be chosen. When we evaluate the verb in the sentence concept, all this information set must be handled with the reification approach in order to concretize this discrete concept with qualitative properties as well as to cite all the on-going attributes (e.g. subject, complement, time and location) hidden in this verb.

For example; when we evaluate the English sentence "John eats an apple" with predicate logic, the representation model will be:

$$eat(John, apple)$$

If we evaluate the same sentence with reification method, with the basic logical operators, the representation model become as follows:

$$\exists m\ (fact\_of\_eating(m) \wedge agent(m, John) \wedge object(m, apple))$$

m: A specific event → The fact that Jean eats

If we evaluate the Turkish translation of the same sentence, "John bir elma yiyor", with the same method, the output will be

$$yemek(John, elma)\ (*)$$

As in this example, the linguistic differences between Turkish and English are not very recognizable; the results are similar to each other. However, when we take into consideration the dominant role of the verb in Turkish sentence, it is obvious that Turkish is closer to be evaluating with reification method. The verbal "X"-me/ma belonging to the verb "X"-mek/mak can be done in English with the transformation: the verb "to X" → the fact of "X"-ing.

Thanks to the existence of these verbals in Turkish language, we obtained the following form (**) if we passed from the previous example (*) to a reified representation format:

$$\exists m\ (yeme(m) \wedge agent(m, Jean) \wedge object(m, elma))\ (**)$$

The analysis realized on the previous example demonstrate that both predicate and reification representation method can be used for Turkish and English.

However, it may be argued that Turkish is more advantageous in terms of reification comparing to English, especially in the following cases:

- The sentences where the subject is changed directly or indirectly
- Turkish sentence where the subject is indefinite while the subject is very explicit in English translation of the same sentence.

3.1 Gerunds

Gerunds are verbal constructions which in Turkish are often used in order to replace the relative and completive proposition. These gerunds may mention more or less complete form of the verbal construction such as person. They can have several grammatical functions and play the roles of nouns, adjectives or adverbs.

The following examples of gerunds introduce briefly our systematic approach by comparing their respective translations into English. The logical and reified logical versions were given as follows;

Araba **sürdüğünde** cebi çaldı. While he was driving his mobile phone rang

∃p1,p2 (fact_of_ringing(p1)∧agent(p1,mobile)∧time(p,t1)) ∧

(fact_of_driving(p2)∧agent(p2,he)∧time(p2,t1)): fact_of_ringing (mobile,time) ∧

fact_of_driving (he,car)

∃p1,p2 (çalma(p1)∧agent(p1,mobile)∧zaman(p,t1)) ∧

(sürme(p2)∧agent(p2,he)∧time(p2,t1)): çalma (cep,zaman) ∧ sürme (o,araba)

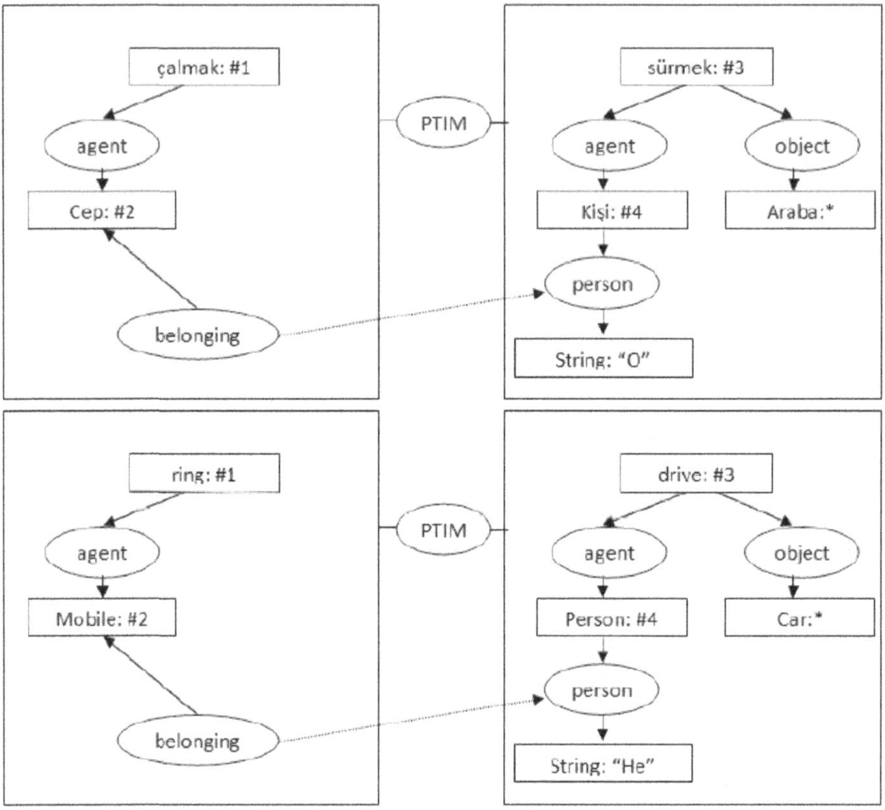

Fig. 1. Representations of *Arabasürdüğündecebi çaldı. While he was driving his mobile phone rang* with conceptual graphs where PTIM represents the concept of time.

Geldiğin zaman çayımı içiyordum. When you arrived I was drinking tea

∃p1,p2 (fact_of_arriving(p1)∧agent(p1,you)∧time(p,t1)) ∧ (fact_of_drinking(p2)∧agent(p2,I)∧object(p2,tea)): fact_of_arriving (you,time) ∧ fact_of_drinking (I,tea)

∃p1,p2 (gelme(p1)∧agent(p1,sen)∧zaman(p,t1)) ∧ (içme(p2)∧agent(p2,ben)∧object(p2,çay)): gelme (sen,zaman) ∧ içme (ben,çay)

3.2 Verbal Nouns and Noun Clauses and Moods

As gerunds these verbal nouns may play the role of relative and completive propositions and can also occur as epithets. The following examples illustrate the comparison between Turkish and English using the similar methodology.

Due to the lack of auxiliary (to be, to have) Turkish language uses noun clauses with the particles (var/yok≈there is/not)

1 Usage of indefinite subject: (The possession association)

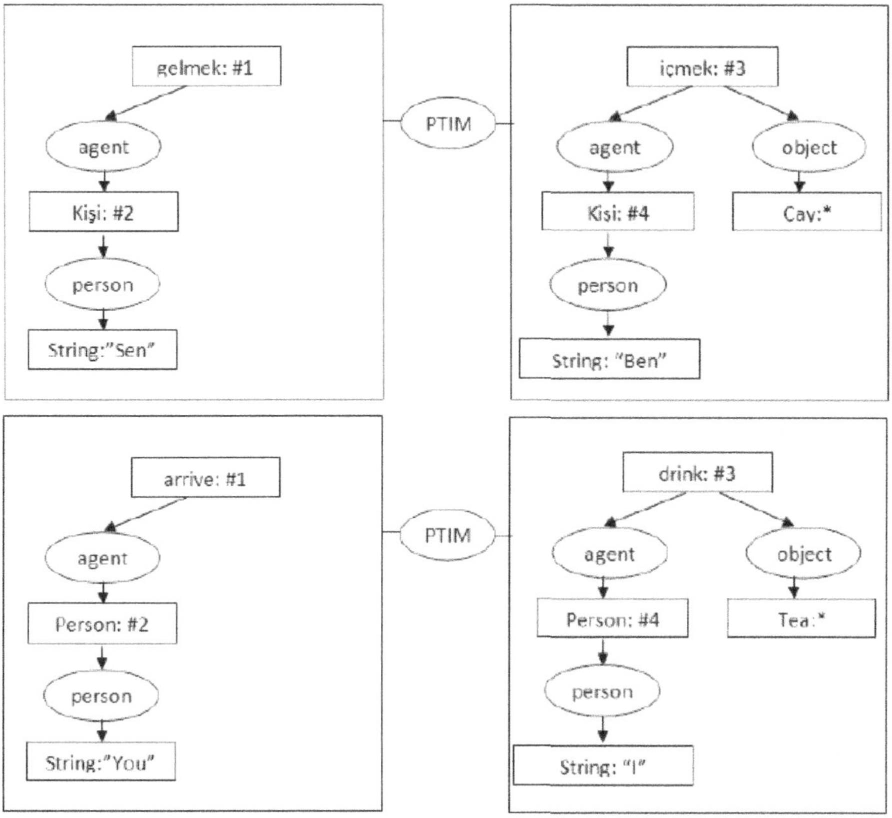

Fig. 2. Representations of ***Geldiğin zaman*** *çayımı içiyordum. When you arrived I was drinking tea* with conceptual graphs where PTIM represents the concept of time.

Mehmet'in arabası var → Mehmet has a car (or There is a car of Mehmet)

∃p(fact_of_having(p)∧agent(p,Mehmet)∧objet(p,car)) : fact_of_having (Mehmet,car)

∃p(sahip_olma(p)∧agent(p,Mehmet)∧ object(p,araba)) : sahip_olma (Mehmet,araba)

2 Usage of subject transformation: English and Turkish thematization greatly differs and this results in having different ways in expliciting agents of actions.
Canım yapmak ist(em)iyor ⇔ I do not want to do

∃ p (fact_of_wanting(p) ∧ agent(p,je) ∧ action(p,faire)) : fact_of_wanting (me,do)

∃ p (isteme(p) ∧ agent(p,can) ∧ action(p,yapmak) ∧ belonging(can,ben)): wanting (can,yapmak) ∧ belonging(can,ben)

Başım ağrıyor ⇔ I have a headache

∃ p (fact_of_having_ache(p) ∧ agent(p,me) ∧ location(p,head)) : fact_of_having_ache (me, head)

∃ p (ağrıma(p) ∧ agent(p, baş) ∧ belonging(baş,ben)) : aching (baş)∧ belonging (baş,ben)

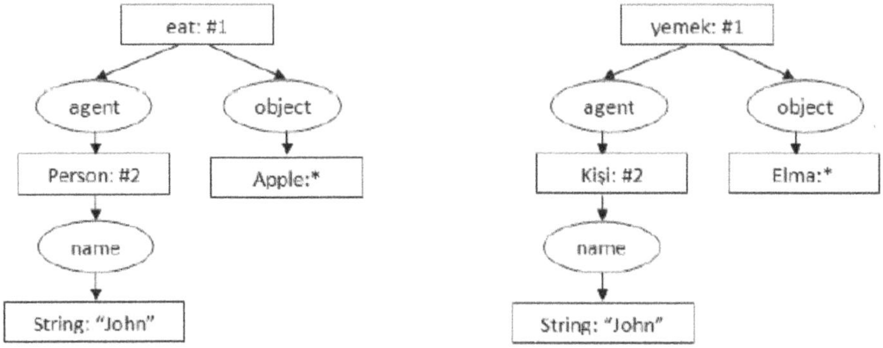

Fig. 3. Representations of *John eats an apple-Jean bir elma yiyor* with conceptual graphs

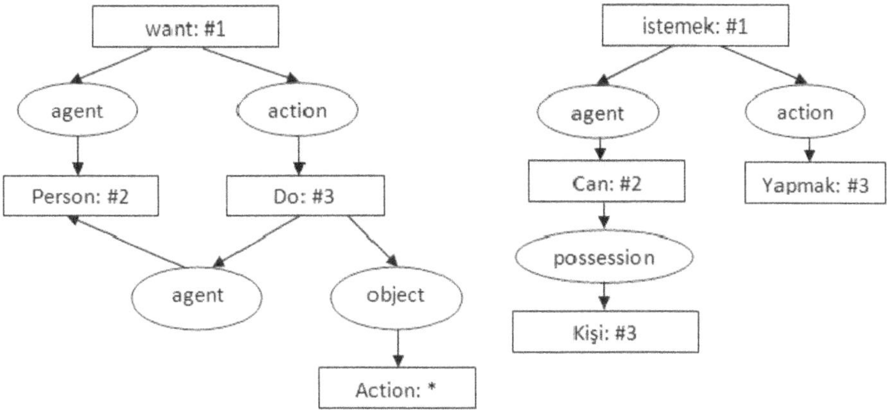

Fig. 4. Representations of *I want to do-Canım yapmak istiyor* with conceptual graphs

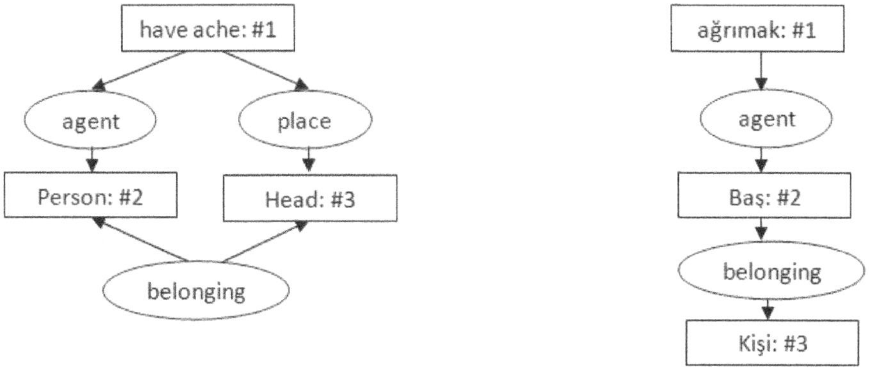

Fig. 5. Representations of *I have an headache - Başım ağrıyor* with conceptual graphs

Geç **kalanlar** sınıfa girdiler. The persons who arrived in late entered into the classroom

∃p1,p2 (fact_of_entering(p1)∧agent(p1,persons)∧location(p1,classroom)) ∧ (fact_of_arriving (p2)∧agent(p2,persons)): fact_of_entering (persons,classroom) ∧ fact_of_arriving (they)

∃p1,p2 (girme(p1)∧agent(p1,kişiler)∧location(p1,sınıf)) ∧ (geç_kalma(p2)∧agent(p2,kişiler)): girme (kişiler,sınıf) ∧ geç_kalma(kişiler)

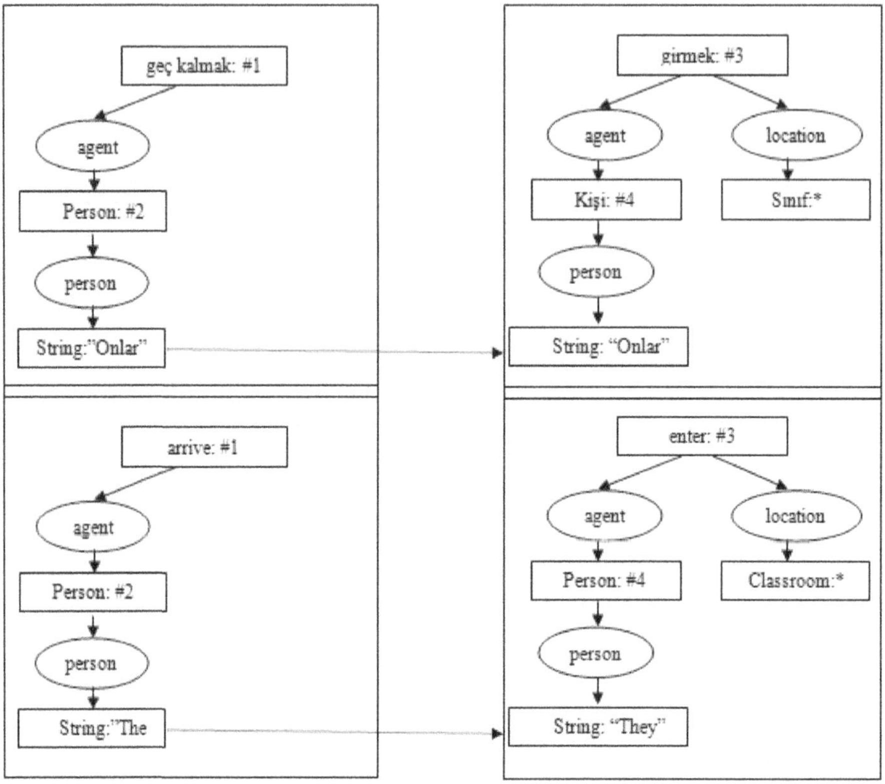

Fig. 6. Representations of Geç **kalanlar** sınıfa girdiler. The persons who arrived in late entered into the classroom with conceptual graphs.

4 Conclusion

In this study, we try to develop a new approach if the representation methods used in linguistic and NLP domains are suitable for Turkish and English which may be generally accepted as equivalent in terms of language representation, in an automatic analysis environment.

We concretize the aim of the study with the utilization differences of the subject in Turkish and English sentences. In an English sentence, the subject which is found at the beginning of the sentence is emphasized significantly, whereas in the Turkish sentence which has the same meaning, the subject will be hidden, and may either be a null subject or may be expressed with an indefinite person.

This property becomes more explicit when a predicate logic representation for English and reification representation for Turkish are preferred and transferred to automatic analysis.

Finally, we indicate how the texts can be transformed to automatic analysis for NLP by applying conceptual graphs to our examples. This graph based representation method barely known may reify naturally the sentence and so the text and render the analysis possible.

References

Sowa, J.F.: Conceptual Structures: Information processing in mind and machine. Addison-Wesley (1984)

Sowa, J.F.: Knowledge representation: logical. MIT Press (2000)

Findler, N.V.: Associative networks: The representation and use of knowledge by computers. Academic Press (1979)

Collins, A.M., Quillian, M.R.: Retrieval time from semantic memory. Journal of Verbal Learning and Verbal Behavior 8, 240–247 (1969)

Information Retrieval with Porter Stemmer: A New Version for English

Wahiba Ben Abdessalem Karaa[1] and Nidhal Gribâa[2]

[1] University of Tunis. Higher Institute of Management, Tunisia
RIADI-GDL laboratory, ENSI, National School of Computer Sciences. Tunisia
wahiba.abdessalem@isg.rnu.tn
[2] University of Tunis, Higher institute of management, Tunis. Tunisia
gribaa.nidhal@yahoo.fr

Abstract. The performance of information retrieval systems can be improved by matching key terms to any morphological variant. A stemming algorithm is a technique for automatically conflating morphologically related terms together. Several stemming algorithms exist with different techniques. The most common algorithm for English is Porter, Porter (1980). It has been widely adopted for information retrieval applications in a wide range of languages. However, it still has several drawbacks, since its simple rules cannot fully describe English morphology.

The present paper develops an improved version of Porter stemming algorithm for the English language. The resultant stemmer was evaluated using the error counting method. With this method, the performance of a stemmer is computed by counting the number of understemming and overstemming errors committed during the stemming process. Results obtained show an improvement in stemming accuracy, compared with the original stemmer.

Keywords: stemming, porter stemmer, Information retrieval.

1 Introduction

Stemming is a technique to identify different inflections and derivations of the same word in order to transform them to one common root called stem. A word's stem is its most basic form which may or may not be a genuine word. Stems are obtained by stripping words of derivational and inflectional affixes to allow matching between the ones having the same semantic interpretation. In text mining applications using stemming, documents are represented by stems rather than by the original words. Thus, the index of a document containing the words "try", "tries" and "tried" will map all these words to one common root which is "try". This means that stemming algorithms can drastically reduce the dictionary size especially for highly inflected languages which leads to significant efficiency benefits in processing time and memory requirements. Porter (1980) estimates that stemming reduces the vocabulary to about one third.

D. Nagamalai et al. (Eds.): *Adv. in Comput. Sci., Eng. & Inf. Technol.*, AISC 225, pp. 243–254.
DOI: 10.1007/978-3-319-00951-3_24

First researches concerning stemming have been done in English which has a simple morphology. A variety of stemming algorithms have been proposed for the English language such as Lovins stemmer, Lovins(1968), Paice/Husk stemmer, Paice (1990), etc. The most common algorithm for English stemming is Porter algorithm, Porter (1980). It is a rule-based and language dependent stemmer that has been widely adopted for information retrieval applications in a wide range of languages. Many studies have shown that Porter stemmer improves greatly retrieval. However, this stemmer makes common errors that may hurt retrieval performance. The present paper proposes an improved version of Porter stemming algorithm for English. The improvements concern essentially the addition of new morphological rules.

This paper is organized as follows: In Section 2, we briefly describes Porter stemmer, we present common errors made by Porter stemmer in section 3. Section 4 presents our approach. Experiments and results are given in section 5. Finally, we conclude our study and discuss future work in Section 6.

2 Porter Stemmer

Porter stemmer was developed by Martin Porter in 1980 at the University of Cambridge. It is a significant contribution to the literature since it is highly cited in diverse journals. Many of these publications appeared in information science journals, computer science literature relating to data and knowledge and some other subjects coming from widely dispersed fields such as Bioinformatics and Neuroinformatics, Willett (2006). Nowadays, Porter stemmer has become the most popular stemmer and the standard approach for stemming.

Porter stemmer is a suffix-removal algorithm. It is known to be very simple and concise. The algorithm is based on the idea that most of English suffixes are composed of other simpler ones. Thus, the stemmer is a linear step algorithm that applies morphological rules sequentially allowing removing suffixes in stages. Specifically, the algorithm has six steps. Each step defines a set of rules. To stem a word, the rules are tested sequentially. If one of these rules matched the current word, then the conditions attached to that rule are tested. Once a rule is accepted, the suffix is removed and the next step is executed. If the rule is not accepted, then the next rule in the same step is tested, until either a rule from that step is accepted or there are no more rules in that step and hence the control passes to the next step. This process continues for all the six steps; in the last step the resultant stem is returned.

The first step of the algorithm is designed to deal with inflectional morphology. It handles plurals, past participles, etc. The second step presents a recoding rule that simply transforms a terminal –y to an –i to solve problems of mismatch between similar words; ones ending in –y; and other in –i. For instance, if "happy" is not converted to "happi", the words "happy" and "happiness" will not be conjoined together after the removal of the suffix –ness from the word "happiness". Step3 and 4 deal with special word endings. They map double suffixes to single ones. So, the suffix –iveness (–ive plus –ness) is mapped to –ive. After dealing with double suffixes, Step 5 removes the remaining suffixes. The last step defines a set of recoding rules to normalize stems.

Implementations of Porter stemmer usually have a routine that computes the m measure each time there is a possible candidate for removal to judge if a suffix must be removed or no. In fact, all rules for suffix removal are of the following form: (condition) S1 ➔ S2. The condition is usually given in terms of m. The value m is the number of vowel consonant sequences in a word or word part.

3 Common Errors Made by Porter Stemmer

Although Porter stemmer is known to be powerful, it still faces many problems. The major ones are Overstemming and Understemming errors. The first concept denotes the case where a word is cut too much which may lead to the point where completely different words are conjoined under the same stem. For instance, the words "general" and "generous" are stemmed under the same stem "gener". The latter describes the opposite case, where a word is not cut enough. This may cause a situation where words derived from the same root cannot be identified as the same stem. This is due essentially to the fact that Porter stemmer ignores many cases. In fact, Porter stemmer does not treat irregular forms such as irregular plural and irregular past participles. Moreover, many suffixes are not handled by Porter such as –ship, –ist, –atory, –ingly, etc. This would decrease the stemming quality, since related words are stemmed to different forms. For example, the words "ability" and "able" are stemmed respectively to "abil" and "abl". In an information retrieval context, such cases reduce recall since some useful documents will not be retrieved. In the previous example, a search for "ability" will not return documents containing the word "able". This would decrease the performance of diverse systems applying Porter stemmer.

4 Contribution

In order to improve the stemmer's performance, we studied the English morphology, and used its characteristics for building the enhanced stemmer. The resultant stemmer to which we will refer as "New Porter" includes five steps. The first one handles inflectional morphology. The second step treats derivational morphology, it maps complex suffixes to single one. The third step deletes simple suffixes. The fourth step defines a set of recoding rules to normalize stems. The last step treats irregular forms that do not follow any pattern. In what follows, we show how we deal with diverse errors:

4.1 Class 1

Porter stemmer ignores irregular forms. These forms can be categorized into two types: forms that do not follow any pattern; for instance "bought" the past participle of "buy". To handle these cases, we inserted a small dictionary in Step 5 containing a list of common irregular forms. The second category concerns forms that follow a general pattern. For instance, words ending in –feet are plural form of words ending in –foot.

We proposed rules to handle this category. Table. 1 presents the proposed rules and results when applying the two approaches. We give also the number of unhandled words for each category:

Table 1. Handling words belonging to class 1

category	Original words	Results using Porter	Rules	Results using New Porter	Number of words
Words ending in -feet are the plural form of words ending in –foot	clubfoot /clubfeet	clubfoot/clubfeet	feet➔-foot	clubfoot/clubfoot	9 words
Words ending in -men are the plural form of words ending in -man	drayman/ dray-men	drayman/ dray-men	men➔-man	drayman/ dray-man	427 words
Words ending in -ci are the plural form of words ending in -cus	abacus/ abaci	abacus/ abaci	ci➔-cus	abacu/ abacu	35 words
Words ending in -eaux are the plural form of words ending in -eau	plateau/ plateaux	plateau/ plateaux	eaux➔-eau	plateau/ plateau	29 words
Words ending in -children are the plural form of words ending in -child	child/children	child/children	children➔-child	child/child	6 words
Words ending in -wives are the plural form of words ending in -wife	farmwife/ farm-wives	farmwif/farmwiv	-wives➔-wife	farmwif/farmwif	12 words
Words ending in -knives are the plural form of words ending in -knife	knife/ knives	knif/ kniv	-knives➔-knife	knif/ knif	5 words

Table 1. (*continued*)

Words ending in –staves are the plural form of words ending in -staff	flagstaff/ flag-staves	flagstaff/ flagsttav	-staves➔-staff	flagstaff/ flag-staff	7 words
Words ending in -wolves are the plural form of words ending in -wolf	werewolf/ werewolves	werewolf/ werewolv	-wolves➔-wolf	werewolf/ werewolf	4 words
Words ending in -trices are the plural form of words ending in -trix	aviatrix/ aviatrices	aviatrix/ aviatric	-trices➔-trix	aviatrix/ aviatrix	18 words
Words ending in -mata are the plural form of words ending in -ma	chiasma/ chias-mata	chiasma/ chias-mata	-mata➔-ma	chiasma/ chi-asma	108 words
Words ending in -ei are the plural form of words ending in -eus	clypeus/ clypei	clypeu/ clypei	-ei➔-eus	clypeu/clypeu	26 words
Words ending in -pi are the plural form of words ending in -pus	carpus /carpi	carpu /carpi	-pi➔-pus	carpu /carpu	17 words
Words ending in -ses are the plural form of words ending in -sis	analysis/analyses	analysi/analys	-sis➔-s	analys/analys	492 words
Words ending in -xes are the plural form of words ending in -xis	praxis/praxes	praxi/prax	-xis➔-x	prax/prax	32 words

Table (1) shows that the proposed approach performs better than the original stemmer. It handles many cases that are not taken into account by Porter. The proposed rules handle about 1200 exceptional cases that were ignored by the original algorithm.

4.2 Class 2

Porter stemmer does not conflate verbs ending in –s (not –ss) with their participle forms. The stem of the infinitive form is obtained by simply removing the terminal –s. For the past or present participle, Porter stemmer removes respectively the suffixes –ed and –ing and keeps the terminal –s. Table 2 presents the proposed rules to handle this category and results when applying the two approaches.

Table 2. Handling words belonging to class 2

Original words	**Results using Porter**	**Rules**	**Results using New Porter**	**Number of words**
focus/focuses/focused/ focusing	focu/focus/focus/focus	-sed and !(-ssed)➔s -sing and !(-ssing)➔s	focu/focu/focu/ focu	28 verbs
chorus/choruses/ chorused /chorusing	choru/chorus/chorus /chorus		choru/choru/ chor/ choru	

4.3 Class 3

Porter stemmer does not conflate words ending in –y and that do not contain a vowel with their derived form. In fact, Porter defines a recoding rule that transforms an –y terminal to –i only if the word contains a vowel. Another rule is defined to handle words ending in –ies which is ies-> i. This will conflate carry-carries, marry-marries, etc. However, words such as try-tries-tried are not conflated since the stem does not contain a vowel. Similarly, verbs ending in –ye are not handled by Porter stemmer. In fact, when a verb ends in –ye, the terminal –e is removed in the last step and hence, the –y is not replaced by –i. To stem the past or present participle of this verb, the suffix –ed or –ing will be removed in the first step, leaving the stem with an –y terminal which will be replaced by –i in the next step and hence, the infinitive form of the verb, its past and present participle are not conflated. To handle these cases, we propose to eliminate the rule defined in the first step that transforms a terminal –y to –i if it contains a vowel. Table. 3 presents results when applying the two approaches on a set of words belonging to this category.

Table 3. Handling words belonging to class 3

Original words	Results using Por-	Results using New	Number of
cry/cries/cried/cryin	cry/cri/cri/cry	cry/cry/cry/cry	About 20 verbs
dye/dyes/dyed/dying	dye/dyes/dyed/dyin	dy/dy/dy/dy	

4.4 Class 4

Porter stemmer makes errors concerning verbs ending in a double consonant and their derivations. Thus, if the stem of the present or past participle form ends in a double consonant and that the consonant is other than 'l', 's' or 'z', then the stemmer removes a letter and keeps the stem with a single consonant. In the last step, the stemmer removes a consonant only for words ending in –ll and for which the m value is greater than 1. This will cause some problems. For instance, "ebbed" is stemmed to "eb". However, "ebb" is stemmed to "ebb". Hence, the two words are not conflated. Exceptional cases are all verbs ending in double consonant other than –l, –s and –z. Verbs ending in –z that double the –z to form the present or past participle are also not treated by Porter stemmer. Thus, these verbs get their past or present participle by doubling the terminal –z. Consequently, these verbs will not be conflated with their infinitive form. For instance, "whizzed" the past participle of "whiz" is stemmed to "whizz"; "whiz" is kept unchanged and hence "whiz" and "whizzed" are not conflated. To resolve these problems, we propose to redefine the recoding rule of the last step. Initially, the rule deletes a double consonant if the consonant in question is –l and that the m value of this stem is greater than 1. The rule will be modified in the way that it deletes the consonant of all the stems ending in a double consonant. For words ending in –ll, It removes a consonant if the stem has an m value greater than 1. Table.4 presents results when applying the two approaches on a set of words belonging to this category.

Table 4. Handling words belonging to class 4

Original words	Results using Porter	Results using New Porter	Number of words
ebb/ebbed/ebbing	ebb/eb/eb	eb/eb/eb	160 verbs
add/added/adding	add/ad/ad	ad/ad/ad	
staff/staffed/staffing	staff/staf/staf	staf/staf/staf	
spaz/spazzes/spazzed	spaz/spazz/spazz	spaz/spaz/spaz	
whiz/whizzes/whizzed	whiz/whizz/whizz	whiz/whiz/whiz	

4.5 Class 5

Porter stemmer does not treat present or past participle derivations. For instance, 'studiedly' is stemmed to 'studiedli' however 'study' is stemmed to 'studi'. Hence, the two forms are not conflated. Table.5 presents the proposed rules to handle this category and results when applying the two approaches on a set of words belonging to this category:

Table 5. Handling words belonging to class 5

category	Original words	Results using Porter	Rules	Results using New Porter	Number of words
Words ending in –iedly or –iedness are related to word ending in -ied	study/studied/ studiedness/ studiedly	studi/studi/studied/studiedli	-ly➔-ied -ss➔-ied	study/study/study/study	13 words
Words ending in –edly or –edness are related to word ending in –ed	amaze/amazed /amazedly/ amazedness	amaz/amaz/ amazedli/ amazed	-ly➔-ed -ss➔-ed	amaz/amaz/amaz/ amaz	439 words
Words ending in –ingly or –ingness are related to word ending in -ing	amaze/amazing /amazingly/ amazingness	amaz/amaz/ amazingli/ amazing	-ly➔-ing -ss➔-ing	amaz/amaz/amaz/ amaz	543 words

4.6 Class 6

Porter stemmer ignores many suffixes such as –est, –ist, –tary, –tor, –sor, –sory, –nor, –ship, –acy, –ee, etc. We propose new rules to handle these suffixes. Many other compound suffixes are also ignored by Porter stemmer. To deal with this problem, we propose to generate all possible compound suffixes derived from each suffix. For instance, suffixes derived from –ate are –ative, –ativist, –ativistic, –ativism, etc. These suffixes will be then mapped to a common suffix (the suffix from which all suffixes were derived). The proposed rules handle an important number of exceptions.

In this section, we presented a new version of Porter. The resultant stemmer handles an important number of errors that were ignored by the original stemmer. Some output from both Porter and New Porter are given in Appendix 1 to demonstrate dissimilarities between the two approaches. In what follows, we propose to evaluate New Porter using a standard method inspired by Paice (1994).

5 Validation

5.1 Paice's Evaluation Method

The motivation for the development of stemmers was to improve information retrieval performance by conflating morphologically related terms to a single stem. This means that an efficacious stemmer should conflate only pairs of words which are semantically equivalent. This definition creates the problem of how the program will judge when two words are semantically equivalent. The solution proposed by Paice (1994) was to provide an input to the program in the form of grouped files. These files contain list of words, alphabetically sorted and any terms that are considered by the evaluator to be semantically equivalent are formed into concept groups. An ideal stemmer should stem words belonging to the same group to a common stem. If a stemmed group includes more than one unique stem, then the stemmer has made understemming errors. However, if a stem of a certain group occurs in other stemmed groups, the stemmer has made overstemming errors. This permits the computation of the Overstemming and Understemming Indexes (UI and OI) and their ratio, the stemming weight (SW) for each stemmer.

The Understemming and Overstemming Indexes are measurements of specific errors that occur during the implementation of a stemming algorithm. According to these measures, a good stemmer should produce as few understemming and overstemming errors as possible. However, they cannot be considered individually during results analysis. To determine the general relative accuracy of the stemmers, Paice defines a measure, called error rate relative to truncation ($ERRT$). It is useful for deciding on the best overall stemmer in cases where one stemmer is better in terms of understemming but worse in terms of overstemming. To calculate the $ERRT$, a baseline is used. The baseline is obtained by performing the process of length truncation which means reducing every word to a given fixed length. Paice estimates that length truncation is the crudest method of stemming, and he expects any other stemmer to do better. To do so, Paice proposed to determine values of UI and OI for a series of truncation lengths. This defines a truncation line against which any stemmer can be assessed. Any reasonable stemmer will give an (UI, OI) point P between the truncation line and the origin. The further away the point is from the truncation line, the better the stemmer is. The $ERRT$ is obtained by extending a line from the origin O through the (UI, OI) point P until it intersects the truncation line at T, as illustrated in FIG. 2. $ERRT$ is then defined as : $ERRT = length(OP) / length(OT)$.

To apply the evaluation method proposed by Paice, lists of grouped word files are required. We used two word lists downloaded from the official website for Paice et Hooper (2005). The first list (Word List A) was initially used by Paice (1994) in his first experiments to evaluate the stemming accuracy of three stemmers, it contains about 10000 words. The sample of words was taken from document abstracts from the CISI test collection which is concerned with Library and Information Science.

The second list (Word List B) refers to a larger grouped word set (about 20000) compiled from word lists used in Scrabble® word checkers. These were chosen as they list a large number of word variants.

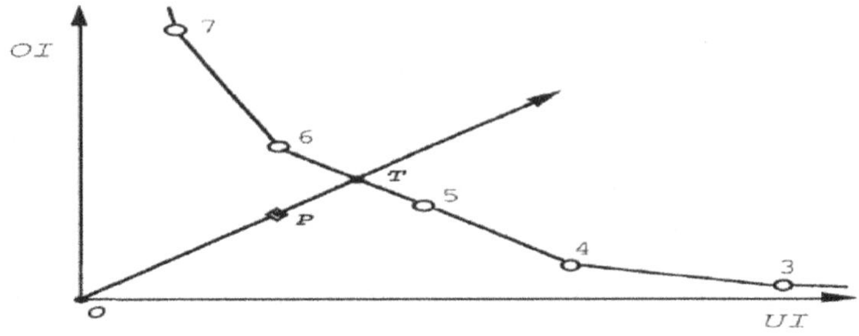

Fig. 1. Computation of $ERRT$ value, Paice (1994) Experiments

In this work, we decided to run tests initially with the Porter's original stemmer and the improved version of the stemmer in order to evaluate our approach. We also ran tests with the Paice/husk and Lovins stemmers. The results of these tests are presented in Table. 6.

Table 6. Stemming results using the two versions of the stemmers

	Word List A				Word List B			
	UI	OI	SW	ERRT	UI	OI	SW	ERRT
New Porter	0.1882	0.0000518	0.0002753	0.59	0.1274	0.0000441	0.0003464	0.44
Porter	0.3667	0.0000262	0.0000715	0.75	0.3029	0.0000193	0.0000638	0.63
Paice/Husk	0.1285	0.0001159	0.0009020	0.57	0.1407	0.0001385	0.0009838	0.73
Lovins	0.3251	0.0000603	0.0001856	0.91	0.2806	0.0000736	0.0002623	0.86

5.2 Discussion

Comparing Porter stemmer to New Porter, the relative values of indexes are summarised as follows:

UI (Porter)> UI (New Porter)
OI (New Porter)> OI (Porter)
$ERRT$ (Porter)> $ERRT$ (New Porter)

The higher value of understemming for Porter indicates that it leaves much more words understemmed than New Porter. For instance words such as "ability" and "able" are not conflated when using Porter stemmer which is not true for New Porter. This will reduce Understemming Index and conflate much more related words than Porter especially that the two word lists contain many words ending in suffixes that are not treated by the original stemmer. Hence, the UI value is improved by the proposed approach. On the other hand the OI value for New Porter is higher than that of the original stemmer. This is not entirely surprising as New Porter removes an important number of suffixes affecting a lower number of words. In fact, these rules tend to improve the UI value which hurt OI and generate more overstemming errors. For instance, the rule (m>=1) est➔' ' which means that if the word to stem ends in the suffix 'est', then the proposed stemmer will remove this suffix. This rule will improve the UI value by conflating more related words such as "fullest-full, furthest-further". These cases are ignored by Porter. However, this rule will hurt other words such as "forest" which is stemmed to "for". Hence "for" and "forest" are reduced to the same stem while they are unrelated. This will increase the OI.

For a more detailed analysis, we analyze the structure of word list A and Word List B. We find that an important number of words are related to the derivational morphology. Porter ignores many derivational suffixes such as –est, –ship, –ist, –tor, –ionally, –antly, –atory, etc. All of the words ending in these suffixes are not conflated with other related forms in the original version of the stemmer. This was resolved in the proposed approach. The proposed stemmer copes very well with derivations and inflections. Concerning inflectional morphology, many words in word list A are not handled by Porter stemmer. These cases concern irregular forms. This was handled by adding a small dictionary containing a list of exceptions. New rules handle also other words ending in suffixes such as –'s, –ingly, –ingness that were ignored by Porter.

The $ERRT$ is the general measure used by Paice (1994) to evaluate the accuracy of a stemmer. According to this value, the best stemmer would have the lowest $ERRT$ value compared to the rest. So, if we take $ERRT$ as a general indicator of performance accuracy, we would have to conclude that New Porter is a better stemmer than Porter. This means that the (OI, UI) point of New Porter is farther than the (OI, UI) point of Porter from the truncated line. Consequently, Porter stemmer generates more errors than the New Porter. We remark also large improvement in the $ERRT$ values, about 27% for word list A and about 43% for Word List B. This means that the proposed approach performs much better than the original version.

Comparing our stemmer to the other approaches (Lovins and Paice/HUSK stemmer). We find that the New stemmer not only performs better than the original version but also it is more accurate than Paice/HUSK and Lovins stemmers. Hence, it is the best stemmer overall. In fact, the differences in error rate values ($ERRT$) are so important (about 66% with Paice and 95% for Lovins). Regarding the stemmer strength, New Porter is lighter than the Paice/Husk stemmer since it has a lower SW value. This is advantageous for the information retrieval task since this would improve precision. Hence, less useless information is retrieved.

6 Conclusion

This paper describes an improved version of Porter stemmer for English. The stemmer was evaluated using the error rate relative to truncation method. In these experiments, we used two grouped word lists of 10000 and 20000 words respectively. Encouraging results are obtained. Thus, the New Porter stemmer performs much better than the original stemmer and other English stemmers for both the two word lists.

The results obtained in this work show that the New Porter stemmer is relatively a light stemmer and hence it seems to be appropriate for the information retrieval task. In order to confirm this observation, a perspective to this work is to evaluate the New Porter stemmer on an information retrieval context. The original Porter stemmer and the New Porter stemmer will be used in order to pre-process a textual corpus. The effectiveness of the two stemmers will be then measured in terms of their effect on retrieval performance. Precision and recall measures will judge which stemmer is the best overall.

References

Lovins, J.B.: Development of a stemming algorithm. Journal of Mechanical Translation and Computational Linguistics 11(1-2), 22–31 (1968)

Paice, C.D.: Another stemmer. SIGIR Forum 24(3), 56–61 (1990)

Paice, C.D.: An evaluation method for stemming algorithms. In: Proceedings of the 7th Annual Intl, ACM-SIGIR Conference, Dublin, Ireland, pp. 42–50 (1994)

Paice, C., Hooper, R.: (2005), `http://www.comp.lancs.ac.uk/computing/research/stemming/Links/program.htm`

Porter, M.F.: An Algorithm for Suffix Stripping. The journal Program 14(3), 130–137 (1980)

Willett, P.: The Porter stemming algorithm: Then and now. The Journal of Program 40(3), 219–223 (2006)

Agglomerative Hierarchical Clustering Techniques for Arabic Documents

Hanane Froud and Abdelmonaime Lachkar

L.S.I.S, E.N.S.A
University Sidi Mohamed Ben Abdellah (USMBA), Fez, Morocco
{hanane_froud,abdelmonaime_lachkar}@yahoo.fr

Abstract. Arabic Documents Clustering is an important task for obtaining good results with Search Engines, Information Retrieval (IR) systems, Text Mining Applications especially with the rapid growth of the number of online documents present in Arabic language. Document clustering is the process of segmenting a particular collection of texts into subgroups including content based similar ones. Clustering algorithms are mainly divided into two categories: Hierarchical algorithms and Partition algorithms. In this paper, we propose to study the most popular approach of Hierarchical algorithms: ***Agglomerative Hierarchical algorithm*** using ***seven linkage techniques*** with a wide variety of distance functions and similarity measures, such as the Euclidean Distance, Cosine Similarity, Jaccard Coefficient, and the Pearson Correlation Coefficient; in order to test their effectiveness on Arabic documents clustering, and finally we recommend the best techniques tested. Furthermore, we propose also to study the effect of using the stemming for the testing dataset to cluster it with the same documents clustering technique and similarity/distance measures cited above. The obtained results show that, on the one hand, the Ward function outperformed the other linkage techniques; on the other hand, the use of the stemming will not yield good results, but makes the representation of the document smaller and the clustering faster.

Keywords: Arabic Text Mining Applications, Arabic Language, Arabic Text Clustering, Hierarchical Clustering, Agglomerative Hierarchical Clustering, Similarity Measures, Stemming.

1 Introduction

Clustering is a crucial area of research, which finds applications in many fields including bioinformatics, pattern recognition, image processing, marketing, data mining, economics, etc. Cluster analysis is one of the primary data analysis tools in data mining. Clustering algorithms are mainly divided into two categories: Hierarchical algorithms and Partition algorithms. A hierarchical clustering algorithm divides the given data set into smaller subsets in hierarchical fashion, they organizes clusters into a tree or a hierarchy that facilitates browsing. A partition clustering algorithm partition the data set into desired number of sets in a single step.

D. Nagamalai et al. (Eds.): *Adv. in Comput. Sci., Eng. & Inf. Technol.*, AISC 225, pp. 255–267.
DOI: 10.1007/978-3-319-00951-3_25

One popular approach in documents clustering is Agglomerative Hierarchical clustering [1]. Algorithms in this family follow a similar template: Compute the similarity between all pairs of clusters and then merge the most similar pair. Different agglomerative algorithms may employ different similarity measuring schemes. Recently, Steinbach et al. [16] shows that UPGMA [1][2] is the most accurate one in its category.

In our days, Arabic Language is used by more than 265 millions of Arabs; also it is understood by more than one billion of Muslims worldwide, as the Muslims' holy book (the Koran) is written in Arabic. Arabic documents became very popular on an electronic format, so the need for documents clustering became very necessary.

This ever-increasing importance of Arabic documents clustering and the expanded range of its applications led us to do an experimental study of the most popular Hierarchical algorithms: Agglomerative Hierarchical algorithm. In particular, we compare seven linkage techniques using a wide variety of distance functions and similarity measures, such as the Euclidean Distance, Cosine Similarity, Jaccard Coefficient, and the Pearson Correlation Coefficient [3], for the Agglomerative Hierarchical algorithm, in order to test their effectiveness on Arabic documents clustering with and without stemming.

The main contribution of this paper is to decide which are the best linkage techniques to produce coherent clusters from a heterogeneous dataset especially for Arabic documents, choosing the best distance functions and similarity measures to define similarity of two documents, and finally we describe the effect of using the stemming to cluster Arabic documents.

The remainder of this paper is organized as follows. The next section describes the Arabic text preprocessing, stemming and document representation used in the experiments. Section 3 presents the different clustering techniques used in this work. Section 4 discusses the similarity measures and their semantics and Section 5 and 6 explain experiment settings, dataset, evaluation approaches, results and analysis. Section 7 concludes and discusses future work.

2 Arabic Text Preprocessing

Prior to applying document clustering techniques to an Arabic document, the latter is typically preprocessed: it is parsed, in order to remove stop words, and then words are stemmed using two stemming algorithms: Morphological Analyzer from Khoja and Garside [4], and the Light Stemmer developed by Larkey [5]. In addition, at this stage in this work, we computed the term-document using *tfidf* weighting scheme.

2.1 Stemming

Stemming aim to find the lexical root or stem for words in natural language, by removing affixes attached to its root, because an Arabic word can have a more complicated form with those affixes. There are four kinds of affixes: antefixes, prefixes, suffixes and postfixes that can be attached to words. An Arabic word can represent a phrase in English, for example the word أَتَتَذكرونَنَا :" **do you remember us?"** is decomposed as follows:

Table 1. Arabic Word Decomposition

Postfix	Suffix	Root	Prefix	Antefix
نا	ون	تذكر	ت	أ
A pronoun meaning "us"	Termination of conjugation	remember	A letter meaning the tense and the person of conjugation	Preposition for asking question

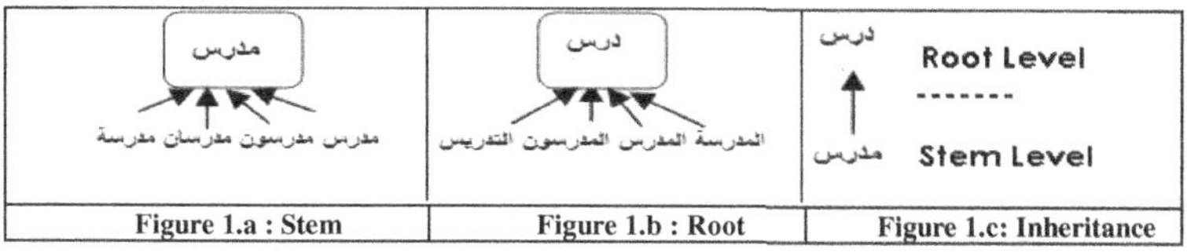

Fig. 1. An Example of Root/Stem Preprocessing

We selected tow famous stemming algorithms for which we had ready access to the implementation and/or results: the Morphological Analyzer from Khoja and Garside [4], and the Light Stemmer developed by Larkey [5] (Fig. 1).

2.2 Document Representation

There are several ways to model a text document. For example, it can be represented as a bag of words, where words are assumed to appear independently and the order is immaterial. This model is widely used in information retrieval and text mining [6].

Each word corresponds to a dimension in the resulting data space and each document then becomes a vector consisting of non-negative values on each dimension. Let $D=\{d_1,...,d_n\}$ be a set of documents and $T=\{t_1,...,t_m\}$ the set of distinct terms occurring in D. A document is then represented as an m-dimensional vector $\overrightarrow{t_d}$. Let $tf(d,t)$ denote the frequency of term $t \in T$ in document $t \in D$. Then the vector representation of a document d is:

$$\overrightarrow{t_d} = (tf(d,t_1),...,tf(d,t_m)) \tag{1}$$

Although more frequent words are assumed to be more important, this is not usually the case in practice (in the Arabic language words like الى that means to and في that means in). In fact, more complicated strategies such as the *tfidf* weighting scheme as described below is normally used instead. So we choose in this work to produce the *tfidf* weighting for each term for the document representation. In the practice terms those appear frequently in a small number of documents but rarely in the other documents tend to be more relevant and specific for that particular group of documents, and therefore more useful for finding similar documents. In order to capture these terms and reflect their importance, we transform the basic term frequencies $tf(d,t)$ into the *tfidf* (term frequency and inversed document frequency) weighting scheme. *Tfidf* weights the frequency of a term t in a document d with a factor that discounts its importance with its appearances in the whole document collection, which is defined as:

$$tfidf(d,t) = tf(d,t) \times \log(\frac{|D|}{df(t)}) \quad (2)$$

Here $df(t)$ is the number of documents in which term t appears, |D| is the numbers of documents in the dataset. We use $w_{t,d}$ to denote the weight of term t in document d in the following sections.

3 Clustering Techniques

The four main classes of clustering algorithms available in the literature are partitioning methods, hierarchical methods, density-based clustering and grid-based clustering (see [19] for an extensive survey). For the purpose of our comparative study we select to study the most popular hierarchical clustering algorithms.

Hierarchical algorithms create decomposition of the database [13]. It is categorized into agglomerative and divisive clustering. Hierarchical clustering builds a tree of clusters, also known as a dendrogram. Every cluster node contains the child cluster. An agglomerative clustering start with a one-point (singleton) cluster and recursively merges two or more most appropriate clusters. A divisive clustering starts with one cluster of all data points and recursively splits into the most appropriate clusters. The process continues until a stopping criterion is achieved.

Agglomerative methods start with an initial clustering of the term space, where all documents are considered to represent a separate cluster. The closest clusters using a given inter-cluster similarity measure are then merged continuously until only 1 cluster or a predefined number of clusters remain.

Simple Agglomerative Clustering Algorithms are

1. Compute the similarity between all pairs of clusters i.e. calculates a similarity matrix whose ijth entry gives the similarity between the ith and jth clusters.
2. Merge the most similar (closest) two clusters.
3. Update the similarity matrix to reflect the pairwise similarity between the new cluster and the original clusters.
4. Repeat steps 2 and 3 until only a single cluster remains.

Divisive clustering algorithms start with a single cluster containing all documents. It then continuously divides clusters until all documents are contained in their own cluster or a redefined number of clusters are found.

4 Similarity Measures

In this section we present the five similarity measures that were tested in [3] and our works [17][18], and we include these five measures in our work to effect the Arabic text document clustering.

4.1 Euclidean Distance

Euclidean distance is widely used in clustering problems, including clustering text. It satisfies all the above four conditions and therefore is a true metric. It is also the default distance measure used with the K-means algorithm.

Measuring distance between text documents, given two documents d_a and d_b represented by their term vectors $\vec{t_a}$ and $\vec{t_b}$ respectively, the Euclidean distance of the two documents is defined as:

$$D_E(\vec{t_a}, \vec{t_b}) = \left(\sum_{t=1}^{m} \left|w_{t,a} - w_{t,b}\right|^2\right)^{1/2}, \tag{3}$$

where the term set is $T = \{t_1, \ldots, t_m\}$. As mentioned previously, we use the *tfidf* value as term weights, that is $w_{t,a} = tfidf(d_a, t)$.

4.2 Cosine Similarity

Cosine similarity is one of the most popular similarity measure applied to text documents, such as in numerous information retrieval applications [6] and clustering too [7]. Given two documents $\vec{t_a}$ and $\vec{t_b}$, their cosine similarity is:

$$SIM_C(\vec{t_a}, \vec{t_b}) = \frac{\vec{t_a} \cdot \vec{t_b}}{\left|\vec{t_a}\right| \times \left|\vec{t_b}\right|}, \tag{4}$$

where $\vec{t_a}$ and $\vec{t_b}$ are m-dimensional vectors over the term set $T = \{t_1, \ldots, t_m\}$. Each dimension represents a term with its weight in the document, which is non-negative. As a result, the cosine similarity is non-negative and bounded between $[0,1]$ An important property of the cosine similarity is its independence of document length. For example, combining two identical copies of a document d to get a new pseudo document d_0, the cosine similarity between d and d_0 is 1, which means that these two documents are regarded to be identical.

4.3 Jaccard Coefficient

The Jaccard coefficient, which is sometimes referred to as the Tanimoto coefficient, measures similarity as the intersection divided by the union of the objects. For text document, the Jaccard coefficient compares the sum weight of shared terms to the sum weight of terms that are present in either of the two documents but are not the shared terms. The formal definition is:

$$SIM_J(\vec{t_a}, \vec{t_b}) = \frac{\vec{t_a} \cdot \vec{t_b}}{\left|\vec{t_a}\right|^2 + \left|\vec{t_b}\right|^2 - \vec{t_a} \cdot \vec{t_b}} \tag{5}$$

The Jaccard coefficient is a similarity measure and ranges between 0 and 1. It is 1 when the $\vec{t_a} = \vec{t_b}$ and 0 when $\vec{t_a}$ and $\vec{t_b}$ are disjoint. The corresponding distance measure is $D_J = 1 - SIM_J$ and we will use D_J instead in subsequent experiments.

4.4 Pearson Correlation Coefficient

Pearson's correlation coefficient is another measure of the extent to which two vectors are related. There are different forms of the Pearson correlation coefficient formula. Given the term set $T = \{t_1, \ldots, t_m\}$, a commonly used form is

$$SIM_P(\vec{t_a}, \vec{t_b}) = \frac{m\sum_{t=1}^{m} w_{t,a} \times w_{t,b} - TF_a \times TF_b}{\sqrt{\left[m\sum_{t=1}^{m} w_{t,a}^2 - TF_a^2\right]\left[m\sum_{t=1}^{m} w_{t,b}^2 - TF_b^2\right]}} \tag{6}$$

where $TF_a = \sum_{t=1}^{m} w_{t,a}$ and $TF_b = \sum_{t=1}^{m} w_{t,b}$, this is also a similarity measure. However, unlike the other measures, it ranges from -1 to +1 and it is 1 when $\vec{t_a} = \vec{t_b}$. In subsequent experiments we use the corresponding distance measure, which is $D_P = 1 - SIM_P$ when $SIM_P \geq 0$ and $D_P = \left|SIM_P\right|$ when $SIM_P \prec 0$.

4.5 Averaged Kullback-Leibler Divergence

In information theory based clustering, a document is considered as a probability distribution of terms. The similarity of two documents is measured as the distance between the two corresponding probability distributions. The Kullback-Leibler divergence (KL divergence), also called the relative entropy, is a widely applied measure for evaluating the differences between two probability distributions. Given two distributions P and Q, the KL divergence from distribution P to distribution Q is defined as

$$D_{KL}(P \parallel Q) = P \log(\frac{P}{Q}) \tag{7}$$

In the document scenario, the divergence between two distributions of words is:

$$D_{KL}(\vec{t_a} \parallel \vec{t_b}) = \sum_{t=1}^{m} w_{t,a} \times \log(\frac{w_{t,a}}{w_{t,b}}). \tag{8}$$

However, unlike the previous measures, the KL divergence is not symmetric, i.e. $D_{KL}(P \parallel Q) \neq D_{KL}(Q \parallel P)$. Therefore it is not a true metric. As a result, we use the averaged KL divergence instead, which is defined as:

$$D_{AvgKL}(P \| Q) = \pi_1 D_{KL}(P \| M) + \pi_2 D_{KL}(Q \| M), \tag{9}$$

where $\pi_1 = \frac{P}{P+Q}, \pi_2 = \frac{Q}{P+Q}$ and $M = \pi_1 P + \pi_2 Q$ For documents, the averaged KL divergence can be computed with the following formula:

$$D_{AvgKL}(\vec{t_a} \| \vec{t_b}) = \sum_{t=1}^{m} (\pi_1 \times D(w_{t,a} \| w_t) + \pi_2 \times D(w_{t,b} \| w_t)), \tag{10}$$

where $\pi_1 = \frac{w_{t,a}}{w_{t,a} + w_{t,b}}, \pi_2 = \frac{w_{t,b}}{w_{t,a} + w_{t,b}}$, and $w_t = \pi_1 \times w_{t,a} + \pi_2 \times w_{t,b}$.

The average weighting between two vectors ensures symmetry, that is, the divergence from document i to document j is the same as the divergence from document j to document i. The averaged KL divergence has recently been applied to clustering text documents, such as in the family of the Information Bottleneck clustering algorithms [8], to good effect.

5 Evaluation of Cluster Quality

The quality of the clustering result was evaluated using two evaluation measures: purity and entropy, which are widely used to evaluate the performance of unsupervised learning algorithms [10], [11].

The Purity measure evaluates the coherence of a cluster, that is, the degree to which a cluster contains documents from a single category. Given a particular cluster C_i of size n_i, the purity of C_i is formally defined as

$$P(C_i) = \frac{1}{n_i} \max_h (n_i^h) \tag{11}$$

where $\max_h (n_i^h)$ is the number of documents that are from the dominant category in cluster C_i and n_i^h represents the number of documents from cluster C_i assigned to category h. Purity can be interpreted as the classification rate under the assumption that all samples of the cluster are predicted to be members of the actual dominant class for the cluster. For an ideal cluster, which only contains documents from a single category, its purity value is 1. In general, the higher the purity value, the better the quality of the cluster is.

The Entropy measure evaluates the distribution of categories in a given cluster. The entropy of a cluster C_i with size n_i is defined to be

$$E(C_i) = -\frac{1}{\log c} \sum_{h=1}^{k} \frac{n_i^h}{n_i} \log\left(\frac{n_i^h}{n_i}\right) \tag{12}$$

where c is the total number of categories in the data set and n_i^h is the number of documents from the hth class that were assigned to cluster C_i. The entropy measure is more comprehensive than purity because rather than just considering the number of objects in the cluster C_i and not in the dominant category, it considers the overall distribution of all the categories in a given cluster. Contrary to the purity measure, for an ideal cluster with documents from only a single category, the entropy of the cluster will be 0. In general, the smaller the entropy value, the better the quality of the cluster is.

The total entropy for a set of clusters is calculated as the sum of the entropies of each cluster weighted by the size of each cluster:

$$Entropy = \sum_{i=1}^{k} \frac{n_i}{n} E(C_i) \tag{13}$$

where n is the total number of documents in the dataset.

6 Experiments and Results

In our experiments, we used the Agglomerative Hierarchical algorithms as documents clustering methods for Arabic documents. The similarity measures do not directly fit into the algorithm, because smaller values indicate dissimilarity [18]. The Euclidean distance and the Averaged KL Divergence are distance measures, while the Cosine Similarity, Jaccard coefficient and Pearson coefficient are similarity measures. We apply a simple transformation to convert the similarity measure to distance values. Because both Cosine Similarity and Jaccard coefficient are bounded in $[0,1]$ and monotonic, we take $D = 1 - SIM$ as the corresponding distance value. For Pearson coefficient, which ranges from −1 to +1, we take $D = 1 - SIM$ when $SIM \geq 0$ and $D = |SIM|$ when $SIM \prec 0$. For the testing dataset, we experimented with different similarity measures for three times: without stemming, and with stemming using the Morphological Analyzer from Khoja and Garside [4] , and the Light Stemmer [5] for the all documents in dataset . Moreover, each experiment was run for many times and the results are the averaged value over many runs. In the total we had 105 experiments for Agglomerative Hierarchical algorithm using 7 techniques for merging the clusters described below in the next section.

6.1 Agglomerative Hierarchical Techniques

Agglomerative algorithms are usually classified according to the inter-cluster similarity measure they use. The most popular of these are [14][15]:

- Single Linkage: minimum distance criterion ,
- Complete Linkage : maximum distance criterion ,
- Average Group : average distance criterion,
- Centroid Distance Criterion ,
- And Ward : minimize variance of the merge cluster.

Jain and Dubes (1988) showed general formula that first proposed by Lance and William (1967) to include most of the most commonly referenced hierarchical clustering called SAHN (Sequential, Agglomerative, Hierarchical and nonoverlapping) clustering method. Distance between existing cluster k with nk objects and newly formed cluster (r,s) with nr and ns objects is given as:

$$d_{k\to(r,s)} = \alpha_r d_{k\to r} + \alpha_s d_{k\to s} + \beta d_{r\to s} + \gamma\left|d_{k\to r} - d_{k\to s}\right| \quad (14)$$

The values of the parameters are given in the in Table 2.

Table 2. The values of the parameters of the general formula of hierarchical clustering SAHN

Clustering Method	α_r	α_s	β	γ
Single Link	1/2	1/2	0	-1/2
Complete Link	1/2	1/2	0	1/2
Unweighted Pair Group Method Average (UPGMA)	$\frac{n_r}{n_r+n_s}$	$\frac{n_s}{n_r+n_s}$	0	0
Weighted Pair Group Method Average (WPGMA)	1/2	1/2	0	0
Unweighted Pair Group Method Centroid (UPGMC)	$\frac{n_r}{n_r+n_s}$	$\frac{n_s}{n_r+n_s}$	$\frac{-n_r n_s}{(n_r+n_s)^2}$	0
Weighted Pair Group Method Centroid (WPGMC)	1/2	1/2	-1/4	0
Ward's Method	$\frac{n_r+n_k}{n_r+n_s+n_k}$	$\frac{n_s+n_k}{n_r+n_s+n_k}$	$\frac{-n_k}{n_r+n_s+n_k}$	0

Table 3. Number of texts and number of Terms in each category of the testing dataset

Text Categories	Number of Texts	Number of Terms
Economics	29	67 478
Education	10	25 574
Health and Medicine	32	40 480
Interviews	24	58 408
Politics	9	46 291
Recipes	9	4 973
Religion	19	111 199
Science	45	104 795
Sociology	30	85 688
Spoken	7	5 605
Sports	3	8 290
Tourist and Travel	61	46 093

6.2 Dataset

The testing dataset [9] (Corpus of Contemporary Arabic (CCA)) is composed of 12 several categories, each latter contains documents from websites and from radio Qatar. A summary of the testing dataset is shown in Table 3. As mentioned previously, we removed stop words and applied stemming, and we ranked terms by their weighting schemes *Tfidf* and use them in our experiments.

6.3 Results

Tables 4-7 show the average purity and entropy results for each similarity/distance measure with the Morphological Analyzer from Khoja and Garside [4], the Light Stemmer [5], and without stemming with document clustering algorithm cited above.

The goal of these experiments is to decide which are the best and appropriate techniques to use for producing consistent clusters for Arabic Documents.

Results without Stemming
The overall entropy and purity values, for our experiments without stemming, are shown in the tables 4 and 5, the obtained results using the Agglomerative Hierarchical algorithm with 7 schemes for merging the clusters show that this document clustering technique performs good using the COMPLETE, UPGMA, WPGMA schemes, and Ward function with the Cosine Similarity, the Jaccard measures and Pearson Correlation.

Results with Stemming
Tables 6 and 7 present the results of using Agglomerative Hierarchical algorithms as document clustering methods with stemming using the Morphological Analyzer from Khoja and Garside [4] , and the Light Stemmer [5] for the all documents in dataset.

A closer look to the Table 5 lead as to observe that the lower entropies are obtained with the Cosine Similarity, the Jaccard and the Pearson Correlation measures using the COMPLETE, UPGMA, WPGMA schemes, and Ward function as linkage techniques.

In Table 6, the use of Larkey's Stemmer for the all documents in dataset clustered by the Agglomerative Hierarchical algorithm produce good entropy and purity scores for the Cosine Similarity, the Jaccard and the Pearson Correlation measures using the COMPLETE, UPGMA, WPGMA schemes, and Ward function as linkage techniques.

The above obtained results (shown in the different Tables) lead us to conclude that:

- For the ***Agglomerative Hierarchical algorithm***, the use of the COMPLETE, UPGMA [16], WPGMA schemes, and Ward function as linkage techniques yield good results.
- Cosine Similarity, Jaccard and Pearson Correlation measures perform better relatively to the other measures for three times: without stemming, and with stemming using the Morphological Analyzer from Khoja and Garside [4], and the Light Stemmer [5].
- The tested documents clustering technique perform well without using the stemming.
- The **Larkey's Stemmer** outperforms the **Khoja's Stemmer** because this later affects the words meanings [17][18].

6.4 Discussion

The above conclusions shows that, in one side, the use of stemming affects negatively the clustering, this is mainly due to the ambiguity created when we applied the stemming (for example, we can obtain two roots that made of the same letters but semantically different); this observation broadly agrees with our previous works [17][18].

In the other side, we can explain the behavior of the all tested linkage techniques as follows:

Table 4. Entropy Results without Stemming using Agglomerative Hierarchical Algorithm

	Euclidean	Cosine	Jaccard	Pearson	KLD
COMPLETE	0.872	**0.219**	**0.136**	**0.109**	0.879
SINGLE	0.881	0.881	0.877	0.877	0.876
UPGMA	0.872	**0.329**	**0.064**	**0.116**	0.879
WPGMA	0.881	**0.255**	**0.131**	**0.367**	0.879
UPGMC	0.872	0.869	0.885	0.877	0.879
WPGMC	0.881	0.885	0.884	0.884	0.879
Ward	0.699	**0.100**	**0.091**	**0.073**	0.707

Table 5. Purity results without stemming using Agglomerative Hierarchical Algorithm

	Euclidean	Cosine	Jaccard	Pearson	KLD
COMPLETE	0.878	**0.556**	**0.558**	**0.535**	0.933
SINGLE	0.933	0.933	0.933	0.933	0.933
UPGMA	0.878	**0.725**	**0.611**	**0.750**	0.933
WPGMA	0.933	**0.749**	**0.748**	**0.823**	0.933
UPGMC	0.878	0.913	0.933	0.892	0.933
WPGMC	0.933	0.933	0.933	0.933	0.933
Ward	0.753	**0.458**	**0.537**	**0.457**	0.818

Table 6. Entropy Results with Khoja's Stemmer, and Larkey's Stemmer using Agglomerative Hierarchical Algorithm

		Euclid	Cosine	Jaccard	Pearson	KLD
Khoja's stemmer	**COMPLETE**	0.878	**0.172**	**0.165**	**0.213**	0.875
	SINGLE	0.882	0.887	0.888	0.889	0.877
	UPGMA	0.881	**0.075**	**0.590**	**0.250**	0.878
	WPGMA	0.882	**0.319**	**0.446**	**0.156**	0.875
	UPGMC	0.882	0.870	0.859	0.794	0.878
	WPGMC	0.882	0.887	0.887	0.881	0.878
	Ward	0.631	**0.127**	**0.117**	**0.109**	0.649
Larkey's stemmer	**COMPLETE**	0.883	**0.051**	**0.157**	**0.057**	0.878
	SINGLE	0.883	0.886	0.886	0.882	0.878
	UPGMA	0.885	**0.561**	**0.654**	**0.354**	0.878
	WPGMA	0.883	**0.223**	**0.064**	**0.260**	0.878
	UPGMC	0.885	0.886	0.869	0.857	0.878
	WPGMC	0.883	0.886	0.886	0.886	0.878
	Ward	0.740	**0.059**	**0.116**	**0.144**	0.648

Table 7. Purity Results with Khoja's Stemmer, and Larkey's Stemmer using Agglomerative Hierarchical Algorithm

		Euclid	Cosine	Jaccard	Pearson	KLD
Khoja's stemmer	**COMPLETE**	0.892	**0.481**	**0.529**	**0.417**	0.892
	SINGLE	0.933	0.933	0.933	0.932	0.933
	UPGMA	0.891	**0.578**	**0.751**	**0.559**	0.933
	WPGMA	0.933	**0.793**	**0.695**	**0.601**	0.892
	UPGMC	0.933	0.919	0.888	0.878	0.933
	WPGMC	0.933	0.933	0.932	0.891	0.933
	Ward	0.765	**0.445**	**0.465**	**0.392**	0.819
Larkey's stemmer	**COMPLETE**	0.933	**0.437**	**0.512**	**0.449**	0.933
	SINGLE	0.933	0.933	0.933	0.933	0.933
	UPGMA	0.933	**0.729**	**0.655**	**0.667**	0.933
	WPGMA	0.933	**0.674**	**0.676**	**0.619**	0.933
	UPGMC	0.933	0.933	0.891	0.878	0.933
	WPGMC	0.933	0.933	0.933	0.933	0.933
	Ward	0.786	**0.415**	**0.423**	**0.440**	0.816

The COMPLETE linkage technique is non-local, the entire structure of the clustering can influence merge decisions. This results in a preference for compact clusters with small diameters, but also causes sensitivity to outliers.

The Ward function allows us to minimize variance of the merge cluster; the variance is a measure of how far a set of data is spread out. So the Ward function is a non-local linkage technique.

With the two techniques described above, a single document far from the center can increase diameters of candidate merge clusters dramatically and completely change the final clustering. That why these techniques produce good results than UPGMA [16], WPGMA schemes and better than the other all tested linkage techniques; because this merge criterion give us local information. We pay attention solely to the area where the two clusters come closest to each other. Other, more distant parts of the cluster and the clusters overall structure are not taken into account.

7 Conclusion

In this paper, we have proposed to study the most popular approach of Hierarchical algorithms: Agglomerative Hierarchical algorithm using seven linkage techniques with five similarity/distance measures with and without stemming for Arabic Documents.

Our results indicate that the Cosine Similarity, the Jaccard and the Pearson Correlation measures have comparable effectiveness and performs better relatively to the other measures for all documents clustering algorithms cited above for finding more coherent clusters in case we didn't use the stemming for the testing dataset.

Furthermore, our experiments with different linkage techniques for yield us to conclude that COMPLETE, UPGMA, WPGMA and Ward produce efficient results than other linkage techniques. A closer look to these results, show that Ward technique is the best in all cases (with and without using the stemming), although the two other techniques are often not much worse.

The main contribution of this paper is three manifolds:

1. The stemming affects negatively the final results, it makes the representation of the document smaller and the clustering faster,
2. Cosine Similarity, Jaccard and Pearson Correlation measures are quite similar for finding more coherent clusters for all documents clustering algorithms,
3. Ward technique is effective than other linkage techniques for producing more coherent clusters using the ***Agglomerative Hierarchical algorithm***.

Finally, we confirm that this comparative study presented in this paper will be very useful for the researchers to support the research in the field any Arabic Text Mining applications.

References

1. Dubes, R.C., Jain, A.K.: Algorithms for Clustering Data. Prentice Hall College Div, Englewood Cliffs (1998)
2. Kaufman, L., Rousseeuw, P.J.: Finding Groups in Data: An Introduction to Cluster Analysis. John Wiley and Sons (March 1990)
3. Huang, A.: Similarity Measures for Text Document Clustering. In: NZCSRSC 2008, Christchurch, New Zealand (April 2008)

4. Khoja, S., Garside, R.: Stemming Arabic Text. Computing Department. Lancaster University, Lancaster (1999)
5. Larkey, L.S., Ballesteros, L., Connell, M.: Improving Stemming for Arabic Information Retrieval: Light Stemming and Co-occurrence Analysis. In: Proceedings of the 25th Annual International Conference on Research and Development in Information Retrieval (SIGIR 2002), Tampere, Finland, August 11-15, pp. 275–282 (2002)
6. Yates, R.B., Neto, B.R.: Modern Information Retrieval. Addison-Wesley, New York (1999)
7. Larsen, B., Aone, C.: Fast and Effective Text Mining using Linear-time Document Clustering. In: Proceedings of the Fifth ACM SIGKDD International Conference on Knowledge Discovery and Data Mining (1999)
8. Tishby, N.Z., Pereira, F., Bialek, W.: The Information Bottleneck Method. In: Proceedings of the 37th Allerton Conference on Communication, Control and Computing (1999)
9. Al-Sulaiti, L., Atwell, E.: The Design of a Corpus of Contemporary Arabic. University of Leeds
10. Zhao, Y., Karypis, G.: Evaluation of Hierarchical Clustering Algorithms for Document Datasets. In: Proceedings of the International Conference on Information and Knowledge Management (2002)
11. Zhao, Y., Karypis, G.: Empirical and Theoretical Comparisons of Selected Criterion Functions for Document Clustering. Machine Learning 55(3) (2004)
12. El-Kourdi, M., Bensaid, A., Rachidi, T.: Automatic Arabic Document Categorization Based on the Naïve Bayes Algorithm. In: 20th International Conference on Computational Linguistics, Geneva (August 2004)
13. Sathiyakumari, K., Manimekalai, G., Preamsudha, V., Phil Scholar, M.: A Survey on Various Approaches in Document Clustering. Int. J. Comp. Tech. Appl., IJCTA 2(5), 1534–1539 (2011)
14. Müllner, D.: Modern hierarchical, agglomerative clustering algorithms, arXiv: 1109.2378 (2011)
15. Teknomo, K.: Hierarchical Clustering Tutorial (2009), `http://people.revoledu.com/kardi/tutorial/clustering/`
16. Steinbach, M., Karypis, G., Kumar, V.: A comparison of document clustering techniques. In: KDD Workshop on Text Mining (2002)
17. Froud, H., Lachkar, A., Alaoui Ouatik, S.: A Comparative Study Of Root-Based And Stem-Based Approaches For Measuring The Similarity Between Arabic Words For Arabic Text Mining Applications. Advanced Computing: An International Journal (ACIJ) 3(6) (November 2012)
18. Froud, H., Lachkar, A., Ouatik, S., Benslimane, R.: Stemming and Similarity Measures for Arabic Documents Clustering. In: 5th International Symposium on I/V Communications and Mobile Networks ISIVC. IEEE Xplore (2010)
19. Berkhin, P.: Survey of clustering data mining techniques. Technical report, Accrue Software, San Jose, California (2002), `http://citeseer.nj.nec.com/berkhin02survey.html`
20. Cutting, D.R., Karger, D.R., Pedersen, J.O., Tukey, J.W.: Scatter/gather: A cluster-based approach to browsing large document collections. In: Proceedings of the Fifteenth Annual International ACM SIGIR Conference on Research and Development in Information Retrieval, pp. 318–329 (1992)
21. Larsen, B., Aone, C.: Fast and effective text mining using linear-time document clustering. In: KDD 1999, pp. 16–22 (1999)

A Slippery Window and VH2D Approach to Recognition Offline Arabic Words

Ahlam Maqqor*, Akram Halli, Khalide Satori, and Hamed Tairi

University Sidi Mohamed Ben Abdellah Faculty of Science Dhar EL Mahraz
Laboratory LIIAN, Fez, Morocco
ahlamhfl@gmail.com,
{akram_halli,khalidsatorim3i,h_tairi}@yahoo.fr

Abstract. In this paper, we propose an analytical approach of an offline recognition of handwritten Arabic. Our method is based on Markov modeling and takes into account the characteristic of the Arabic script.

The objective is to propose a methodology for rapid implementation of our approach. To this end, a preprocessing phase that can prepare the data was introduced. These data are then used by two methods of feature extraction: The first is the use of a sliding window on binary images to recognize words from right to left. The second is the use of the. The results of this step are converted to data sequence as vectors that are combined through a multi-stream.

Keywords: Cursive Arabic, Hidden Markov Models, sliding window, multi-stream approach, VH2D approach, Extraction of primitives.

1 Introduction

The recognition of Arabic handwriting is an active field in the pattern recognition domain [15]. Several solutions have been proposed for this recognition, quoting among others [9] [10] [11] [12] [13], these systems based MMCs have been developed for the recognition of cursive Arabic words.

Constructing off-line recognition systems is a challenging task because of the variability and the cursive nature of the Arabic handwriting ie the segmentation of handwritten words, tilt, overlap, the spaces between and within words are of varying lengths, words contain dots and diacritical marks that can change the meaning of a word.

To overcome these problems, an analytical approach has been proposed to require the inclusion of a large number of variability.

The main objective of our system is to design and implement a multi-stream approach of two types of feature extraction. Characteristics based on local densities and configurations of pixels and features a projection based on vertical, horizontal and diagonal 45 °, 135 ° (VH2D approach). these characteristics is considered independent of the others and the combination is in a space of probability that is to say, combine the outputs of classifiers with a creation of a system of higher reliability, [16] [17] showed the combination of classifiers for the recognition of handwriting.

* Corresponding author.

D. Nagamalai et al. (Eds.): *Adv. in Comput. Sci., Eng. & Inf. Technol.*, AISC 225, pp. 269–279.
DOI: 10.1007/978-3-319-00951-3_26

In this paper, we discuss the first the characteristics of a handwritten Arabic script. Then, the image preprocessing to recognize the word, the segmentation of text, then the feature extraction with the use: the technique of sliding windows VH2D and approach. Finally we describe the modeling MMC with a combination multi-stream.

2 Characteristic of Arabic Script

Arabic script is different compared to other types of writing Latin, Chinese.... By their own structure and binding mode to form a word.

The difficulties related to the morphology of writing:

- Arabic script written from right to left.
- The Arabic alphabet is richer than its Latin equivalent, it contains 28 letters.
- Arabic script is inherently cursive that is to say that the letters are usually related to each other.
- Depending on its position in the word each character can take four different forms (beginning, middle, end, isolated) (Fig. 1).
- Change in calligraphic styles, six different graphic styles (Fig. 2).
- An Arabic word usually consists of one or more connected components (pseudo-word) each containing one or more characters (Fig. 3).
- Some characters in a word can be overlapped vertically without contact (Fig. 4).
- Multiple characters can be combined vertically to form a ligature (Fig.5).
- Some characters have the same body, but the presence and position of a point or group of points, the features are critical to distinguish these characters (Fig. 6).

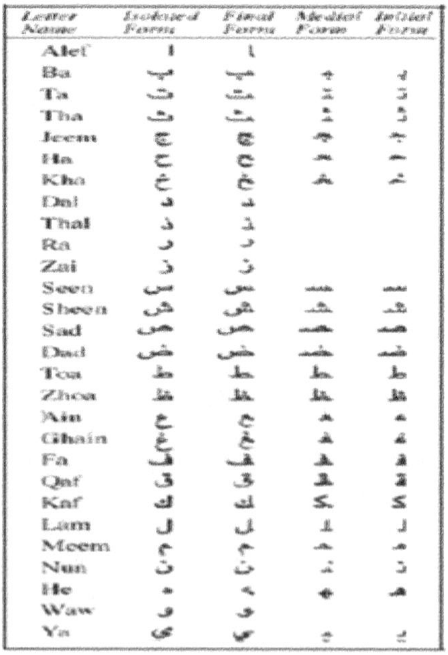

Fig. 1. Different forms of the letters according to their position

Fig. 2. Variation of calligraphic style

Fig. 3. Example of Arabic word trios with connected components

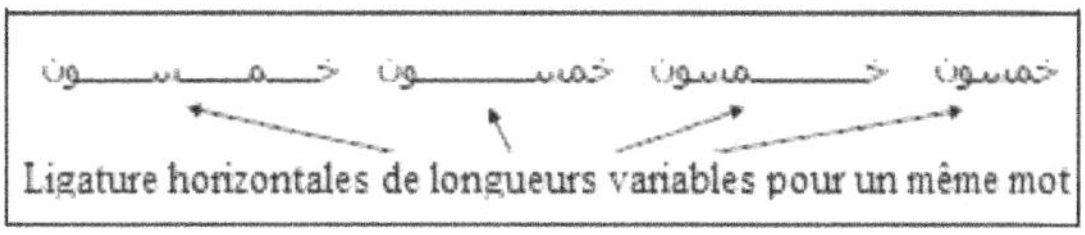

Fig. 4. A sample of Ligature horizontal

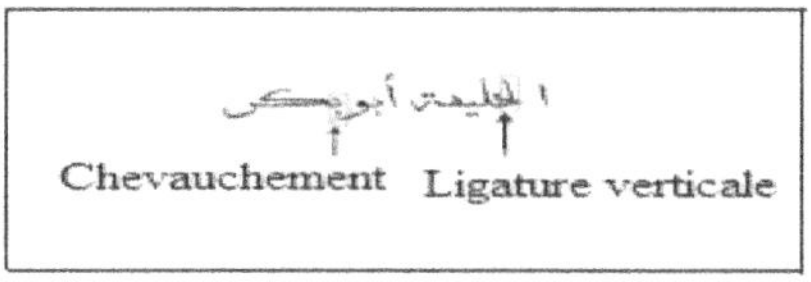

Fig. 5. A sample of the Arabic script illustrating some of these characteristics

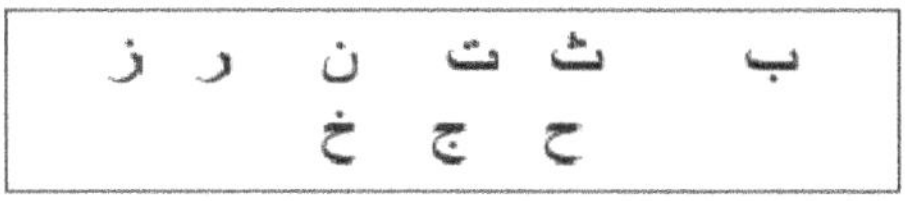

Fig. 6. Characters depending on the diacritical points characteristics

3 Recognition System

The proposed recognition system is based on a multi-band model. The system input is a word. The first step is to apply a series of preprocessing to the image. The next step in the segmentation of the image of the text lines after the separation of lines, we focus on the separation of words for each line. Then the most important step of the recognition feature extraction. Each word is shown in two sets of characteristics with the use of two different methods (Sliding Window approach and VH2D). Both sets used by the following two MMCs. The combination of these two sources is studied through the Multi-flow models.

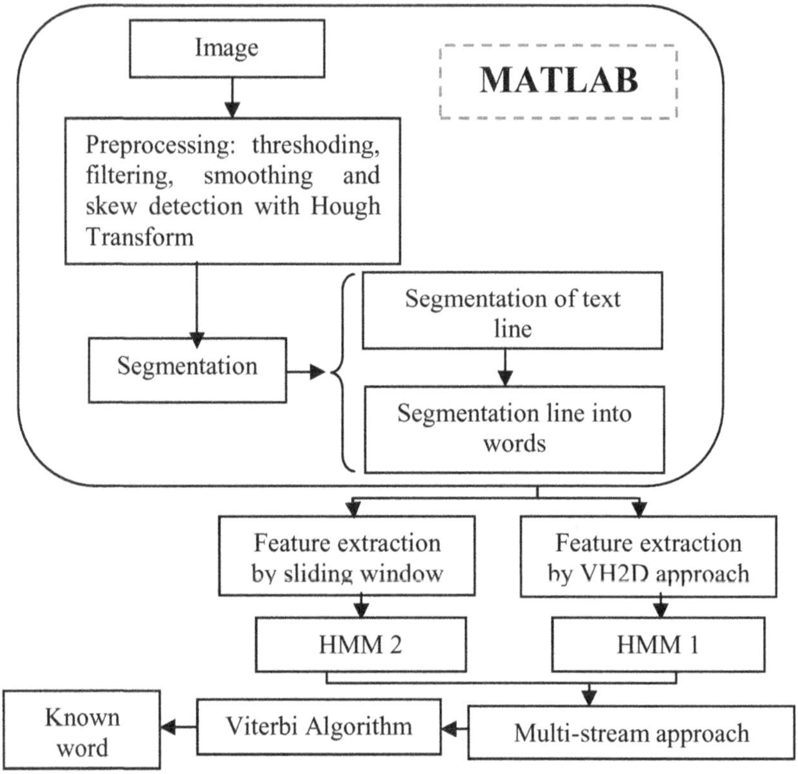

Fig. 7. Description of the recognition system

4 Pre-processing

The text is scanned and stored as a binary image. The preprocessing applied to the image of the word can, firstly, to eliminate or reduce noise in the image. Another important function is to align the text image in true horizontal.

To reach this purpose, in this paper a set of operations are applied to the text image including binarization or thresholding , filtering, smoothing, normalization and skew detection it to simplify the feature extraction.

In our approach the median filter is applied to reduce the image noise. Simply, median filter calculates the median value of all pixels in the N x N windows, and replaces the windows centre pixel by the median value. See Figure 8.

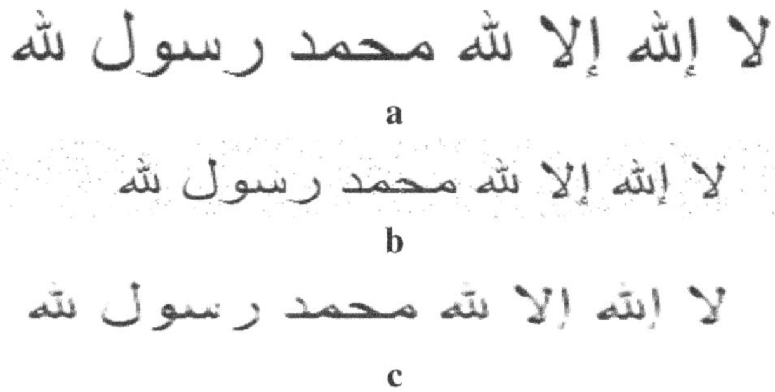

Fig. 8. The median filter effect. a) original image without noise. b) image after adding a desired noise of type "salt & pepper". c) image after using median filter.

The data corpus in scanned page by page, sometimes the text lines are not aligned to the true horizontal axis. The most effective approach to estimate the angle of inclination is to use the Hough transform [14]. Hough transform is a method of detecting fragmented lines in a binary image. Let us consider a group of black pixels, one can find the imaginary line or lines that go through the maximum number of these pixels.

a). A skewed image

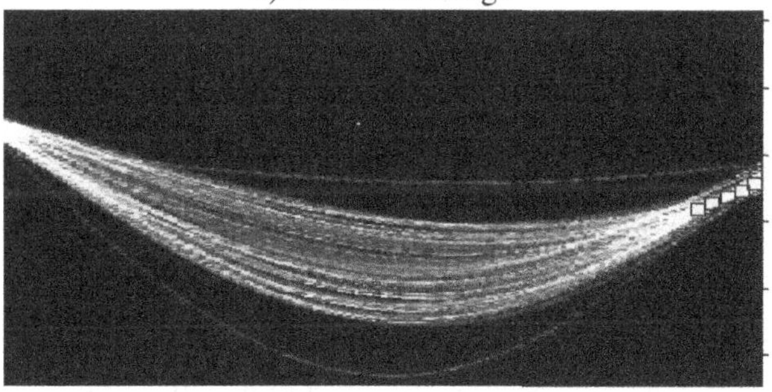

b). Hough Transform of a)

Fig. 9. Skew detection a) the text in the image is skewed by 8° from the true horizontal axis. b) the Hough Transform of image in a) detects the skewing angle.

Once the skew angle is determined, the text image is then rotated with the skewing angle but in the reverse direction as show in Figure 9.

In our approach the extraction of the base line of writing the word is based on the method described in [1] It is based on the analysis of the histogram of horizontal projection.

5 Segmentation

The most important step of automatic segmentation of documents is the image of the text lines in the future studies conducted based on a decomposition of the image connected components [1] [2] [3].

We used in our system a segmentation of the image of the text in line with the trivial use of the method of separation of lines uses the horizontal projection that is nothing more than a simple sum of the number of bits on each line. We can say the beginning or end of a text line is detected, if the value of horizontal projection is below a threshold (the threshold is obtained at least in the histogram).

After separation of the lines, we focus on the separation of words for each line.

6 Feature Extraction

We used two methods in our system for the extraction of features:

6.1 Sliding Window

Each picture of the word is transformed into a sequence of feature vectors are extracted from right to left on binary images of words from a sliding window of size N are successively offset pixel and pixels of £ (£ parameter that takes values between 1 and N-1). Each window is divided vertically into a number of fixed cells.

These extracted features are divided into two families: the characteristics of local densities and configuration of the pixels.

The advantage of using this type of feature they are independent of the language used and also can be used for any type of cursive including poles and legs like writing Arabic and Latin writing.

The characteristics of the densities of pixels:

- F1: density of black pixels in the window.
- F2: density of white pixels in the window.
- F3: Number of black / white transitions between cells.
- F4: difference in position between the center of gravity g of pixels to write in two consecutive windows.
- F5 to F12 are the densities of pixels in each writing column of the window.
- F13: center of gravity of the pixels of writing.

The characteristics of the local configurations of pixels:

- F14 to F18: The number of white pixels that belong to one of five configurations of the figure in each window.

The result of this step is a feature vector with 18 features.

6.2 VH2D Approach

The VH2D approach proposed in (Xia and Cheng, 1996) consists in projecting every character on the abscissa, on the ordinate and the diagonals 45° and 135°.

The projections take place while calculating the sum of the values of the pixels i_{xy} according to a given direction.

Presentation of the VH2D

1. **Vertical projection :** The vertical projection of an image I=Ixy (of dimension NxN) representing a character C is indicated by :

$$P^v = [P_1^v, P_2^v, p_3^v \ldots, P_{N-1}^v, P_N^v] \text{ where } P_y^v = \sum_{x=1}^{N} i_{xy}$$

2. **Horizontal projection :** Flat projection of an image I=Ixy (of dimension NxN) representing a character C is indicated by:

$$P^h = [P_1^h, P_2^h, \ldots, P_{N-1}^h, P_N^h] \text{ where } P_y^h = \sum_{y=1}^{N} i_{xy}$$

3. **Diagonal projection (45°) :** Flat projection of an image I=I_{xy} (of dimension N x N) representing a character C is indicated by :

$$P^{d1} = [P_1^{d1}, P_2^{d1}, p_3^{d1}, \ldots, P_{N-1}^{d1}, P_N^{d1}] \text{ where :}$$

$$P_m^{d1} = \begin{cases} \displaystyle\sum_{l=N-m+1}^{N} \sum_{k=1}^{m} i_{lk} & 1 \le m \le N \quad et\ l = k + N - m \\ \displaystyle\sum_{l=1}^{2N-m} \sum_{k=m-N+1}^{N} i_{lk} & N + 1 \le m \le 2N - 1\ et\ l = k + N - m \end{cases}$$

4. **Projection on the diagonal 135° :** Projection on the diagonal 135° of an image I=Ixy (of dimension N x N) representing a character C is indicated by :

$$P_m^{d2} = \begin{cases} \displaystyle\sum_{l=1}^{m} \sum_{k=1}^{m} i_{lk} & 1 \le m \le N \quad et\ k = m - l + 1 \\ \displaystyle\sum_{l=m-N+1}^{2N-m} \sum_{k=m-N+1}^{N} i_{lk} & N + 1 \le m \le 2N - 1 \quad et\ k = m - l + 1 \end{cases}$$

Therefore, the process of the VH2D approach consists in calculating the feature vectors of each image of the handwritten word.

7 HMM Recogniser

Methods to recognize handwritten words are well known and widely used for many different languages. In opposition to printed text in most languages, the characters in cursive handwritten words are connected. In recent years, methods based on Hidden Markov Models (HMM) particularly, have been very successfully used for recognizing cursively handwritten words.

The training of the HMM-parameters is done by means of the Viterbi algorithm using a segmental k-means algorithm. The initial codebook is incorporated into the training procedure. At each iteration, only the state-vector assignment resulting from best path, obtained from applying the Viterbi algorithm, is used to re-estimate model parameters.

The system models words and characters in the form of models of Markov Hidden. The system is analytical: the models words are built by concatenation of models of type characters. that are left-right as shown in Figure 10 [4][5][6] gives an example of the training, showing that each character shape of the same type, independent of the word where it was written, contributes to the statistical character shape model. This enables a statistical training with less training data than in the case of word based models.

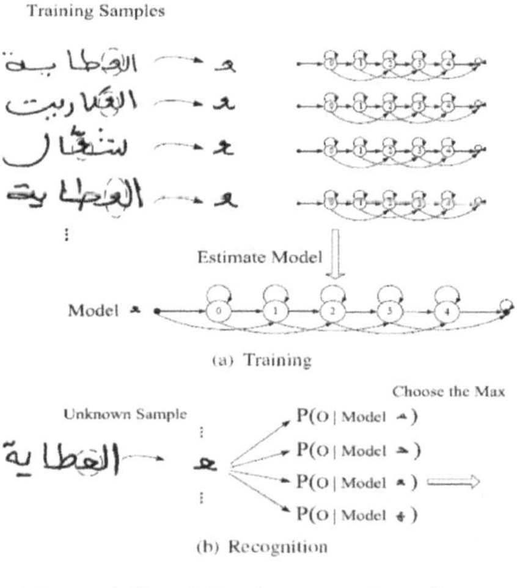

Fig. 10. HMM is trained for each "mode" using a number of examples of that "mode" from the training set, and (b) to recognize some unknown sequence of "modes", the likelihood of each model generating that sequence is calculated and the most likely (maximum) model identifies the sequence.

8 Multi-stream Model

The multi-stream initially proposed in [7] [8] is an adaptive method for combining different sources of information using Markov models. The multi-flow method is intended to fill these gaps. It includes three types of fusion:

1. Multi-Classifiers

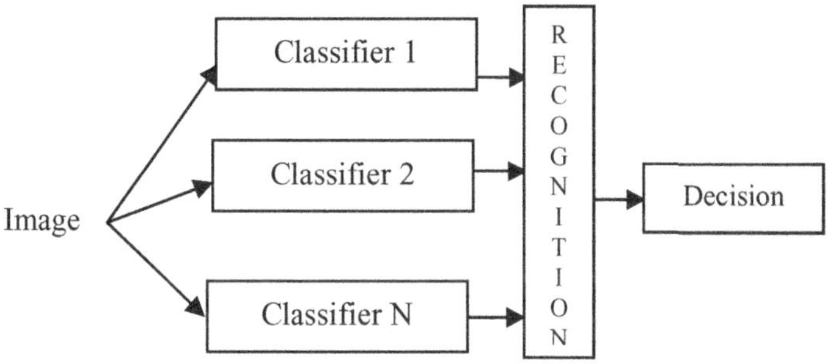

2. Multi-modal approach

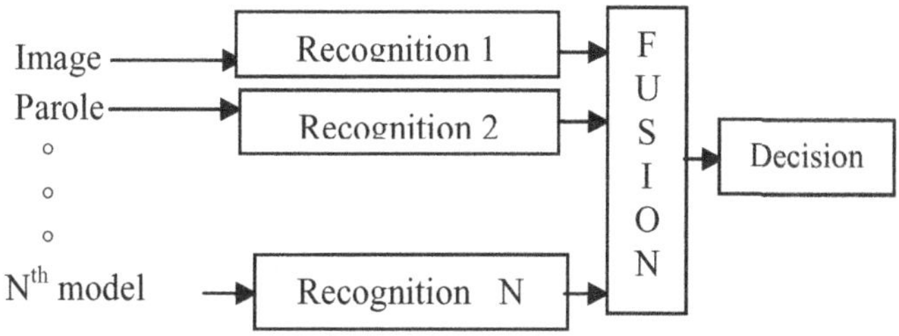

3. Partial recognition approach or multi-band

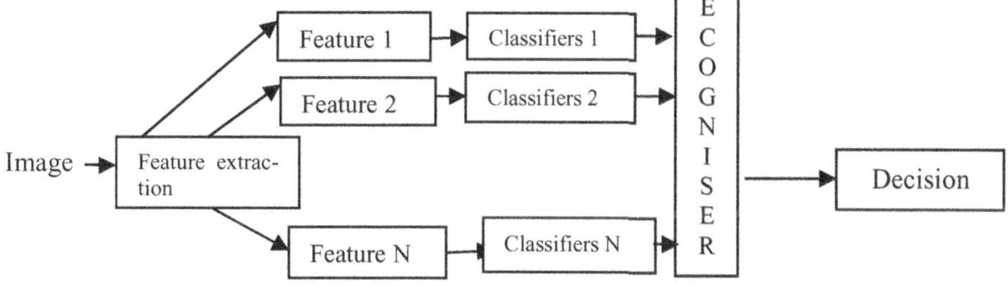

Fig. 11. The method consists of three multi-stream fusions: Multi-classifier approach, the multi-modal approach and partial recognition or multi-band

In our case we chose the multi-flow multi-fusion by dander that there are two types of data from respectively the image of the word.

8.1 Multi-stream Formalism

K is the number of information sources and M is the model consists of a sequence of J (eg letters) in models that correspond to lexical items in Mj (j = 1,2, ... N). Each sub model M consists of K Markov models (HMM) independent M_j^k(k=1,2,...A).

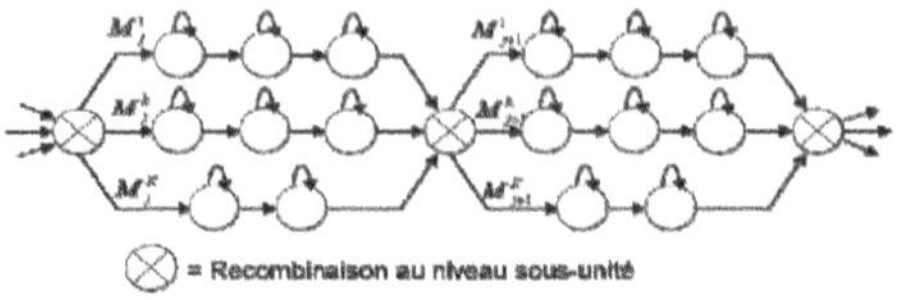

Fig. 12. General structure of a multi-stream

Definition of the Multi-stream Model:

C_j is sequence of states associated with multi-stream sequence of multi-stream observations Xj, the probability P (X_j, C_j | M_j) is calculated on the basis of the likelihoods associated with each of the sources of information according to the following:

$$P(X_j, C_j|M_j) = f(\{P(X_j^k, C_j^k|M_j^k), k = 1,2, \dots \dots \dots k\})$$

f Function of linear combination

X_j^k Sequence of observation vectors associated with the flow k

C_j^k Sub-sequence of states associated with the flow pattern in the k

M_j^k The likelihood of the subsequence from the model X_j subunit M_j and C_j the path is written

$$\log P(X|M) = \sum_{j=1}^{N} \sum_{k=1}^{A} W_j^k \log P(X_j^k|M_j^k)$$

W_j^k Represents the reliability of the flow k.

The main role of multi-stream model is to calculate the probability of the sequence of vectors of observations of different vectors of the primitive image of a handwritten word.

9 Conclusion

We have presented a recognition system off-line handwritten Arabic script based on a multi-stream with a Markov model hidden. The multi-stream models proposed method is illustrated the advantage of extracting primitive vectors by vertical windows and approach VH2D, Our system considers that the VH2D is sufficient because the recognition will be confirmed with two classifiers.

References

1. Pechwitz, M., Maergner, V.: HMM Based Approach for Handwritten Arabic Word Recognition Using the IFN/ENIT- Database. In: ICDAR (Proceedings of the Seventh International Conference on Document Analysis and Recognition), pp. 890–894 (2003)
2. Likforman, S., Faure, L.: Une methode de resolution des conflits d'alignements pour la segmentation des documents manuscrits. In: CNED 1994, 3eme Colloque National sur l'ecrit et le Document. Rouen 6,7et 8 Juillet 1994, pp. 265–272 (1994)
3. Pal, U., Roy, P.P.: Multioriented and Curved Text Lines Extraction From Indian Documents. IEEE Transactions on Systems, Man, and Cybernetic B 34(4) (August 2004)
4. El-Hajj, R., Likforman-Sulem, L., Mokbel, C.: Arabic Handwriting Recognition Using Baseline Dependant Features and Hidden Markov Modeling. In: ICDAR, pp. 893–897 (2005)
5. Makhoul, J., Schwartz, R.M., Lapre, C., Bazzi, I.: A Script-Independent Methodology For Optical Character Recognition. Pattern Recognition 31(9), 1285–1294 (1998)
6. Märgner, V., Pechwitz, M., El-Abed, H.: ICDAR 2005 Arabic Handwriting Recognition Competition. In: Eighth International Conference on Document Analysis and Recognition (ICDAR 2005), vol. 1, pp. 70–74 (2005)
7. Wellekens, C.J., Kangasharju, J., Milesi, C.: The use of meta-HMM in multiflux HMM training for automatic speech recognition. In: Proc. of Intl. Conference on Spoken Language Processing (Sydney), pp. 2991–2994 (December 1998)
8. Bourlard, H., Dupont, S.: Sub-band-based Speech Recognition. In: IEEE Int. Conf. on Acoust., Speech, and Signal Processing, pp. 1251–1254 (1997)
9. Ben Amara, N., Belaïd, A., Ellouze, N.: Utilisation des modèles markoviens en reconnaissance de l'écriture arabe état de l'art. In: Colloque International Francophone sur l'Ecrit et le Document (CIFED 2000), Lyon, France (2000)
10. Khorsheed, M.S.: Recognizing Arabic manuscripts using a single hidden Markov model. Pattern Recognition Letters 24, 2235–2242 (2003)
11. Makhoul, J., Schwartz, R., Lapre, C., Bazzi, I.: A script independent methodology for optical character recognition. Pattern Recognition 31(9), 1285–1294 (1998)
12. Miled, H., Olivier, C., Cheriet, M., Lecourtier, Y.: Coupling observation/letter for a Markovian modelization applied to the recognition of Arabic handwriting. In: ICDAR 1997, Ulm, pp. 580–583 (1997)
13. Pechwitz, M., Märgner, V.: HMM Based approach for handwritten Arabic Word Recognition Using the IFN/ENIT– DataBase. In: ICDAR 2003, Edinburgh, pp. 890–894 (2003)
14. Parker, J.R.: Algorithms for Image Processing and Computer Vision. John Wiley and Sons, Inc. (1997)
15. Larigo, L.M., Govindaraju, V.: Off-line Handwriting Recognition: A survey. IEEE PAMI, 712–724 (2006)

Confirming the Design Gap

Benjamin Menhorn and Frank Slomka

Institute for Embedded Systems/Real-Time Systems
Ulm University
Albert-Einstein-Allee 11, Ulm, Germany
{benjamin.menhorn,frank.slomka}@uni-ulm.de

Abstract. The design gap is the divergence between technology capabilities and design capabilities. This work uses a complexity measure to calculate complexity figures for highly regular and irregular structures. This measurement allows to assess complexity without the need of empirical data to chart the design gap. This will demonstrate that the design gap is caused by the structure itself.

1 Introduction

The design gap is well-known to digital hardware designers and others. It shows the spread between technology capabilities and hardware design capabilities. Among others, one explanation lists the structure of designs as cause for the design gap. For now, there are only reasonable explanations for the structure as source. The aim of this work is to show, that the design gap is caused by the structure itself. Therefore a complexity measurement method is applied to designs with different degrees of regularity.

This approach faces two main challenges. On the one hand two different designs have to reflect the technology capabilities and hardware design capabilities. On the other hand an adequate measurement method for complexity has to be found. The complexity measure has to provide figures as mathematical statements about the designs.

This work is organized as follows: First, the design gap is discussed with its possible explanations in related work. Then, it will be discussed why a memory as a highly regular structure and a processor as a highly irregular structure can reflect technology capabilities and hardware design capabilities. The next part introduces a hardware design measurement which is used to calculate complexity. With all presuppositions the last part introduces and analyzes the designs and reveals the design gap. The work closes with the conclusion.

2 Related Work

2.1 Moore's Law

In context with integrated circuits, Moore's Law describes the long-term trend for the amount of transistors which can be placed on a chip inexpensively. Depending on the source, the amount doubles every 18 to 24 months [1] [2] [3].

D. Nagamalai et al. (Eds.): *Adv. in Comput. Sci., Eng. & Inf. Technol.*, AISC 225, pp. 281–292.
DOI: 10.1007/978-3-319-00951-3_27

Figure 1 charts the trend of integrating transistors on a single chip over the past two decades. Even though Moore's Law is not a scientific natural law it is widely accepted as an observation prediction of integrated circuit development. At the same time one can speak of a "self-fulfilling prophecy" since various industries are involved in the development of better microchips [4].

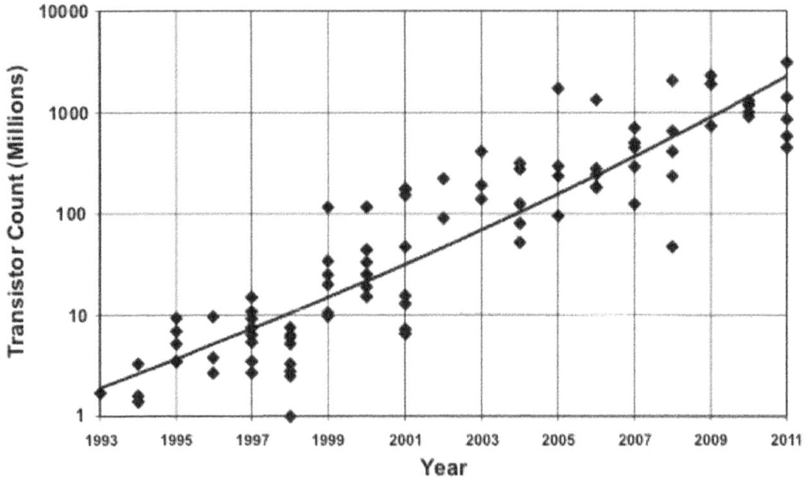

Fig. 1. Chip complexity development (from [5])

Originally Gordon Moore observed the number of components per integrated function [1]. Later, transistors per chip were counted instead of components. Even though Moore's Law counts transistors over time, it is often used to describe the development of complexity over time [6]. The figures are based on observations and forecasts. In order to compare the complexity from Moore's Law with the productivity, the following section will introduce the design gap.

2.2 The Design Gap

The design gap is the divergence between technology and design capabilities. Design capabilities (also called productivity) are measured in transistors per day [7]. Technology capabilities (also called complexity) are measured in transistors per chip as in Moore's Law [7]. **Figure 2** plots productivity and complexity over time. The design gap can be clearly identified. Complexity has an annual growth rate of 58% while productivity has an annual growth rate of 21% [8]. A common opinion for the slower growing productivity found in literature bases the design gap on missing tools, abstraction layers and design possibilities [9] [10]. Another work states, that the "designers appear limited by the very tools and processes that made those billion transistors a reality" [11]. As well as that the "design gap

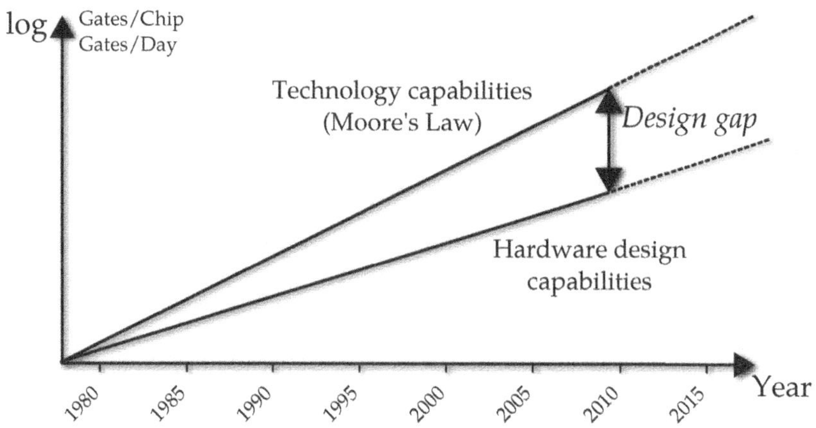

Fig. 2. Design gap (Data from [13])

might be avoided if more ESL (**E**lectronic **S**ystem **L**evel) methodologies would be deployed" [12].

This works advances the view that the design gap is originally caused by the difference between regular and irregular structures itself. Therefore the following section will identify designs representing highly regular and highly irregular structures as well as introduce a complexity measure.

3 Presuppositions

In order to calculated complexity for charting the design gap, two challenges appear. The first challenge is to identify a design which represents the technology capabilities (Moore's Law) and another design representing the design capabilities (productivity). The second challenge is to find an adequate complexity measure for hardware designs.

3.1 Structure Regularity

The design gap is the disparity between transistors available to a designer and the ability to use them effectively in a design [9]. The technology capabilities in Moore's Law represent the amount of transistors which can be placed on a chip inexpensively. The highest gate density can be found in memories, a highly regular structure. Breaking it down to the basic components, a memory consists of single memory cells and an inverse multiplexer as control unit. With the development of one (elementary) memory cell the basic component for a memory is designed. This cell can be copied in any quantity and put together to a grid. This grid and a separately developed control logic compose a whole memory. The reusability of single memory cells and the grid arrangement reduce the design complexity drastically.

On the other hand, a design representing the productivity can be found in a processor. Processors are highly irregular structures. Their control logic needs most of their chip area. This is a highly irregular structure which is more complicated to design. There are no elementary cells which can be copied in any quantity. But parts of the processor such as caches or registers also consist of single, reusable components. Therefore processors are not completely irregular structures.

In contrast to memories, reusability in processors is very limited. This increases design complexity. By adding regular structures, for example by increasing the processor's memory (cache), the throughput can be increased. By reducing the complexity of irregular structures at the same time, e.g. by using RISC architectures, the ratio between regular and irregular structures can be enhanced in favor of less complex designs.

The design gap is caused by this difference in structure. Additional abstraction layers and design tools shorten the gap between technology capabilities and design capabilities. Those approaches reduce the decision possibilities for designers and engineers. Thereby the amount of possible states of a system is reduced. This is especially interesting in design automation processes.

3.2 Complexity Measure

The determination of complexity makes it possible to give system related statements and allows to compare different systems [14] [15]. In order to develop a complexity model, complexity itself needs to be understood [16]. But today, most methods for estimating system size use empirical data by analyzing previous projects [17]. In [18] a measurement, called design entropy concept, was proposed which doesn't rely on empirical data. This concept bases its calculations on an abstract variable: states. "A state is a situation in which a system or system's component may be at a certain point of time. It refers to the interior of a system and ignores external influences such as input and output. The set of states is the abstraction of a real system" [19].

The main statements of the design entropy concept are summarized in the following in order to calculate complexity for the different designs in the next section. The approach of the design entropy bases on Shannon's information entropy. In order to give mathematical statements about transmitted information, Claude Elwood Shannon developed a model which became famous as Shannon's information theory [20]. The design entropy concept identifies the single components of a design as sources and drains of information. Connections between components are channels and information are symbols transmitted from a pool of available symbols. In digital hardware connections are normally implemented by wires. The available symbols are "high" and "low" signals in the simplest model. For instance an assignment `a:=b` between two components would be identified as: The information *(=signal level)* from component *(=source/sender)* `b` is transmitted *(=assigned)* to component *(=drain/receiver)* `a`.

Complexity can be considered to be a measure of a system's disorder which is a property of a system's state. Complexity varies with changes made at the amount of possible states of a system. An entropy measurement can be used to measure project size by using states. States are abstract variables which depend on the analyzed property of a project. It becomes possible to calculate the effect of introducing design tools to a development process by calculating complexity on different abstraction levels and comparing them.

$$H = -K \sum_{i=1}^{N} p_\alpha \log p_\alpha \tag{1}$$

The form of Shannon's theorem (see equation (1)) is recognized as that of entropy as defined in certain formulations of statistical mechanics (e.g. [21]), where p_α is the probability of a system being in cell α of its phase space. H is from Boltzmann's famous H theorem and the constant K merely amounts to a choice of unit of measure [20]. Equation (1) can be rewritten as (2) and leads to definition 1 [18] [22].

Definition 1 (Behavior Entropy) *Let c be a component with inputs (component's sources) $n_i(c)$, $i = \{1, \ldots, n(c)\}$ and outputs (component's drains) $m_j(c)$, $j = \{1, \ldots, m(c)\}$, where $n(c)$ is the amount of inputs of c and $m(c)$ the amount of outputs of c. Let $z(n_i(c))$, $i = \{1, \ldots, n(c)\}$ be the amount of possible states for the inputs $n_1(c) \ldots n_{n(c)}(c)$ and $z(m_j(c))$, $j = \{1, \ldots, m(c)\}$ be the amounts of possible states for the outputs $m_1(c) \ldots m_{m(c)}(c)$. Then the behavior entropy $H_B(c) \in \mathbb{R}_0^+$ of component c is given by:*

$$H_B(c) = log\left(\prod_{i=1}^{n(c)} z(n_i(c)) \cdot \prod_{j=1}^{m(c)} z(m_j(c))\right) \tag{2}$$

The behavior entropy gives a statement about the (usage) complexity of a component. The behavior complexity does not allow for the actual implementation complexity of a component. It only provides complexity information about the usage of a component. It can be compared to an outer or black box view on a component. In contrast to the behavior entropy the structure entropy in definition 2 provides information how complex the implementation/realization of a component is. This is similar to an inner or white box view on a component.

Definition 2 (Structure Entropy) *Let c be a component with instances c_b and implemented sub-components c_s. Then the structure entropy $H_S(c) \in \mathbb{R}_0^+$ for component c is given by the sum of all behavior entropies of all instances c_b and the structure entropy of all implemented sub-components c_s:*

$$H_S(c) = \sum_{i \in c_b} H_B(i) + \sum_{j \in c_s} H_S(j) \tag{3}$$

The structure entropy allows to add up the entropies from the single sub-components. If these sub-components have also been implemented, their structure complexity needs to be considered, too.

In the following section the measurement will be applied on a well-known effect, the design gap. This will demonstrate on the one hand, that the source for complexity lays in the structure itself and on the other hand that the design entropy model is capable to give mathematical statements about the complexity of designs.

4 Explaining the Design Gap

In order to demonstrate that the design gap is caused by the structure itself, we implemented a memory and a processor in VHDL. For both implementations, only basic gates (such as AND, OR, XOR, NOR etc.) were used in order to make both designs comparable. In order to demonstrate the design gap, we need to calculate the complexity. Because both implementations consist of several single components, we will show the complexity calculation with a memory cell. Therefore, the structure of the memory will be described in the following and the VHDL code will be given.

4.1 Memory

As discussed before, this work considers a memory as a design where transistors can be placed on a chip inexpensively. The memory is linked to Moore's Law and the technology capabilities. The fundamental structure of a memory in shown in **figure 3**. The main parts are an address decoder and the storage matrix. The storage matrix consists of single memory cells. The structure of a memory cell in shown in **figure 4**. The core is a D-flip-flop, which can hold one bit. The VHDL implementation of such a memory cell is given in **listing 1.1**.

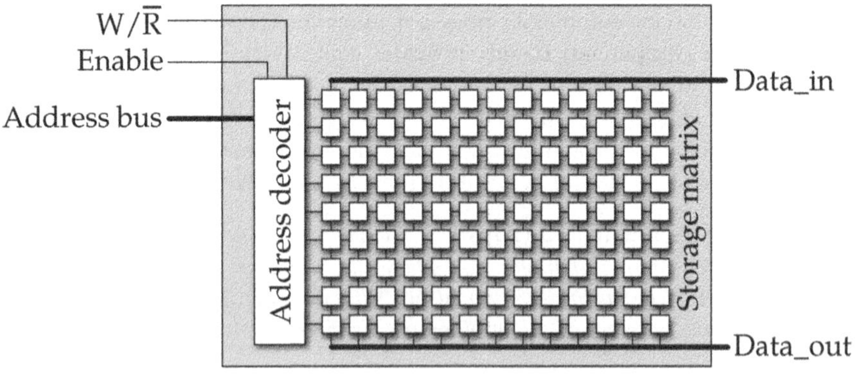

Fig. 3. Fundamental structure of a memory

```
entity memory_cell is
port (
     I           :in  std_logic; --input bit
     W           :in  std_logic; --write_enable
     S           :in  std_logic; --select_enable
     O           :out std_logic  --output bit
);
end memory_cell;

architecture structural of memory_cell is
     signal E,D_g,E_g,Qa,Qb,NotD :std_logic;

begin
     E <= W AND S;
     NotD <= Not I;
     D_g <= NotD AND E;
     E_g <= I AND E;
     Qa <= Qb NOR D_g;
     Qb <= Qa NOR E_g;
     O <= Qa and S and not W;
end structural;
```

Listing 1.1. VDHL code for a memory cell

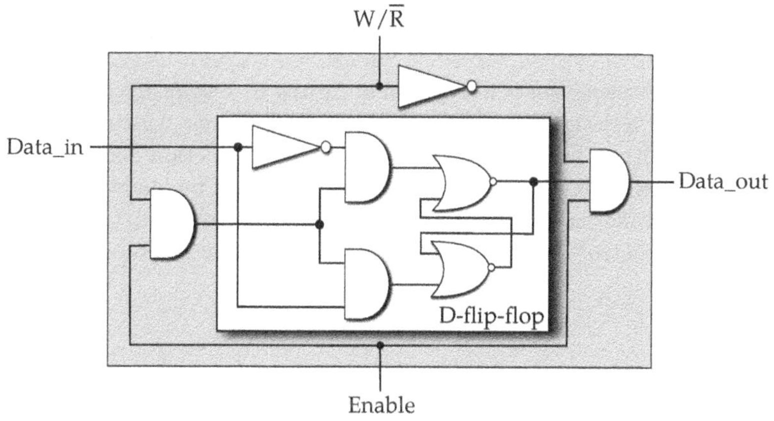

Fig. 4. Memory Cell

In order to calculate the complexity for the memory equations (2) and (3) are used. The calculation starts with the complexity of a single memory cell:

$$H_B(\text{memory\_cell}) = 4 \cdot \log(2) \tag{4}$$

$$H_S(\text{memory\_cell}) = 2 \cdot H_B(\text{Inverter}) + 3 \cdot H_B(\text{AND}) + 2 \cdot H_B(\text{NOR}) + H_B(\text{AND\_2}) = 23 \cdot \log(2) \tag{5}$$

The memory cells are used to build up 1-bit word arrays which compose m-bit word arrays. These word arrays are the storage matrix of the memory. The address decoders consists of a n-bit wide inverse multiplexer which itself consists of single 1-to-2 inverse multiplexers. The whole structural design hierarchy can be found in **figure 5**. The complexity for the whole memory is calculated in equation (6)

$$H_S(\text{memory}) = H_B(address\_decoder) + H_S(address\_decoder) + H_B(storage\_matrix) + H_S(storage\_matrix) = ((m+4)2^n + m^2 + 2n + m + 33) \cdot \log(2) \tag{6}$$

The next part will calculate the complexity for the processor. Because of its more heterogeneous design there are more components to consider.

4.2 Processor

The design for the processor is illustrated in **figure 6**. The design of the processor is based on a MIPS design from [23]. We use a fixed address width of 32-bit and a variable data width. The processor has a separate instruction and data memory. The program counter is a m-bit wide register which can be loaded with the next address from the adder or directly with an address from an (un-)conditional jump from the control unit. The ALU can perform five basic operations (add, sub, and, or, slt).

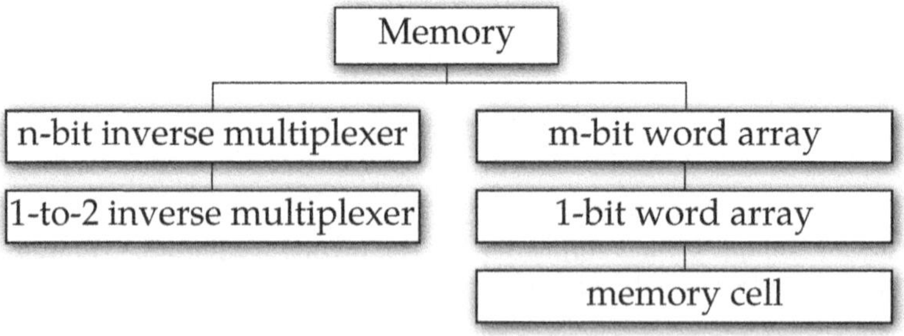

Fig. 5. Structural design hierarchy of a memory

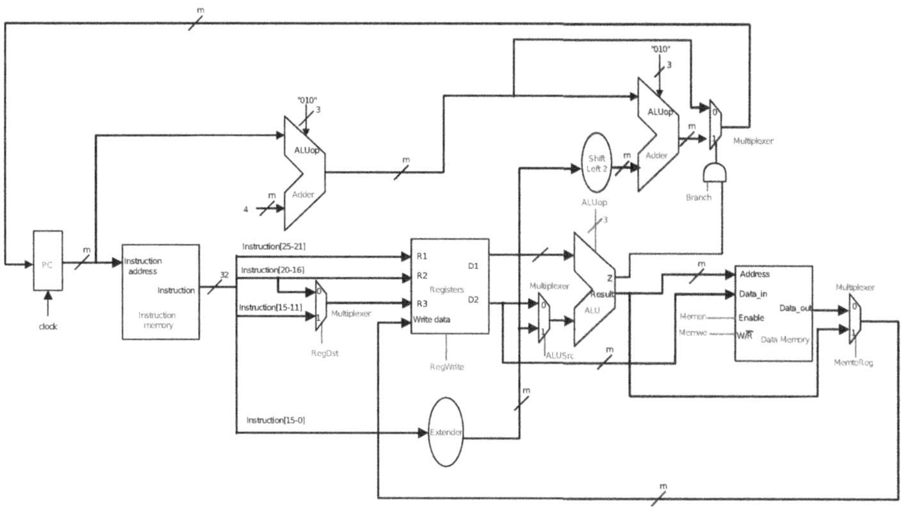

Fig. 6. Basic processor architecture

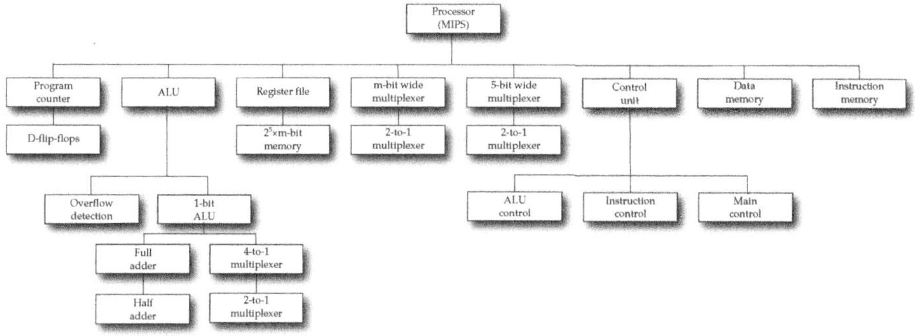

Fig. 7. Structural design hierarchy of a processor

In order to calculate the design entropy of our MIPS, the behavior entropy and structure entropy of every single component has to be calculated. As figure 6 shows, the design is very heterogeneous. Therefore, the structural design hierarchy of this processor becomes large, as shown in **figure 7**. The ALU is composed by single 1-bit ALUs. Since we already built a memory, we used this memory for the data and the instruction memory of the processor. Due to the huge amount of components and sub components, we only provide the complete complexity for the whole processor in equation (7).

$$H_S(\text{processor}) = \\ ((m+40)2^m + 2m^2 + 100m + 1748) \cdot \log(2) \tag{7}$$

4.3 The Design Gap

With equation (6) for the memory's entropy and (7) for the processor's entropy, both designs can be plotted as a function of the data width m. The address width for the memory is also 32-bit. As the minimum data width for the processor is 32-bit, we chart values starting with 32-bit. This leads to the complexity figures in **table 1**. The figures can then be plotted in **figure 8**.

Table 1. Entropy for different data widths

Width	32	64	128	256
Processor	$4,3 \cdot 10^{11}$	$2,7 \cdot 10^{21}$	$8,1 \cdot 10^{40}$	$8,4 \cdot 10^{97}$
Memory	$1,5 \cdot 10^{11}$	$2,9 \cdot 10^{11}$	$5,7 \cdot 10^{11}$	$1,1 \cdot 10^{12}$

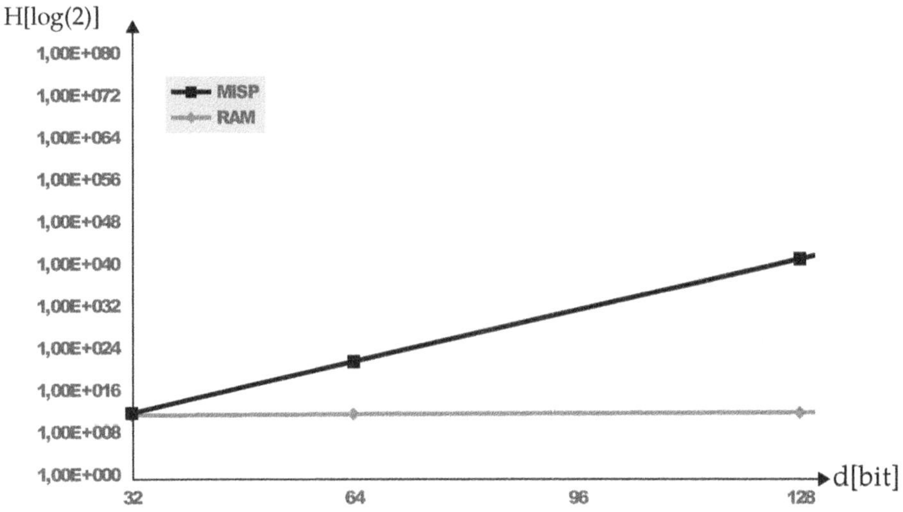

Fig. 8. Entropy of processor and memory

The divergence between the processor and the memory can clearly be identified.

5 Conclusion

As figure 8 shows, the design of a processor is more complex than the design of a memory. A memory was chosen to represent highly regular structures while a processor represents highly irregular structures. Therefore the results from the complexity calculation were expected to show that highly irregular structures are more complex than highly regular structures. In contrast to the chart in figure 2, the values for figure 8 were calculated by the complexity measurement method, which was introduced earlier.

Figures for Moore's Law and for the productivity rely on empirical data. Explanations for the slower growing productivity are missing tools, abstraction layers and design possibilities. But as this work shows, the gap is caused by the structure itself. It also shows that complexity can be calculated by relying only on key figures of a design.

It wasn't necessary to take empirical data into account for the calculation. For the complexity calculation a measurement was chosen which depends on states. With this complexity calculation method, it can also be explained why the introduction of further tools, abstraction layers and design possibilities shorten the gap. Those approaches reduce the decision possibilities for designers and engineers and therefore reduces the amount of possible states.

With this work, it was demonstrated that complexity figures can be calculated. This calculation reflects empirical data known for 30 years in semiconductor industry. Therefore it reflects the effort of complexity in future designs. Concluding, that the different complexities depend on the designs' structure. Therefore the cause for the design gap also depends on the structure.

References

1. Moore, G.E.: Cramming more components onto integrated circuits. Electronics 38(8) (April 1965)
2. Intel Corporation: Excerpts from a conversation with gordon moore: Moore's law (2005), `ftp://download.intel.com/museum/Moores_Law/Videotranscripts/Excepts_A_Conversation_with_Gordon_Moore.pdf`
3. Kanellos, M.: Moore's law to roll on for another decade (2003), `http://news.cnet.com/2100-1001-984051.html`
4. The INQUIRER: Gordon moore says aloha to moore's law (2005), `www.theinquirer.net/inquirer/news/1014782/gordon-moore-aloha-moore-law`
5. Conference, I.S.S.C.: Isscc 2011 trends report (2011), `www.isscc.org/doc/2011/2011_Trends.pdf`
6. Mollick, E.: Establishing moore's law. IEEE Annals of the History of Computing 28(3), 62–75 (2006)
7. Semiconductor Industry Association: International technology roadmap for semiconductors (2007), `www.itrs.net/Links/2007ITRS/Home2007.htm`
8. Semiconductor Industry Association: International technology roadmap for semiconductors (1999), `public.itrs.net/files/1999_SIA_Roadmap/Home.htm`
9. Sedcole, N.P.: Reconfigurable Platform-Based Design in FPGAs for Video Image Processing. PhD thesis (2006)
10. Henkel, J.: Closing the soc design gap. Computer 36(9), 119–121 (2003)
11. Hamilton, S.: Semiconductor research corporation: Taking moore's law into the next century. Computer 32(1), 43–48 (1999)
12. Henkel, J., Wolf, W., Chakradhar, S.: On-chip networks: A scalable, communication-centric embedded system design paradigm. In: International Conference on VLSI Design, p. 845 (2004)

13. Ecker, W., Müller, W., Dömer, R.: Hardware-dependent Software - Principles and Practice. Springer (2006)
14. DeMarco, T.: Controlling Software Projects: Management, Measurement, and Estimates. Prentice Hall PTR, Upper Saddle River (1986)
15. Fenton, N.E., Pfleeger, S.L.: Software Metrics: A Rigorous and Practical Approach, Revised. Course Technology (February 1998)
16. Calvano, C.N., John, P.: Systems engineering in an age of complexity: Regular paper. Syst. Eng. 7, 25–34 (2004)
17. Hinrichs, N., Leppelt, P., Barke, E.: Building up a performance measurement system to determine productivity metrics of semiconductor design projects. In: IEEE International Engineering Management Conference (IEMC), Austin TexasErmolayev 2007. CD–ROM Proceedings. IEEE (2007)
18. Menhorn, B., Slomka, F.: Design entropy concept. In: ESWEEK 2011 Compilation Proceedings (October 2011)
19. Menhorn, B., Slomka, F.: States and complexity. In: Dumitrescu, D., Lung, R.I., Cremene, L. (eds.) Coping with Complexity COPCOM 2011, pp. 68–88 (October 2011)
20. Shannon, C.E.: A mathematical theory of communication. Bell System Technical Journal 27, 379–423, 623–656 (1948)
21. Tolman, R.C.: The principles of statistical mechanics. Oxford Univ. Pr., London (1938)
22. Menhorn, B., Slomka, F.: Project Management Through States. In: IEMS 2009: International Conference on Engineering Management and Service Sciences (2009)
23. Patterson, D.A., Hennessy, J.L.: Computer Organization and Design: The Hardware/software Interface, 3rd edn. Morgan Kaufmann (2005)

A Digital Forensic Investigation Model for Insider Misuse

Ikuesan R. Adeyemi, Shukor Abd Razak, Anazida Zainal, and Nor Amira Nor Azhan

Information Assurance and Security Research Group, Faculty of Computing,
University Teknologi Malaysia, 81310 Johor, Malaysia
{raikuesan2,namira5}@live.utm.my, shukorar@utm.my,
anazida@gmail.com

Abstract. It is no longer a hidden fact, that insider misuse, either intentional of unintentional, constitutes grave consequence to business continuity. Detection and prediction of such misuse are however facing practical setbacks, due in part to the relative proximity of an insider to organizational assets, as well as human dynamics in relation to societal dynamics. The Saying of "prevention is better than cure" thus becomes the best option for such misuse mitigation. One way of prevention is deterrence, through investigative capability. This research therefore presents an investigation model for insider misuse mitigation. This model can be strictly applied for identification of the insider emergence, as well as for identification of misuse activities from an insider action. Implementing this model in forensic process can be a breakthrough for digital forensics in insider misuse occurrences.

Keywords: Digital forensics, investigation, insider definition, insider classification, insider misuse investigation, insider misuse prevention, investigation model, BellLa-Padula Model, Biba Model.

1 Introduction

The internet is a world of virtual reality that aids criminal activities as against the rigors and impossibilities attributed to real world crime. Unfortunately, this is also applicable to insider misuse of organization information technology (IT) infrastructure. An insider to an organization (as expanded in section 3) is a trusted subject that have physical as well as logical access privilege to organization infrastructure. Insider misuse forensics is an under developed (Hunker and Probst, 2011) research field. Research work in insider misuse detection, prevention, protection, assessment and avoidance is gradually curbing the "would have been disaster" caused by malicious user whose clearance level is used as the key instrument for misdemeanor. Insider misuse can be generally classified into masquerader and misfeasors. A masquerader is an unauthorized user with illegal access to a node, while a misfeasor is a legitimate user who deliberately or ignorantly violates the system use, or performs an operation that is capable of affecting the confidentiality, integrity and availability of information system. However, insider misuse has been researched and solution proffered, yet, surveys (Magklaras and Furnell 2004; Kowalski et. al. 2008) and investigation results

D. Nagamalai et al. (Eds.): *Adv. in Comput. Sci., Eng. & Inf. Technol.*, AISC 225, pp. 293–305.
DOI: 10.1007/978-3-319-00951-3_28

(Verizon 2012), insider misuse still lingers; with a more disastrous effect on organization. The paradigm of detection, protection and assessment of insider misuse (human centric problems) faces serious inherent challenges, as an insider may be aware of such mechanism. The definition of "whom" and "what" an insider is, is a major concern to the above-mentioned paradigm. Thus the need for a complementary paradigm that can be viewed as a deterrent for insider misuse. Forensic investigation is perceived as a major deterrent to criminal activities. Insider misuse investigation is a relatively new field of research in digital forensic discipline; a paradigm which helps to proffer suspect-pool list in event of crime occurrence. However, pinpointing a malicious insider in a seemingly virtual environment is a complex challenge with various unquantifiable variables/features. Therefore, establishing such a paradigm will serve as a milestone in the quest for a deterrent mechanism to insider misuse/threat. This research therefore aims to introduce a new paradigm to insider misuse prevention through an investigative model. The rest of this paper is organized as follows: section 2 gives an elucidatory insight into related research in insider misuse. In section, the paper presents findings on proposed investigation model. Conclusion and future works is presented in section 4.

2 Related Works on Insider Misuse Investigation

Relative to other types of investigation, insider attacks/misuse investigation with respect to information technology is an emerging field with a "less than hand full" research works. As against information security paradigm, investigating insider misuse of IT infrastructure is a post mortem process, which could leave substantial or no evidence, complex privacy concerns, and "by the book" process before initiation. Few researchers have worked on the insider investigation from computer forensics perspective, as well as from the network security perspective. Magklaras and Furnell (2004) presented a prediction tool for insider misuse evaluation using probabilistic process. The Authors identified and presented a four taxonomical classification of insider misuse. As a way to identify and hence measure insider level of sophistication, Magklaras,(2011) designed a mechanism which considers user breath of knowledge, depth of knowledge, finesse, and effect on IT infrastructure as a measure of user sophistication. The classification of user breadth into high, medium, and low with arithmetic equivalent of 6, 3, and 1 respectively, yields a distinctive level of sophistication. However, classifying user into three classes for investigation purpose is shallow, and could lead to longer period of investigation with higher approximation.

An investigative classification should consider among other things the need for granularity and discreteness. Thus challenges as to the level of expected granularity desired, especially in bigger organization where end-users falls within different range of sophistication (as against a concluded: {advance user} = 2{normal user} = 4{novice user}), becomes more difficult. Popovsky et. al. (2007) developed a network forensics development life cycle using an embedded cognitive digital forensic investigation methodology. They identified the lack of conceptual digital forensic framework from user perspective as the major challenge sinking insider investigation. However, they failed to propose, develop and implement any augmented cognitive forensics approach, though mentioned. Additionally, their proposed life cycle process was not implemented. Similarly, Adeyemi, Razak, and Azhan (2012) presented an investigation framework for military, industry, and law enforcement paradigm, indicating the critical features required for each perspective and the sequential procedure for effective investigation. Dehkordi and Carr (2011) presents insider misuse detection from multi-perspective user profile paradigm. They argued that existing insider detection algorithms and metrics are user-centric, thus, cannot yield accuracy in a collaborative-user threat scenario. Using a one-class support vector machine (SVM) detection engine, in a controlled environment of 20 users, they developed a user feature vector, file server feature vector, and feature vector of database server. Their approach however did not offer detection process for impersonation and compromised node.

Furthermore, the feature vector selection, which determines the result of the LIBSVM engine, did not consider situational influence, hence devoid of possible consistency and accuracy, which are the key factors for evidence admissibility in investigation. Stiawan et. al. (2012) proposed an insider threat prediction adopting a habitual profile activity methodology, as users profiling process. Four classes of possible misdemeanor: remote login, slammer, DOS/DDOS, ragnawk; were identified. Thus, a new architecture for IPS was proposed. The defined user habitual behavior as the feature vector, which can be used to classify abnormal and normal user activities. Moreover, a behavioral learning system for insider attack requires self-learning, self-updating process, which can reveal user complexity, as well as identify unique changes. Additionally, users' inconsistency and shared-common-expression are factors in user behavior, which are not taken into consideration. Research on malicious insider investigation in Shaw (2006), identified lack of empirical data and accepted typologies on insider activities as a major challenge to insider misuse investigation. Shaw (2006) further identified deductive and inductive profiling techniques as a case-based investigation and general conclusion base on knowledge respectively.

Techniques such as remote psychological profiling for content analysis, which comprises quantitative assessment distribution of psycholinguistic characteristics, and qualitative content analysis; psychoanalytic theory are identified for insider investigation. An insider attack prediction framework presented in Schultz (2002) identified capability, motive, and opportunity as a model for understanding insider attack source. Using multiple regression equations methodology, Schultz (2002) presents a quantitative measure for evaluating attacks. Bhukya and Bariothu (2011), presents a one-class SVM approach for monitoring, analyzing and reporting graphic user interface of user behavior, as an investigative forensic methodology, for computer related insider misuse investigation. Velpula and Gudipudi (2009) approached insider misuse investigation using statistical analysis of system logs, and application servers. The Authors presents a behavior anomaly-based investigative methodology, which profiles users based on their resource access behavior that exceeds their necessary need to know. Using "peer-group access pattern, user behavior statistics, job experience, job-related information, geographical location and personnel rating" as independent variable, they gathered "access to record, time between access of particular application, number of sensitive data accessed, login characteristics, sequential access pattern, and location-base behavior" as dependent variable. It suffix to note that the various variables (both dependent and independent) gathered by the Authors are enterprise specific, and may not necessarily describe the user. Additionally, such investigative methodology requires integration of various evidence sources, which may not be originally structured for such use. Thus, would require high precision and expertise, which could be a scarce resource. According to Greitzer and Frincke, (2010), "the insider threat is manifested when human behavior deviates from compliance with established policies, regardless of whether it result from malice or a disregard for security policy".

This is contingent on physical measurement of human behavior, which can be bias, yield false positive outputs, or even based on false positive assumptions that is bound to fail. The Authors adopted knowledge-based component, which comprise template of malicious behavior. By adopting a template instead of the traditional pattern recognition and anomaly detection, the Authors used sophisticated reasoning approach, which utilizes cognitive inference of meaningful state through Bayesian network, which calculates the probability of event. However, it is arguably insightful to mention that pure template recognition models would generate more authentic, functional and realistic template than a situational pre-defined template, which is subject to the sophistication, disposition, and vastness of the designer, as well as the knowledge of the system under view. This therefore introduces subjectivity in term of the expertise of the designer, robustness of the template, and efficiency of reasoning input variable. Off course, if the input is questionable, irrespective of the procedure and technique used, the output will also be subjective.

3 Insider Misuse Investigation

Insider misuse investigation is a human centric investigation process, which requires a thorough physics of the attributes of users, in relation to their responsibility and organization. Insider (human) can be identified through their personality trait, verbal behavior, expected vastness, skill and acquired knowledge, access privilege, and psychological make-up reflected in their communication over the network. An insider is a subject (S) with legitimately adequate clearance level (C_L) to perform certain function or process ($F_{s(0)}$, $P_{s(0)}$) on an object (O) with classification level C_{La}. Defining insider misuse can be subjective, but a general perspective is described in equations 1, 2, 3.

$$C_L\,(F_{s(0)}, P_{s(0)})\,S \neq C_{La}\,(F_{s(0)}, P_{s(0)})\,O \quad \cdots(1)$$

$$\text{Where, } C_L\,(F_{s(0)}) \neq 0 \quad \cdots(2)$$

$$\&\ C_L\,(P_{s(0)}) \neq 0 \quad \cdots(3)$$

When a subject performs an access ($F_{(s,0)}$, $P_{(s,0)}$) on an object, a negative equivalent maker is tagged on the subject that counteracts the value on the object, as long as they both have same clearance level. If however they do not have same clearance level, the access privilege is denied. On the other hand, if they subject violates the appropriate access right to an object, the access privilege will be automatically revoked/marked

As shown in equations 1, 2, and 3, the legitimately adequate clearance level (C_L) can be described as the position of category to which a particular user otherwise called subject, possesses. For example, an organization with varying degree of staff position; staffs can be tagged with a numerical equivalent of 1, 2, 3, 4 ,5 and so on, depending on the size of the organization and its hierarchical distribution adopted. $F_{s(0)}$, $P_{s(0)}$ are the various function which a subject can perform on an object. The apprehension of these underlying features forms the basis for investigative research on insider misuse.

Moreover, thorough investigation involves logical and evidentiary analysis of the victim behavior, offender behavior and the environmental circumstances. To understand insider in the paradigm of investigation, the above definition of insider misuse, elucidated in equations 1, 2, 3; is further restructured to reflect the influence of the surrounding environment, and victimology, using existing information security models.

3.1 Who an Insider Is

Several definitions of who an insider is, has trailed the stage of insider research community. Shatnawi et. al. (2011), defined insider as subject with appropriate access to system under view: an individual possessing both access knowledge and authorization access. Thus, an insider is a subject who in addition to knowledge, possesses access right privilege and knowledge of infrastructural vulnerabilities:

- knowledge of other insider schedule, position and timing
- knowledge of different kinds of dependencies style menu
- Knowledge of organization schedule and routines

of an organization. Magklaras (2011) further identified the importance of "the process" adopted. The Author defined insider as a subject with appropriate and adequate access right through interaction with requisite mechanism. Thus, an insider is a subject with substantial level of trust viewed from the microscope of implemented security. Shaw (2006), gives an abstractive definition of an insider based on their action, motivation and modus operandi, resulting in classification of subjects. An interesting descriptive definition of insider is also given by Neumann (1999), as illustrated in Table 1.

Table 1. Classification of insider, based on location around IT infrastructure

Classification of insider based on presence		Physical	Logical
Type 'A'	Inside/present	x	
	Outside/absent		x
Type 'B'	Inside/present		x
	Outside/absent	x	
Type 'C'	Inside/present	x	x
	Outside/absent		
Type 'D'	Inside/present		
	Outside/absent	x	x

Types 'A', 'B', 'C' and 'D' describes the classification of user based on their location, with respect to insider definition in Neumann (1999). The classifications represents insider who is physically present in the location of the IT infrastructure, but logically absent; insider with logical presence but without physical presence; insider without physical and logical presence; and insider with both physical and logical presence respectively. Neumann (1999) suggests that insider definition is relative to the terms of reference, and at such, cannot be definitively implied, in its pure ideal state. Hence, the definition of an insider can be viewed both from an outsider-insider, and from insider-outsider perspective.

Defining an insider encompasses the perceived knowledge of organization IT system an individual posses. Sarkar (2010) categorized insider into "pure insider, insider affiliates, outside affiliates, and insider associates". This categorization is vague, but can be used as a channel for a more appropriate definition. Thus this paper coins-out suitable interpretation to 'who an insider really is'. An insider could be a current employee, laid-off employee, contract workers/staffs, or affiliates of the three-mentioned classes. It suffix to note that all existing definition and classification of insider have been abstractive, consequently, insufficient for investigative purposes.

Investigative classification encompasses the relativity between various classes of subject, environmental and or circumstantial occurrence. Therefore, there is a need for an insider classification or definition system which takes into consideration the position of a subject, the possible influence to a subject, overall access knowledge skill required by a subject to access an object. Kandias et. al. (2010), identified the use of indicators such as personality trait, and verbal behavior as a good feature, but difficult to quantify. This study is directed towards defining insider from the holistic perspective, relevant for investigative description.

Moreover, we group subject into current employee, laid-off staff, contract staff, and affiliates to either current employee, contract staff, or laid-off staff, but particularly, with respect to current employee. A suitable illustration of an insider is the interpretation using crisp set, where the organization under view is universe of discourse. The following dcfinition therefore holds:

1. A current employee {CE} is defined by the properties which satisfies the rule

$$CE = \{x | P(x)\} + \{\hat{r} | P(\hat{r})\} \tag{4}$$

where 'x' is the access right of CE, and P(x) is the property of 'x' which, at any given condition of x belongs to X, P(x) is either true of false. '$\hat{r}$' is the access knowledge of CE, and P($\hat{r}$) is the property of '$\hat{r}$' which at any given condition, '$\hat{r}$' is required in X. Denoting this in the characteristics function, we have:

$$X_{CE}(x, \hat{r}) = \begin{cases} 1 & \text{for } x \in CE, \hat{r} \in CE \\ 0 & \text{for } x \notin CE, \hat{r} \notin CE \end{cases} \tag{5}$$

2. A laid-off employee {LE} is defined by the properties which satisfies the rule:

$$\text{LE} \subseteq \text{CE, and LE} \neq \text{CE} \tag{6}$$

Such that the knowledge possessed about the access right is the commonality while the access right ($\boldsymbol{x}$) is no longer valid for LE.

3. Contract Staff {CS} which include outsourcee, technical and non-technical consultants, cleaner; is defined by the property which satisfies the rule

$$X_{CS}(\boldsymbol{x}) = X - X_{CE}(\boldsymbol{x}) \tag{7}$$

Such that the knowledge possessed about the access right of CE is not disbursed/disclosed to CS. However, there could be some exception and commonality, but not involving classified files, and only for a limited duration.

4. An affiliate of either 'CE', 'CS', or 'LE' is defined by the properties which satisfies the rule of absorption by 'X' given by:

$$CE \cup X = X \tag{8}$$

Such that the personal/special knowledge possesses by CE is the key distinguishing factor between the affiliate and the CE. The access right however, remains common to both the affiliate and the CE. Thus from definitions 1, 2, 3, and 4, we can infer on the level of interaction between the various properties of 'who an insider to an organization is'; which this research clarifies, as who is an insider.

3.2 What Misuse Is

The term insider misuse in IT world refers to conglomerates of threat, posed by an insider, hence, mitigating such threat requires conglomerates of factors such as technical, and psychosocial approaches (Karagianmis et. al. 2004). Challenges such as environmental influence, psychosocial dynamics of human, as well as technical deficiencies (which cannot be wholly erased), further reveal the elasticity in the meaning of insider misuse. Existing IT policies and policy languages are abstractive and often times differential (De-jury policy and De-facto policy); without a detail subject to object relationship. IT policies or information security management system are anchored on certain models, which clearly delineate misuse and use-cases. While these can be satisfactorily applied to the technicalities of IT outsider misuse, they offer little or no measure to insider misuse protection, much less then would such policies be applicable or relevant for investigative purpose. Lack of stringent definition(s) which can clearly articulate use-cases constitutes one of the banes of insider misuse investigation.

Like an outsider who would consume more time and effort to obtain access, a malicious insider would also consume more time and effort to perpetuate malicious act, which could be from the deviation or contradiction to established behavior pattern, without a necessary precursor. For instance, Pramanik et. al.,(2009) defined IT insider misuse with respect to conflict of interest within organization under view. Various taxonomies for insider misuse definitions is discussed in Hunker and Probst (2011). Moreover, the modus ponens of IT misuse is anchored on the existing IT security models. Such models include the Bell-La Padula confidentiality model [24],

Biba Integrity model (Busko 2010), Lattice Model of secure information flow (Denning 1976), Chinese Wall Security model and conflict of interest (Breweer and Nash, 1989; Lin 2000), and Clark-Wilson security model (Clark and Wilson, 1987). Insider misuses are reflected against the triad of information security: confidentiality, integrity and availability. According to (Computer Economics, 2010; Keeney et. al. 2005), insider misuse is mostly attributed to information confidentiality and integrity attack. Thus in defining insider misuse for investigative purpose, an extract from existing security model for instauration model of use-case definition could be term a starting point.

3.2.1 Bell-La Padula (BLP) Confidentiality Use-Case Model

It is a no write down, No read up model. It adopts a state machine in defining acceptable (which defines use-case and the converse) access control system. Access operation is defined by (S, O, A) triple. A set of all current access operation P(s, o a), such that s∈S, o∈O, and a∈A; maximum security level (f_s), current security level (f_c), classification of object (f_o), such that the security level assignment f = (f_o, f_s, f_c); subject classification level is described as the clearance level of the subject, while object classification, object security level. Clearance level of s∈S must be greater than the security level of o∈O meant to read, and s∈S can write to an object o∈O in higher security level; modus tollens.

Therefore, BLP Model for use case = $s \in S \geq o \in O$, *and* $s \in S \leq o \in O$

$$\Rightarrow \therefore s \in S \geq o \in O = \textit{read access, else misuse: } s \in S \leq o \in O = \textit{write access, else misuse}$$

However in some cases, confidentiality is defined in the constraint of time (described as a function of time (f(t)). Therefore a read or write operation with respect to time is defined by the following equation.

$$\text{BLP investigative Model} = \frac{d}{dt}\left(s_t \in S \geq o_t \in O \textit{ read, else misuse: } s_t \in S \leq o_t \in O \textit{ write, else misuse}\right) \quad (9)$$

such that the summation of the current read and write operation of a subject over a period of time could be investigated, either as a possible misuse, or in conformity with benign operation. Suppose read write operation is referred to as "access" denoted by α,

$$\Rightarrow \therefore \alpha_{(t)} = \frac{d}{dt}(s_t \in S \geq o_t \in O \textit{ read, else, misuse : } s_t \in S \leq o_t \in O \textit{ write, else, misuse})$$

However, s∈S is a function of access knowledge possessed ($\hat{r}$) and access right ($\boldsymbol{x}$) of the subject.

$$\Rightarrow \therefore \alpha_{(t)} = \frac{d}{dt}(s\{\boldsymbol{x}, \hat{r}\}_t \in S \geq o_t \in O \textit{ read, else, misuse : } s\{\boldsymbol{x}, \hat{r}\}_t \in S \leq o_t \in O \textit{ write, else, misuse}) \quad (10)$$

Thus, a simple investigation process of subject, which has the capacity to write, read as well as absorb information contained in an object, is given by a distribution bounded by time, information flow and knowledge as explicated in the equation below.

$$\sum \alpha_{(t)}\, dt = \sum d(s\{x, \hat{r}\}_t \in S \geq o_t \in O \text{ read, else, misuse} : \\ s\{x, \hat{r}\}_t \in S \leq o_t \in O \text{ write, else, misuse}) \quad (11)$$

Information flow in misuse cases is also a key factor in confidentiality breach. Lattice model of access security is an information flow model, which complements Bell-la Padula confidential model. Lattice model of secure information flow adopts (⊕, SC, ⟶, N, P) tuple which represents class combining operator, security class corresponding to disjoint information, flow operation, sets of logical storage object, and sets of processes respectively; which defines sequentially secured information flow. Information theory introduces a measure of average uncertainty (also referred to as entropy) in random variable. The term random variable is described as the measure of unexpectedness of a probabilistic event, which also defines Shannon's measure of entropy (Mu and Clark, 2007) which is the measure of entropy based on logarithmic measure of unexpectedness with non-zero probability defined as;

$$\mathrm{P} = \log_2 1 / P \quad (12)$$

Hence, the total information carried by a set of event (information flow) can be expressed as the summation of their individual unexpectedness given in the equation below;

$$\sum P = \sum_{i=1}^{n} Pi \log_2 1 / Pi \quad (13)$$

The expression in Eqn XIII denote the entropy of a discrete random variable [31], generally represented as

$$\mathrm{H(x)} = \sum_x P(x) \log_2 1 / P(x) \quad (14)$$

A flow model (FM) is said to be secure iff the execution of a sequence of operation cannot give rise to a flow that violates the relation "⟶ ". The class of an SC ($a_{1...}a_n$ ∈ A) can flow into the class of another SC ($b_1 \ldots b_n$ ∈ B), if the combining operator of all the objects in "A" is permitted and the entropy (H) is non-negative. It follows therefore that all permissible flow, must be permitted by the flow relation: a sequence of operation is secure if individual operation is secure. The following thus holds for information flow use-case:

- A⟶A (reflective/associative)
- A ⟶ B, & B ⟶ C, then, A ⟶ C (transitive)
- A ⟶ B, & B ⟶A (redundancy, a possible cause for misuse)

The security structure of Lattice model describes typical secure information flow, and a converse flow would be termed misuse, especially in instances where a secure information flow have been defined and implemented. Such flow however, is limited to machine, which is relatively static (preprogrammed/programmed to follow pre-defined and recognizable pattern) compared to insider misuse which is human centered, thus dynamic. Adopting such security model in human control, will require the integration of stringent stable human dependent factors, such as time with respect to intuitive description of workflow and information-flow, impact of the environment, and psychosocial influence. Therefore secure information flow for insider misuse investigation should encompass knowledge of subject, psychosocial features, in addition to the Lattice flow model parameters. Suppose P_i, H_{df} and t represents psychosocial influence, human dependent factors, and time respectively; then defining use-case for investigative purpose for a savvy malicious insider would adopt the following tuple:

$$\text{Information flow Tuple} = (\oplus, SC, \rightarrow, N, P, H_{df}, P_i, t). \quad (15)$$

Unlike the Bell-La Padula and Lattice model, Chinese wall model (CWM) constrained access to data based on the prior access knowledge of the subject, and not on the object attributes alone. Thus, data are classified into group of "conflict of interest". A conflict of interest model such as the CWM adopts access right, which comprises read and write access. The CWM use-case states that subject is allowed access right to object if and only if the object it does not fall into different class of conflict of interest, to previously accessed object or knowledge of object. Table 2 illustrates 4-different classes of subject, belonging to 2-different conflict of interest group and 16 objects: each belonging to one of the conflict of interest class.

Table 2. Conflict of Interest Table

Conflict of Interest	Subject Class	Object 1	Object 2	Object 3	Object 4
CoI 1	Class A	X_1	X_2	X_3	X_4
	Class B	Y_1	Y_2	Y_3	Y_4
CoI 2	Class C	X_1	X_2	X_3	X_4
	Class D	Y_1	Y_2	Y_3	Y_4

Use-case rule states that any subject in class (say "A"), can access any object in another class as long as the subject have not access object in same conflict of interest. As shown in Table 2, Object 1,X_1; Object 2, X_2; Object 3, X_3 and Object 4 X_4 are in same CoI class. If Subject in class-A had viewed or has knowledge-of X_4, subject will be prohibited from viewing either X_1, X_2 or X_3 but could view object in other class, as long it has never viewed any other object from that group of conflict of interest. Violation of these use-cases can therefore be defined, misuse.

3.2.2 Biba Integrity Use-Case Model

This is a No read down, No write up model. "It describes the set of access control rules for data integrity. Thus it addresses the need to protect unauthorized data modification as well as internal consistency of data". Un-authorization could be in the form of unauthorized entity, or unauthorized process. Biba model adopts integrity classification level for object (I_o) and subject (I_s). The rule state thus:

Write access (W_a) = use-case Iff $I_s \geq I_o$
Read access (R_a) = use-case Iff $I_s \leq I_o$.
$\Rightarrow \therefore W_a = I_s - I_o$ = non-negative output, and $R_a = I_s - I_o$ = negative output.

Thus, an IT integrity misuse/violation occurs when a particular write access within information flow yields negative result, irrespective of the summation of the output from the summation of the write access of the information flow.

Similarly, when a read action within an information flow yields positive output, irrespective of the overall output of the summation or read access in the information flow, misuse is established.

4 Conclusion and Future Works

In this paper, the Authors present the need for an investigation model for insider misuse investigation, as a complement to detection and protection methods. The authors identified the challenges in insider misuse detection from which investigation model can be grafted as suitable deterrent to misuse. Furthermore, the Authors presents a holistic taxonomy of who should be referred to as an insider and what should be formulated as misuse based on confidentiality and integrity preservation. As a further research focus, the authors intend to fully develop the investigation model to incorporate human psychosocial attributes, example of which is the personality classification for insider identification. A thorough model of insider investigation model, which includes the dynamics of human in relation to its environment (organization in view).

References

Adeyemi, I.R., Razak, S.A., Azhan, N.A.: Identifying critical features for netowrk forensics investigation perspective critical. International Journal of computer science and information Seucity, 1–23 (2012), http://arxiv.org/ftp/arxiv/papers/1210/1210.1645.pdf (retrieved October 20, 2012 йил)

Breweer, D.F., Nash, M.J.: The Chinese Wall Security Policy. In: IEEE Symposium on research in Security and Privacy, pp. 206–214. IEEE, Oakland (1989)

Buško, V.: Measuring individual differences in psychological attributes: A psychometric view of contextual effects. Review of Psychology, 43–46 (2010)

Clark, D.G., Wilson, D.R.: A comparison of Commercial and Military Computer Security Policies, pp. 184–194. IEEE (1987)

Dehkordi, M.R., Carr, D.: A multi-perspective approach to insider threat detection. Military comnnunication conference-Track 3- cyber ssecurity and netorkl operation, pp. 1164–1169. IEEE (2011)

Denning, D.E.: A Lattice Model of Secure Information flow. Communication of the ACM, 236–243 (1976)

Economics, C.: Computer EEconomics: Metric for IT Management (May 2010), https://www.computereconomics.com/custom.cfm?name=postPaymentGateway.cfm&id=1435 (retrieved November 17, 2012)

Greitzer, F.L., Frincke, D.A.: Combining Traditional Cyber Security Audit Data with Psychosocial Data: Towards predictive Modelling for insider Threat Mitigation, pp. 85–112. Spinger Science + Business Media (2010)

Hunker, J., Probst, C.W.: Insiders and Insider Threats An Overview of Definitions and Mitigation Techniques. Journal of Wireless Mobile Networks, Ubiquitous Computing, and Dependable Applications 2(1), 4–27 (2011)

Kandias, M., Mylonas, A., Virvilis, N., Theoharidou, M., Gritzalis, D.: An insider threat prediction model. In: Katsikas, S., Lopez, J., Soriano, M. (eds.) TrustBus 2010. LNCS, vol. 6264, pp. 26–37. Springer, Heidelberg (2010)

Karagiannis, T., Molle, M., Faloutsos, M.: Long-Range Dependence Ten years of Internet traffic modeling, pp. 2–9. IEEE Computer Society (2004)

Keeney, M., Kowalski, E., Cappelli, D., Moore, A., Shimeall, T., Rogers, S.: Insider Threat Study: Computer System Sabotage in Critical Infrastructure Sectors. Software Engineering Institute, Carnegie Mellon University, Pittsburgh, PA (2005)

Kowalski, E., Conway, T., Keverline, S.P., Williams, M., Cappelli, D., Willke, B., Moore, A.: Insider Threat Study: Illicit Cyber Activity in the Government Sector. United State secret service, and CERT Software engineering Institute. Carnegie Mellon (2008)

Lin, T.Y.: Chinese Wall Security Model and Conflict Analysis. In: The 24th Annual International Computer Software and Applications Conference, COMPSAC 2000, pp. 122–127. IEEE Society, Taipei (2000), doi:10.1109/CMPSAC.2000.884701

Magklaras, G.B., Furnell, S.M.: A priliminary Model of end user sophistication for insder threat prediction in IT system, pp. 371–280. Elsevier Computer nad Security (2004)

Magklaras, G.B., Furnell, S.M.: The insider misuse threat survey: investyigating IT misuse from legitimate users. We-B Centre & Edith Cowan University, 1–8 (2004)

Magklaras, G. V.: An Insider Misuse Threat Detection and Prediction Language. Faculty of Science and Technology, School of Computing and Mathematics. PhD Thesis. University of Plymouth, Plymouth (2011), http://pearl.plymouth.ac.uk:8080/pearl_xmlui/handle/10026.1/1024?show=full (retrieved December 08, 2012)

Mu, C., Clark, D.: Quantitative analysis of secure information flow via probabilistic semantics. King's College London, London (2007)

Neumann, G.P.: The challenges of Insider Misuse. WorkShop on Preventing, Detecting, and Response to Malicious Insider Misuse. RAND, August 16-18, pp. 1–23. Computer Science lab, SRI Intenational EL-234, Santa Monica (1999), http://www.csl.sri.com/users/neumann/pgn-misuse.html (retrieved December 15, 2012)

Popovsky, B.E., Frincke, D.A., Taylor, C.A.: A Theoretical Framework for Organizational Network Forensic Readiness. Journal of Computers 2(3), 1–11 (2007)

Pramanik, S., Sankaranayanan, V., Upadhyaya, S.: Security policies to mitigate insider threat in the Document Control domain. In: 20th Annual Computer Security Application Conference, pp. 1–10. IEEE xplore (2004)

Rezau, K.M., Grout, V.: On Reducing the Degree of Long-range Dependent Network Traffic Using the CoLoRaDe Algorithm. IJCSNS International Journal of Computer Science and Network Security, 80–86 (2007)

Sakar, K.R.: Assessing insider threat to information security using technical, behavioral and organizational measures. Elsevier Information Security Technical reprot (2010)

Schultz, E.E.: A framework for understanding and predicting insider attacks. Computer Security, pp. 526–531. Elsevier Science Ltd., London (2002), doi:0167-4048/02

Shatnawi, N., Althebyan, Q., Mardini, W.: Detection of Insiders Misuse in Database System. In: International Multiconference of Engineers and Computer Scientist, Honk Kong, pp. 1–6 (2011)

Shaw, E.: The role of behavior research in Malaicious cyber insider investigation. Science Direct- Digital investigation, 20–31 (2006)

Stiawan, D., Idris, M.Y., Salam, M.S., Abdullah, A.H.: Intrusion threat detection from insider attack using learning behavior-base. International Journal of the Physical Science, 624–637 (2012)

Velpula, V.B., Gudipudi, D.: Behavior-Anomaly-Based System for Detecting Insider Attacks and Data Mining. International Journal of Recent Trends in Engineering 2(1), 261–266 (2009)

Verizon. 2012 Data Breach Investigation Report. e Australian Federal Police, Dutch National High Tech Crime Unit, Irish Reporting and Information Security Service, Police Central e-Crime Unit, and United States Secret Service, Verizon RISK Team. verizon (2012), http://www.wired.com/images_blogs/threatlevel/2012/03/Verizon-Data-Breach-Report-2012.pdf (retrieved December 12, 2012)

Comparative Analysis of Gravitational Search Algorithm and K-Means Clustering Algorithm for Intrusion Detection System

Bibi Masoomeh Aslahi Shahri, Saeed Khorashadi Zadeh, Ikuesan R. Adeyemi, and Anazida Zainal

Information Assurance and Security Research Group,
Faculty of Computing,
Universiti Teknologi Malaysia
m_aslahi@yahoo.com, saeedkh@live.com,
{rikuesan,anazida}@gmail.com

Abstract. Intrusion Detection System (IDS) is an active defense technology. Many clustering algorithms are used to improve the performance of accuracy and hit rate and reduce False Alarm Rate (FAR). Conventional k-Means is the most popular clustering algorithms due to its simplicity and efficiency. However, its performance is highly dependent on the initial centroid and may trap in local optima. In recent years, heuristic algorithms have been applied to solve clustering problems. Gravitational Search Algorithm which is one of the newest swarm intelligent provides a prototype classifier to address the classification of instances in multiclass datasets. This paper used KDD Cup 1999 dataset to evaluate the performance of the baseline k-Means and GSA-based classifier in terms of accuracy, FAR and hit rate. The results show that GSA has a capability in order to improve the performance of the system.

Keywords: Gravitational Search Algorithm, K-Means,performance, accuracy, False Error rate, False Alarm Rate, IDS, search algorithm.

1 Introduction

Alongside the growing application of computers and computer networks, computer security system has turned into mainstream. Several mitigation to the attacks that hinders the development of intrusion detection system using clustering algorithms to identify normal and abnormal behavior in the network [1] have been proposed. In data mining and intelligent computing, k-Means is the most popular algorithm that has been applied successfully to large datasets. K-Means is simple to implement and efficient in most cases [2, 3]. However, the performance of k-Means is highly dependent on the initial state of centroids and may converge to the local optima rather than global optima. K-Meansalgorithm tries to minimize the intra-cluster variance, but it does not ensure that the result has a global minimum variance [4, 5].In recent years, interest focuses have been towards optimization algorithms, which have been applied in classification problems. The main reason for that, perhaps, comes from the uncertainty nature of machine

D. Nagamalai et al. (Eds.): *Adv. in Comput. Sci., Eng. & Inf. Technol.*, AISC 225, pp. 307–316.
DOI: 10.1007/978-3-319-00951-3_29

learning techniques that have uncertainty within themselves. Since the accuracy of such techniques largely depends on the qualification of training process, thus, theneed for a well-trained process. In this paper, the performance of application of IDS will be addressed based on conventional k-Means and Gravitational Search Algorithm (GSA) on cluster analysis. The performance of conventional k-Means and GSA has beentested on KDD Cup 1999 dataset. The rest of the paper is organized as follows: Section 2 provides a brief background on clustering problems and k-means algorithms. In section 3 the structure of GSA is described. In section 4 the modified GSA is described so that it conducts a classification task. In section 5, the experimental results areexplained and comparison is made, in terms of accuracy and False Alarm Rate. And finally, in Section 6 the conclusion is stated.

2 Background on Clustering and K-Means Algorithm

K-Means is one of the most efficient clustering algorithms in terms of execution time relative to several other existing clustering algorithms [6]. It has also been reported that k-Means is one of the few algorithms that have been applied successfully to large data sets; most of the existing algorithms cannot handle large datasets. K-Means has a rich anddiverse history as it was independently discovered in different scientific fields by Steinhaus (1956) [7], Lloyd (proposed in 1957, published in 1982) [8], Ball & Hall (1965) [9] and McQueen (1967) [10]. K-Means clustering is a clustering analysis algorithm that groups objects based on their feature values into k disjoint clusters. Objects that are classified into the same cluster have similar feature values. k is a positive integer number specifying the number of clusters, and has to be defined in advance. However, an inherent limitation to k-Means clustering algorithms is, in determining the initial state of centroids, which may converge to the local optima rather than global optima [11]. Figure 1 exemplifies k-Means algorithm. As the algorithm iterates through the training data, each cluster's architecture is updated. In updating clusters, elements are removed from one cluster to another. The updating of clusters causes the values of the centroids to change. This change is a reflection of the current cluster elements. Once there are no changes to any cluster, the training of the k-Means algorithm is complete.

K-Means Algorithm:
Input: The numbers of clusters k and a dataset for intrusion detection.
Output: A set of k clusters that minimize the squared-error criterion.
Algorithm:

1. Initialize k clusters (Randomly select k elements for the data)
2. While cluster structure changes, repeat from2.
3. Determine the cluster to which source data belongs. Use Euclidean distance formula. Add element to cluster with min (Distance (X_i, Y_j))
4. Calculate the means of the cluster.
5. Change cluster centroids to means obtained using Step 3.

Fig. 1. k-Means pseudo code [12]

3 Gravitational Search Algorithm

Gravitational Search Algorithm is one of the newest optimization algorithms based on the law of gravity [13]. The mechanism of GSA is based on the interaction of masses in the universe via Newtonian gravity law. It is defined by Newton as, "every particle in the universe attracts every other particle with a force that is directly proportional to the product of the masses of the particles and inversely proportional to the square of the distance between them."

To describe the GSA, consider a system with N masses (agents) in which the position of the *ith* mass is defined as follows:

$$X_i = (x_i^1, \ldots, x_i^d, \ldots, x_i^n),\ i= 1, 2,\ldots, N \tag{1}$$

Where x_i^d is the position of the *ith* mass in the *dth* dimension and n is the total number of dimensions in the search space. The positions of the masses correspond to the solutions of the problem. Based on [14], the mass of each agent is calculated after computing a current population's fitness as follows:

$$M_i(t) = \frac{fit_i(t) - worst(t)}{\sum_{j=1}^{N}\left(fit_j(t) - worst(t)\right)} \tag{2}$$

Where $M_i(t)$ and $fit_i(t)$ represent the mass and the fitness value of the agent i at t, respectively, and $worst(t)$ is defined as follows (for a minimization problem):

$$worst(t) = \max fit_j(t) j \in \{1, \ldots, N\} \tag{3}$$

To compute the acceleration of an agent, the total forces from a set of heavier masses that act on it should be considered based on the law of gravity (Eq. (4)), followed by the calculation of an agent acceleration using a law of motion (Eq. (5)). After that, the next velocity of an agent is calculated as a fraction of its current velocity added to its acceleration (Eq. (6)). Then, its next position can be calculated using Eq. (7).

$$F_i^d(t) = \sum_{j \in kbest, j \neq i} rand_j G(t) \frac{M_j(t) M_i(t)}{R_{ij}(t) + \varepsilon}\left(x_j^d(t) - x_i^d(t)\right) \tag{4}$$

$$a_i^d(t) = \frac{F_i^d(t)}{M_i(t)} = \sum_{j \in kbest, j \neq i} rand_j G(t) \frac{M_j(t) M_i(t)}{R_{ij}(t) + \varepsilon}\left(x_j^d(t) - x_i^d(t)\right) \tag{5}$$

$$v_i^d(t+1) = rand_i \times v_i^d(t) + a_i^d(t) \tag{6}$$

$$x_i^d(t+1) = x_i^d(t) + v_i^d(t+1) \tag{7}$$

Where $rand_i$ and $rand_j$ are two uniformly distributed random numbers in the range of [0, 1], ε is a small value to avoid division by zero, and $R_{ij}(t)$ is the Euclidean distance between two agents i and j defined as Euclidean distance formula. The set of first k agents with the best fitness value and biggest mass is *kbest. kbest* is a function of time, initialized to *K0* at the beginning and decreasing with time. Here, *K0* is set to N (total number of agents) and is linearly decreased to 1. G is a decreasing function of time initially set to 1, and decreased linearly towards zero at the last iteration. The principal of the GSA is shown in Figure 2.

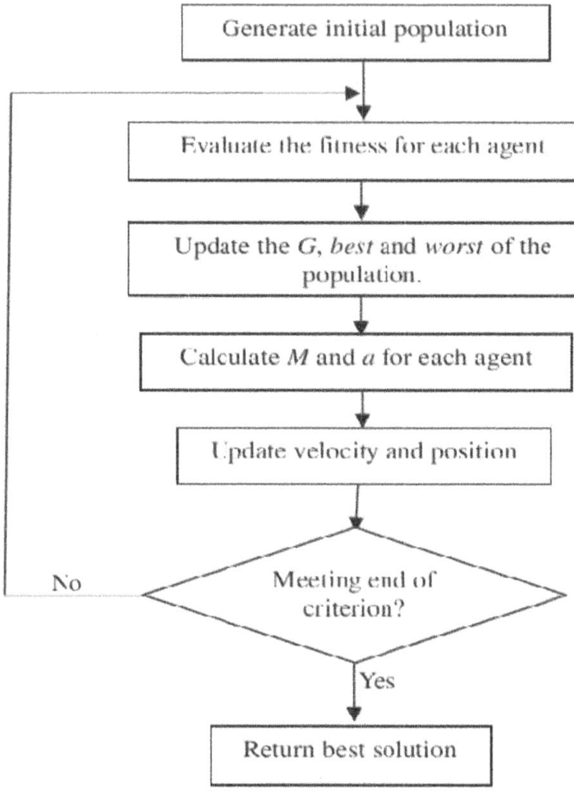

Fig. 2. The principal of GSA [13]

4 The GSA Prototype Classifier

In this section, a prototype based classification approach that uses GSA is used. The issue of finding the appropriate position of each prototype (class representative) is the main objective of this approach. After finding the class representatives, the instance is classified by the class representative that is at the closest distance (i.e. using Euclidean distance). Since GSA is a heuristic based optimization algorithm, it requires some fitness functions which must be defined to measure the quality of the cluster obtained. The most widely used and famous function that is used to specify the goodness of a clustering is the total mean-square quantization error (MSE)[15], which is defined as follows:

$$f(O,C) = \sum_{l=1}^{k} \sum_{O_i \in C_l} d(O_i, Z_l)^2 \tag{8}$$

Where $d(O_i, Z_l)$ specifies the dissimilarity between object O_i and the centroid of cluster C_l (Z_l) to be found by calculating the mean value of objects within the respective cluster. To calculate the dissimilarity between objects, Euclidean distance which is the most popular distance function is chosen. The distance is calculated as in Eq (9):

$$d(X_i, X_j) = \sqrt{\sum_{p=1}^{d} \left(x_i^p - x_j^p\right)^2} \tag{9}$$

Having defined the fitness function, the classification task is shaped into a minimization problem. The performance of GSA then is calculated according to the misclassification percentages of instances by the best found agent. The process of using GSA prototype for classification comprises of two phases, as shown in Figure 3.

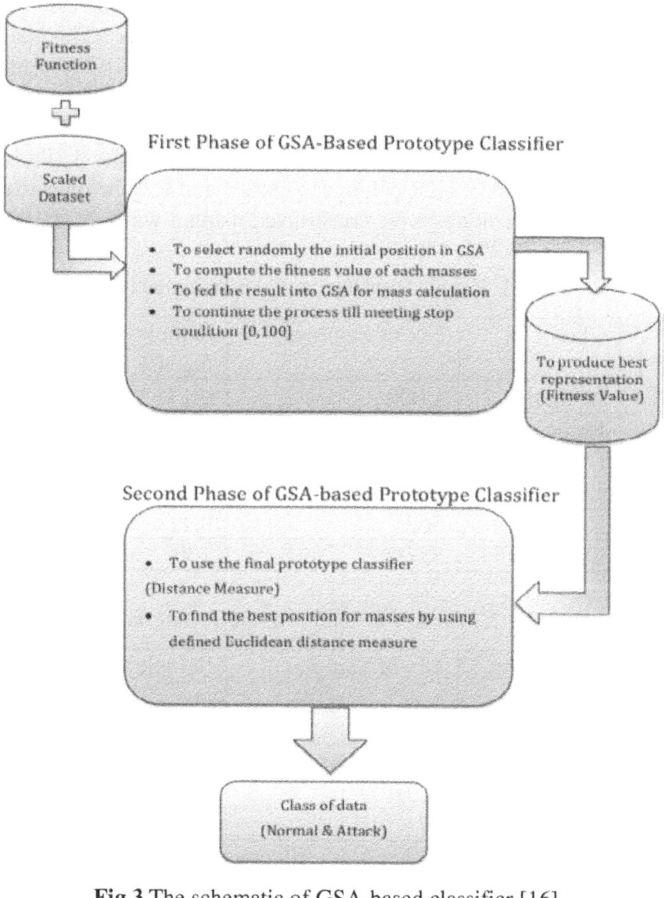

Fig.3. The schematic of GSA-based classifier [16]

5 Experimental and Results

A. Data Description

One of the most popular data set which is common among researchers for (the design and testing) of intrusion detection systems is KDD Cup 1999. The dataset used in this experiment was obtained from 1998 DARPA Intrusion Detection Evaluation Program prepared by MIT Labs and popularly known as KDD Cup 1999 dataset. These datasets were used in many IDS works such as in [17, 18] amongst others. This dataset contains nine weeks of raw tcpdump data from simulated U.S Air Force Local Area Network and injected with multiple attacks. Seven weeks of raw training yielded five millions connection records and two weeks of testing data containing two millions connection records. A connection is a sequence of tcp packets having starting and ending points at some well-defined times. These connections involve data transmitted to and from a source IP address to target IP address under some protocols. Each connection is labeled as either normal or as an attack with exactly one specific attack type. Each connection record consists of around 100 bytes in the dataset. Each TCP connection has 41 features with a label which specifies the status of the connection as either normal or specific attack type. The dataset for this experiment contains randomly generated 11,982 records having 41 features (KDD Cup 1999 intrusion detection dataset). This dataset has five different classes namely Normal, DoS, R2L, U2R and Probes. The training and test comprises of 5092 and 6890 records, respectively. All IDS models were trained and tested with the same set of data. As the data set has five various classes we performed a five-class binary classification. The Normal data belongs to class 1, Probe belongs to class2, DoS belongs to class 3, U2R belongs to class 4 and R2L belongs to class 5 Host-based traffic features: 10 features were constructed using a window of 100 connections

Table 1. Sample Distribution of the Training Dataset

Class	No. of Samples	Sample Percentage (%)
Normal	1,000	0.19
Probe	500	0.1
DoS	3,004	0.59
U2R	26	0.005
R2L	562	0.11
Total	5,092	100

Table 2. Sample Distribution of the TestingDataset

Class	No. of Samples	Sample Percentage (%)
Normal	1,400	0.2
Probe	700	0.1
DoS	4,201	0.61
U2R	27	0.004
R2L	562	0.082
Total	6.890	100

to the same host instead of a time window, because slow scan attacks may occupy a much larger time interval than two seconds. Table 1 and Table 2 summarize the distribution records for training and testing dataset according to class type.

B. Evaluation Measurements

The effectiveness of IDS is evaluated for classification accuracy, false positive rate and hit rate [50]. Table 3 shows a general template for an IDS two-class problem confusion matrix and its evaluation metrics.

Table 3. Typical IDS evaluation metrics [2]

		Actual Classification	
		Normal	Attack
Accuracy= $\frac{a+d}{a+b+c+d}$	Normal	a	b
False alarm rate= $\frac{b}{a+b}$	Attack	c	d
Hit rate= $\frac{d}{c+d}$			

Table 4 shows the categories of data behavior in intrusion detection for binary category classes (Normal and Attacks) in term of true negative, true positive, false positive and false negative.

Table 4. General Behavior of Intrusion Detection Data

Actual	Predicted Normal	Predicted Attack
Normal	TP	FP
Attack	FN	TN

- *True Positive (TP): When Normal data detected as Normal*
- *True Negative (TN) : When Attack data detected as Attack*
- *False Positive (FP) : When Normal data detected as Attack*
- *False Negative (FN) : When Attack data detected as Normal*

C. Results and Discussion

The samples used during the implementation of k-Means and GSA-Based classifier are training and testing dataset which include 5092 and 6890 data respectively. The outputs of clustering algorithms are two Normal and Attack clusters. The standard k-Means is tested on KDD Cup 1999 dataset and its classification performance is summarized in the confusion matrix as shown in Table 5 and Table 6.

Table 5. ConfusionMatrix for Standard k-Means on Training Dataset

	Normal	Attack	Total	Accuracy
Normal	946	1,364	2,310	40.952%
	40.95%	59.05%		
Attack	54	2,728	2,782	98.058%
	2%	98.05%		

Overall Accuracy: 72.15%

False Alarm Rate: 33.33%

Hit rate: 98.06%

Table 6. Confusion Matrix for Standard k-Means on Testing dataset

	Normal	Attack	Total	Accuracy
Normal	1,396	1,741	3,137	44.501%
	44.50%	55.5%		
Attack	54	3,665	3,719	97.655%
	1.5%	97.55%		

Overall Accuracy: 74.67%

False Alarm Rate: 32.2%

Hit rate: 98.55%

In this study, there are two Normal and Attack clusters; therefore the confusion matrix is a 2*2 matrix consists of TP, FP, FN and TN tested on KDD Cup 1999. It can be seen from Tables 5 and 6 that the first row indicates the number of Normal traffic correctly classified as Normal (40.95% and 44.50%) and number of Normal traffic wrongly classified as Attack (59.05%and 55.5%) respectively. Meanwhile, the second row of Tables 5 and 6 shows the number of Attack traffic wrongly classified as Normal (2% and 1.5%), correctly classified as Attack (98.05 and 97.55%) respectively.Also, the confusion matrix for GSA-Based Classifier is shown in Table 7 and 8.It can be seen from Tables 5 and 6 that the first row indicates the number of Normal traffic correctly classified as Normal (40.95% and 44.50%) and number of Normal

Table 7. ConfusionMatrix for GSA-Based Classifier on Training Dataset

	Normal	Attack	Total	Accuracy
Normal	590	707	1,297	45.49%
	45.49%	54.51%		
Attack	410	3,385	3,795	89.2%
	10.8%	89.2%		

Overall Accuracy: 74.67%

False Alarm Rate: 17.28%

Hit rate: 89.2%

traffic wrongly classified as Attack (59.05%and 55.5%) respectively. Meanwhile, the second row of Tables 5 and 6 shows the number of Attack traffic wrongly classified as Normal (2% and 1.5%), correctly classified as Attack (98.05 and 97.55%) respectively.Also, the confusion matrix for GSA-Based Classifier is shown in Table 7 and 8.

Table 8. Confusion Matrix for GSA-Based Classifier on Testing Dataset

	Normal	Attack	Total	Accuracy
Normal	903	1,049	1,952	46.26%
	46.26%	53.74%		
Attack	497	4,441	4,938	89.93%
	10.06%	89.93%		

Overall Accuracy: 77.56%

False Alarm Rate: 19.1%

Hit rate: 89.93%

As shown in Tables 7 and 8 the first row shows the number of Normal traffic correctly classified as Normal (45.49% and 46.26%) and number of Normal traffic wrongly classified as Attack (54.51%and 53.74%) respectively. Meanwhile, the second row of Tables 7 and 8 shows the number of Attack traffic wrongly classified as Normal (10.08% and 10.06%), correctly classified as Attack (89.2 and 89.93%) respectively. Table 9 represents the results across all category classes obtained from k-Means and GSA-Based Classifier using the training and testing sets.

Table 9. The classification result for each class by k-Means and GSA

Dataset	Training		Testing	
Methods	k-Means	GSA	k-Means	GSA
Normal	59.05%	54.51%	55.5%	53.74%
Probe	3.4%	14.2%	80%	56.6%
DoS	26.32%	8%	39.64%	13.32%
U2R	92.59%	77.78%	0%	0%
R2L	94.49%	66.43%	0%	0%

As it is shown in Table 9, the results suggest that traffics of type U2R and R2L are of concern because one possible reason for the lower recall rate is due to the imbalanced data issue. This False Negative Alarm is highly undesirable because the system compromised malicious traffic. Meanwhile, the traffic of type DoS are mostly similar in both clustering using training and testing dataset. This may be because DoS is more consistent than other traffic types and also has a strong characteristic which mostly prevent from being misclassified.

6 Conclusion

False alarm rate in IDS attack identification constitutes the bane of accuracy prediction. Clustering algorithms forms the cardinal for effective attack identification and classification –prediction process. This paper presents a comparative experimental analysis of two clustering algorithm for attack detection in IDS. From the results obtained, k-Means algorithm is discovered to perform relatively stable than GSA using a cluster size of two classes. Further research is however ongoing, on the integration of these two algorithm for better efficiency on IDS attack detection process.

References

1. Fredrik, V.: Real-Time Intrusion Detection Alert Correlation (2006)
2. Jain, A.K.: Data clustering: 50 years beyond K-means. Pattern Recognition Letters 31(8), 651–666 (2010)
3. Kaufman, L., Rousseeuw, P.J.: Finding Groups in Data: An Introduction to Cluster Analysis. John Wiley & Sons, New York (1990)
4. Kao, Y.-T., Zahara, E., Kao, I.W.: A hybridized approach to data clustering. Expert Systems with Applications 34(3), 1754–1762 (2008)
5. Selim, S.Z., Ismail, M.A.: K-means-type algorithms: a generalized convergence theorem and characterization of local optimality. IEEE Transactions on Pattern Analysis and Machine Intelligence PAMI 6(1), 81–87 (1984)
6. Jain, A.K., Topchy, A., Law, M.H.C., Buhmann, J.M.: Landscape of clustering algorithms. In: Proceedings of the International Conference on Pattern Recognition, vol. 1, pp. 260–263 (2004)
7. Steinhaus, H.: Sur la division des corp materiels en parties. Bulletin of Acad. Polon. Sci. IV(C1. III), 801–804 (1956)
8. Lloyd, S.: Least squares quantization in PCM. IEEE Transactions on Information Theory 28, 129–137 (1982)
9. Ball, G., Hall, D.: ISODATA, a novel method of data anlysis and pattern classification. Tech. rept. NTIS AD 699616. Stanford Research Institute, Stanford, CA (1965)
10. Macqueen, J.: Some methods for classification and analysis of multivariate observations. In: Fifth Berkeley Symposium on Mathematics, Statistics and Probability, pp. 281–297. University of California Press (1967)
11. Hatamlou, A., Abdullah, S., Nezamabadi-pour, H.: A combined approach for clustering based on K-Means and gravitational search algorithms. Swarm and Evolutionary Computation (2012)
12. Manas Ranjan Patra, M.P.: Some Clustering Algorithms to enhance the performance of the network intrusion detection system (2005)
13. Rashedi, E., Nezamabadi-pour, H., Saryazdi, S.: GSA: A Gravitational Search Algorithm. Information Sciences 179, 2232–2248 (2009)
14. Holliday, D., Resnick, R., Walker, J.: Fundamentals of physics. John Wiley and Sons
15. Yang, S., et al.: Evolutionary clustering based vector quantization and SPIHT coding for image compression. Pattern Recognition Letters 31(13), 1773–1780 (2010)
16. Bahrololoum, A., Nezamabadi-pour, H., Bahrololoum, H., Saeed, M.: A prototype classifier based on gravitational search algorithm. Applied Soft Computing 12, 819–825 (2012)
17. Mukkamala, S., Sung, A.H., Abraham, A.: Vitorino Ramos Intrusion detection systems using adaptive regression splines. In: Seruca, I., Filipe, J., Hammoudi, S., Cordeiro, J. (eds.) Sixth International Conference on Enterprise Information Systems, ICEIS 2004, Portugal, vol. 3, pp. 26–33 (2004b) ISBN 972-8865-00-7
18. Lee, W., Stolfo, S.J., Mok, K.W.: Adaptive intrusion detection: a data mining approach. Artif. Intell. Rev. 14(6), 533–567 (2000)

Encryption Time Comparison of AES on FPGA and Computer

Yasin Akman[1] and Tarık Yerlikaya[2]

[1] Computer Programming Department, Selcuk University, Turkey
yasinakman@selcuk.edu.tr
[2] Computer Engineering Department, Trakya University, Turkey
tarikyer@trakya.edu.tr

Abstract. Advanced Encryption Standard (AES), which is approved and published by Federal Information Processing Standard (FIPS), is a cryptographic algorithm that can be used to protect electronic data. The AES algorithm can be programmed in software or hardware. This paper presents encryption time comparison of the AES algorithm on FPGA and computer. In the study, Verilog HDL and C programming language is used on the FPGA and computer, respectively. The AES algorithm with 128-bit input and key length 128-bit (AES-128) was simulated on Xilinx ISE Design Suite 13.3. It was observed that, the AES algorithm runs on the FPGA faster than on a computer. We measured the time of encryption on FPGA and computer. Encryption time is 390ns of AES on FPGA and 11 μs of AES on a computer.

Keywords: Advanced Encryption Standard (AES), FPGA, Encryption, Performance Analysis, AES-128.

1 Introduction

With the growth of information and communication technology, the processing of data and transferring them through different media requires high security. [1]

The National Institute of Standards and Technology (NIST) of the United States announced in 1997 an Advanced Encryption Standard (AES) development effort to replace the Digital Encryption Standard (DES). There were five candidates in the last round of the AES algorithm selection process: MARS, RC6, Rijndael, Serpent and Twofish. The Rijndael algorithm, developed by Joan Daemen and Vincent Rijmen [2], was selected as the AES algorithm. The AES algorithm specification is documented in the National Institute of Standards and Technology's (NIST) FIPS 197 publication. [3]

Computers and re-programmable hardware devices are used for the implementation of encryption algorithms. Firstly, we measured encryption time the AES which has coded by verilog HDL is simulated on the Xilinx ISE Design Suite 13.3, and then

D. Nagamalai et al. (Eds.): *Adv. in Comput. Sci., Eng. & Inf. Technol.*, AISC 225, pp. 317–324.
DOI: 10.1007/978-3-319-00951-3_30

we coded the AES by a software using C language and we also measured encryption time the AES on a computer. Accordingly we compared encryption time of FPGA and computer.

2 The Advanced Encryption Standard (AES)

The AES algorithm is a symmetric-key block cipher. It operates on 128-bit data blocks and accepts key sizes of 128, 192, 256 bits and consists of 10, 12 or 14 iteration rounds, respectively. It is given in Table 1. The AES standard is round based and operates on a thirty-two bit by thirty-two bit State array. The array is divided into sixteen bytes as shown in Fig.1. It should be noted that the indexes into the array are row then column. The indexes correspond to the byte sequence; every four bytes form a new row. In this paper, we will present the 128-bit version of AES encryption (AES-128) with 10 rounds.

The number of rounds depends on the length of the key used for the encryption process. For a key length of 128 bits, the required number of iterations is 10 rounds (Nr = 10). The AES structure is shown in Fig.2. As shown in Fig. 2, the first round consists of only AddRoundKey() and each of the Nr-1 rounds consists of 4 transformations: SubBytes(), ShiftRows(), MixColumns(), AddRoundKey(). The last round consists of 3 transformations: SubBytes(), ShiftRows(), AddRoundKey().

Table 1. Comparison of block size, key length and number of rounds of AES keys

Type	Block Size N_b Words	Key Length N_k Words	Number of Rounds N_r
AES-128 bits key	4	4	10
AES-192 bits key	4	6	12
AES-256 bits key	4	8	14

$S_{0,0}$	$S_{0,1}$	$S_{0,2}$	$S_{0,3}$
$S_{1,0}$	$S_{1,1}$	$S_{1,2}$	$S_{1,3}$
$S_{2,0}$	$S_{2,1}$	$S_{2,2}$	$S_{2,3}$
$S_{3,0}$	$S_{3,1}$	$S_{3,2}$	$S_{3,3}$

Fig. 1. AES state array

Table 2. S-Box

	x0	x1	x2	x3	x4	x5	x6	x7	x8	x9	xa	xb	xc	xd	xe	xf
0x	63	7c	77	7b	f2	6b	6f	c5	30	01	67	2b	fe	d7	ab	76
1x	ca	82	c9	7d	fa	59	47	f0	ad	d4	a2	af	9c	a4	72	c0
2x	b7	fd	93	26	36	3f	f7	cc	34	a5	e5	f1	71	d8	31	15
3x	04	c7	23	c3	18	96	05	9a	07	12	80	e2	eb	27	b2	75
4x	09	83	2c	1a	1b	6e	5a	a0	52	3b	d6	b3	29	e3	2f	84
5x	53	d1	00	ed	20	fc	b1	5b	6a	cb	be	39	4a	4c	58	cf
6x	d0	ef	aa	fb	43	4d	33	85	45	f9	02	7f	50	3c	9f	a8
7x	51	a3	40	8f	92	9d	38	f5	bc	b6	da	21	10	ff	f3	d2
8x	cd	0c	13	ec	5f	97	44	17	c4	a7	7e	3d	64	5d	19	73
9x	60	81	4f	dc	22	2a	90	88	46	ee	b8	14	de	5e	0b	db
ax	e0	32	3a	0a	49	06	24	5c	c2	d3	ac	62	91	95	e4	79
bx	e7	c8	37	6d	8d	d5	4e	a9	6c	56	f4	ea	65	7a	ae	08
cx	ba	78	25	2e	1c	a6	b4	c6	e8	dd	74	1f	4b	bd	8b	8a
dx	70	3e	b5	66	48	03	f6	0e	61	35	57	b9	86	c1	1d	9e
ex	e1	f8	98	11	69	d9	8e	94	9b	1e	87	e9	ce	55	28	df
fx	8c	a1	89	0d	bf	e6	42	68	41	99	2d	0f	b0	54	bb	16

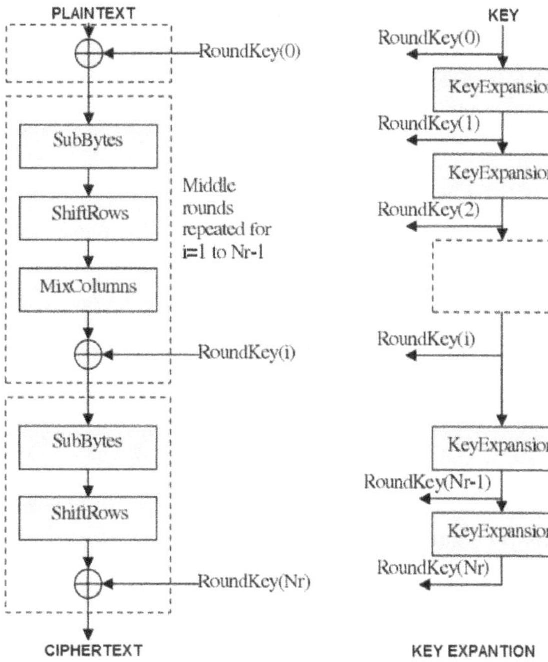

Fig. 2. AES Structure [4]

The four different transformations are described in detail below.

2.1 SubBytes Transformation

The transformation uses 16 identical 256-byte substitution table called S-box as shown in Table 2. It is a non-linear substitution of bytes that operates on each byte of the State using a substitution table (S-box). SubBytes transformation shown in Fig. 3.

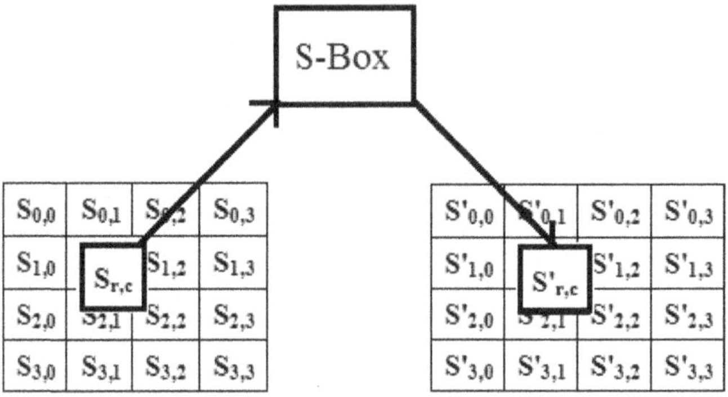

Fig. 3. SubBytes Transformation

2.2 ShiftRows Transformation

In the ShiftRows transformation, the bytes in the last three rows of the State are cyclically shifted over different numbers of bytes (offsets). The first row, r = 0, is not shifted. ShiftRows transformation shown in Fig.4.

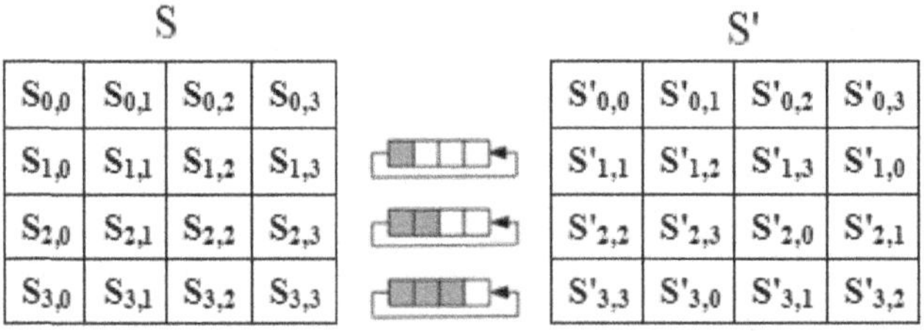

Fig. 4. ShiftRows Transformation

2.3 MixColumns Transformation

Mixing operation which operates separately on every columns of the State using a linear transformation. MixColumns transformation shown in Fig.5.

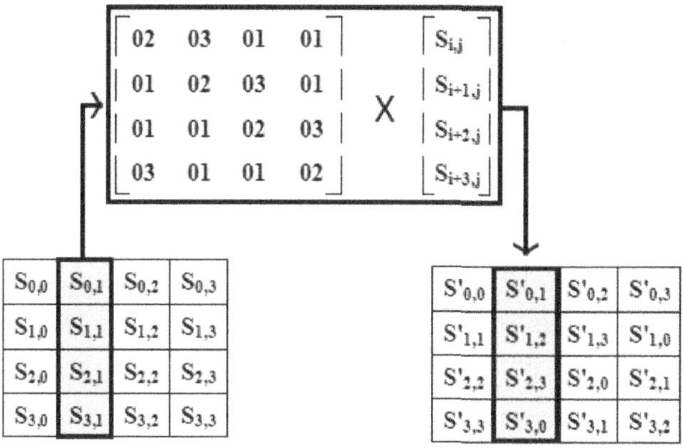

Fig. 5. MixColumns Transformation

2.4 AddRoundKey Transformation

In this transformation, a Round Key is added to the State by a simple bitwise XOR operation. Each round key consists of Nb words from the key expansion. Those Nb words are each added into the columns of the State. AddRoundKey transformation shown in Fig.6. AddRoundKey operation is using a round key which is produced by the key expansion operation.

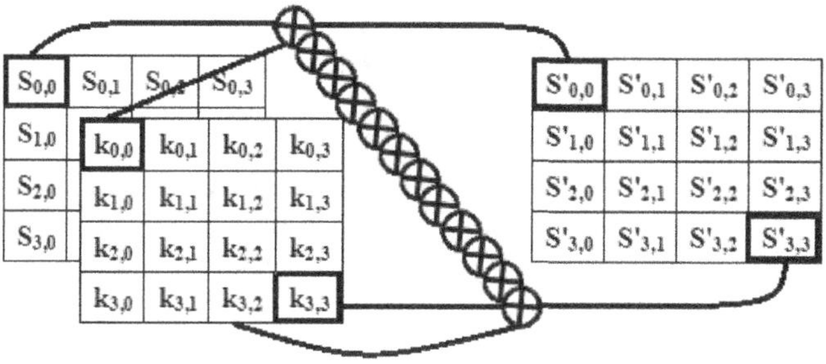

Fig. 6. AddRoundKey Transformation

Key Expansion Phase

The key expansion operation generates a key schedule. Each of four consecutive bytes form a word, denoted w_i as shown in Fig.7, taking into account that the first round-key is the initial key. To generate every w_i (except w_0 - w_3) the routine uses the previous w_{i-1} XOR w_{i-4} (except i mod 4=0). To get the wi, when the i mod 4=0, the operation has two stages, g operation, XOR w_{i-4} as shown in Fig.8. The g operation follows RotWord, SubWord, XOR Rcon[i / 4]. The function RotWord takes a word[a0, a1, a2, a3] as input, performs a cyclic permutation, and returns the word [a1, a2, a3, a0]. SubWord is a function that takes a four-byte input word and applies the

S-box to each of the four bytes to produce an output word. Rcon[i / 4] , contains the values given by [x(i / 4) - 1 , {00}, {00}, {00}], with x(i / 4) - 1 being powers of x (x is denoted as {02}).[5]

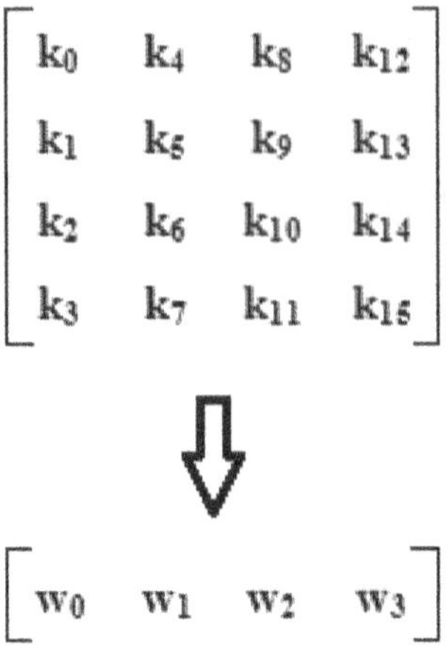

Fig. 7. wi form

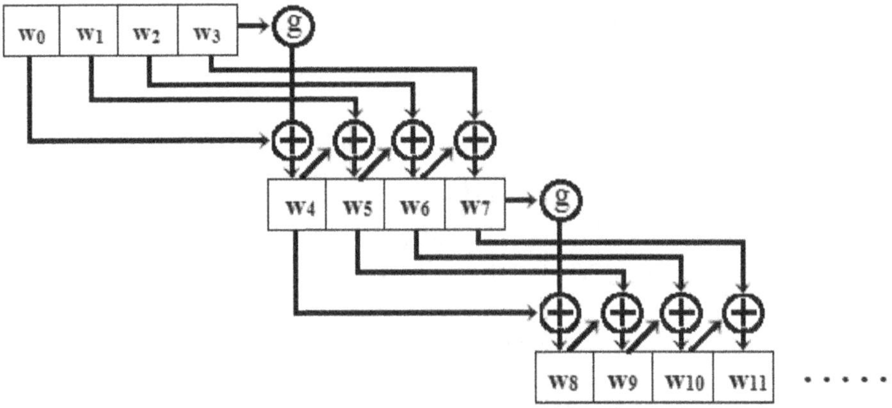

Fig. 8. Key Expansion Phase

3 Encryption Time of AES on FPGA

FPGAs are considered one of the important hardware platforms and integral part for the cryptographic algorithm implementations.[6] FPGAs offer much easier and reasonably cheap solution for the implementation of cryptographic algorithms [7].

The design of the AES encryption algorithm is synthesized and simulated on Xilinx ISE Design Suite 13.3.

Fig.9. illustrates the timing simulation of AES. Encryption time is 390ns of AES on FPGA.

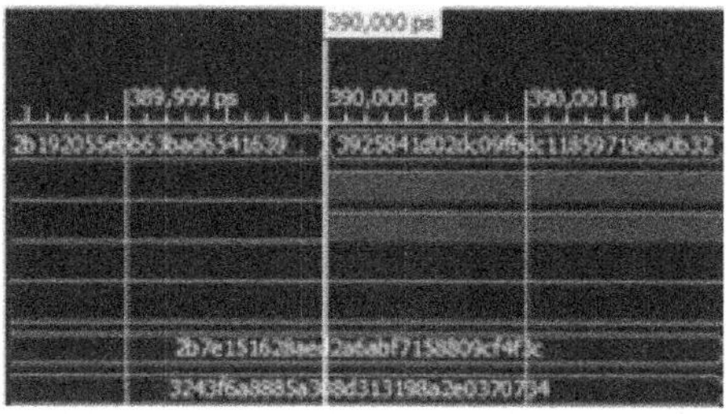

Fig. 9. Timing simulation of AES

Input (text_in) and cipher key (key) values are extracted from FIPS 197. The same output (text_out) values are obtained, as shown in Table 3.

Table 3. Input, cipher key, output values

	Values
Input	3243f6a8885a308d313198a2e0370734
Cipher Key	2b7e151628aed2a6abf7158809cf4f3c
Output	3925841d02dc09fbdc118597196a0b32

4 Encryption Time of AES on Computer

A lot of cryptology algorithms are coded in various software languages to be used on computers. The AES encryption algorithm is coded by a software using C language and GCC is used as the compiler.

The qualifications of the operating computer on which the time is measured are shown in Table 4.

Table 4. Qualifications of computer

	Qualifications
Processor	Intel Core i7-2670 (6MB Cache, up to 3.1 GHz)
Operating System	Ubuntu 12.04 LTS
Memory	6GB DDR3 SDRAM at 1333Mhz
Hard Drive	5400 Rpm SATA (650GB)

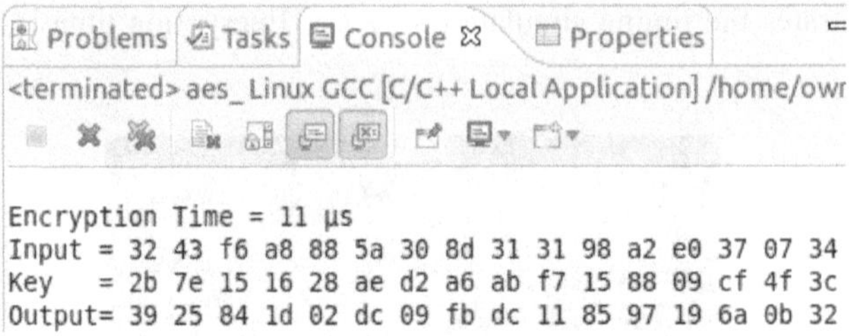

Fig. 10. Encryption time of AES on computer

The time measurement results are shown in Fig.10.

5 Conclusions

In this work, we have presented encryption times of two implementations of the AES-128 encryption algorithm. We aimed to measure the encryption time of AES 128 encryption algorithm on an FPGA and on a computer, and determine which of them would encrypt at the shortest time. Initially we measured the time of encryption on FPGA. Encryption time is 390ns of AES on FPGA.

Then, we measured the encryption time on the computer of which the qualifications are given in Table 4. During the encryption process the best performance of the computer was 11 µs.

Consequently, we reached the conclusion that FPGA that we use is 28 times faster than the operating computer of which the qualifications are shown in the Table 4. This result shows the superiority of hardware implementation in this field.

References

1. Stallings, W.: Cryptography and Network Security: Principles and Practices, 4th edn., pp. 63–173. Pearson Education, Inc. (2006)
2. Daemen, J., Rijmen, V.: The Design of Rijndael. Springer, Heidelberg (2002)
3. FIPS 197: Advanced Encryption Standard. National Institute of Standards and Technology (2001), `http://csrc.nist.gov/publications/fips/fips197/fips-197.pdf`
4. Good, T., Benaissa, M.: AES as Stream Cipher on a Small FPGA In: Circuits and Systems. In: IEEE International Symposium (ISCAS 2006), pp. 529–532 (2006)
5. Zhang, Y., Wang, X.: Pipelined Implementation of AES Encryption Based on FPGA. In: IEEE Conference on Information Theory and Information Security, pp. 170–173 (December 2010)
6. Rais, M.H., Qasim, S.M.: Efficient Hardware Realization of Advanced Encryption Standard Algorithm using Virtex-5 FPGA. International Journal of Computer Science and Network Security (IJCSNS) 9, 59–63 (2009)
7. Järvinen, K., Tommiska, M., Skyttä, J.: Comparative survey of high-performance cryptographic algorithm implementations on FPGAs. In: IEE Proc. Information Security, vol. 152, pp. 3–12 (2005)

Erratum: Design of Optimal Digital Fir Filter Using Particle Swarm Optimization Algorithm

Pavani Uday Kumar, G.R.C. Kaladhara Sarma, S. Mohan Das, and M.A.V. Kamalnath

Department of ECE,
AVR & SVR College of Engineering & Technology, Nandyal, A.P, India
udayaccess@yahoo.com, {grcksrgm,mohantech418,kamalmav}@gmail.com

D. Nagamalai et al. (Eds.): *Adv. in Comput. Sci., Eng. & Inf. Technol.*, AISC 225, pp. 187–196.
DOI: 10.1007/978-3-319-00951-3_19

DOI 10.1007/978-3-319-00951-3_31

In the original version, the first author name was spelt as "Pavani Uday Kumar" but it should be read as "P. Uday Kumar".

The original online version for this chapter can be found at
http://dx.doi.org/10.1007/978-3-319-00951-3_19

Author Index

The manufacturer's authorised representative in the EU is Springer Nature Customer Service Centre GmbH, Europaplatz 3, 69115 Heidelberg, Germany. If you have any concerns regarding our products, please contact ProductSafety@springernature.com

Printed and bound by CPI Group (UK) Ltd, Croydon, CR0 4YY
30/09/2026
02225561-0010